Optimization Methods in Finance

Optimization methods play a central role in financial modeling. This textbook is devoted to explaining how state-of-the-art optimization theory, algorithms, and software can be used to efficiently solve problems in computational finance. It discusses some classical mean–variance portfolio optimization models as well as more modern developments such as models for optimal trade execution and dynamic portfolio allocation with transaction costs and taxes. Chapters discussing the theory and efficient solution methods for the main classes of optimization problems alternate with chapters discussing their use in the modeling and solution of central problems in mathematical finance.

This book will be interesting and useful for students, academics, and practitioners with a background in mathematics, operations research, or financial engineering.

The second edition includes new examples and exercises as well as a more detailed discussion of mean–variance optimization, multi-period models, and additional material to highlight the relevance to finance.

Gérard Cornuéjols is a Professor of Operations Research at the Tepper School of Business, Carnegie Mellon University. He is a member of the National Academy of Engineering and has received numerous prizes for his research contributions in integer programming and combinatorial optimization, including the Lanchester Prize, the Fulkerson Prize, the Dantzig Prize, and the von Neumann Theory Prize.

Javier Peña is a Professor of Operations Research at the Tepper School of Business, Carnegie Mellon University. His research explores the myriad of challenges associated with large-scale optimization models and he has published numerous articles on optimization, machine learning, financial engineering, and computational game theory. His research has been supported by grants from the National Science Foundation, including a prestigious CAREER award.

Reha Tütüncü is the Chief Risk Officer at SECOR Asset Management and an adjunct professor at Carnegie Mellon University. He has previously held senior positions at Goldman Sachs Asset Management and AQR Capital Management focusing on quantitative portfolio construction, equity portfolio management, and risk management.

Optimization Methods in Finance

Second Edition

GÉRARD CORNUÉJOLS
Carnegie Mellon University, Pennsylvania

JAVIER PEÑA
Carnegie Mellon University, Pennsylvania

REHA TÜTÜNCÜ
SECOR Asset Management

CAMBRIDGE
UNIVERSITY PRESS

CAMBRIDGE
UNIVERSITY PRESS

University Printing House, Cambridge CB2 8BS, United Kingdom

One Liberty Plaza, 20th Floor, New York, NY 10006, USA

477 Williamstown Road, Port Melbourne, VIC 3207, Australia

314–321, 3rd Floor, Plot 3, Splendor Forum, Jasola District Centre, New Delhi – 110025, India

79 Anson Road, #06–04/06, Singapore 079906

Cambridge University Press is part of the University of Cambridge.

It furthers the University's mission by disseminating knowledge in the pursuit of education, learning, and research at the highest international levels of excellence.

www.cambridge.org
Information on this title: www.cambridge.org/9781107056749
DOI: 10.1017/9781107297340

First edition © Gérard Cornuéjols and Reha Tütüncü 2007
Second edition © Gérard Cornuéjols, Javier Peña and Reha Tütüncü 2018

First published 2007
Second edition 2018

A catalogue record for this publication is available from the British Library.

ISBN 978-1-107-05674-9 Hardback

Additional resources for this publication at www.cambridge.org/9781107056749

Contents

Preface   *page* xi

Part I   Introduction   1

1   Overview of Optimization Models   3
   1.1   Types of Optimization Models   4
   1.2   Solution to Optimization Problems   7
   1.3   Financial Optimization Models   8
   1.4   Notes   10

2   Linear Programming: Theory and Algorithms   11
   2.1   Linear Programming   11
   2.2   Graphical Interpretation of a Two-Variable Example   15
   2.3   Numerical Linear Programming Solvers   16
   2.4   Sensitivity Analysis   17
   2.5   *Duality   20
   2.6   *Optimality Conditions   23
   2.7   *Algorithms for Linear Programming   24
   2.8   Notes   30
   2.9   Exercises   31

3   Linear Programming Models: Asset–Liability Management   35
   3.1   Dedication   35
   3.2   Sensitivity Analysis   38
   3.3   Immunization   38
   3.4   Some Practical Details about Bonds   41
   3.5   Other Cash Flow Problems   44
   3.6   Exercises   47
   3.7   Case Study   51

4   Linear Programming Models: Arbitrage and Asset Pricing   53
   4.1   Arbitrage Detection in the Foreign Exchange Market   53
   4.2   The Fundamental Theorem of Asset Pricing   55
   4.3   One-Period Binomial Pricing Model   56

4.4	Static Arbitrage Bounds	59
4.5	Tax Clientele Effects in Bond Portfolio Management	63
4.6	Notes	65
4.7	Exercises	65

Part II   Single-Period Models   69

5   Quadratic Programming: Theory and Algorithms   71

5.1	Quadratic Programming	71
5.2	Numerical Quadratic Programming Solvers	74
5.3	Sensitivity Analysis	75
5.4	*Duality and Optimality Conditions	76
5.5	*Algorithms	81
5.6	Applications to Machine Learning	84
5.7	Exercises	87

6   Quadratic Programming Models: Mean–Variance Optimization   90

6.1	Portfolio Return	90
6.2	Markowitz Mean–Variance (Basic Model)	91
6.3	Analytical Solutions to Basic Mean–Variance Models	95
6.4	More General Mean–Variance Models	99
6.5	Portfolio Management Relative to a Benchmark	103
6.6	Estimation of Inputs to Mean–Variance Models	106
6.7	Performance Analysis	112
6.8	Notes	115
6.9	Exercises	115
6.10	Case Studies	121

7   Sensitivity of Mean–Variance Models to Input Estimation   124

7.1	Black–Litterman Model	126
7.2	Shrinkage Estimation	129
7.3	Resampled Efficiency	131
7.4	Robust Optimization	132
7.5	Other Diversification Approaches	133
7.6	Exercises	135

8   Mixed Integer Programming: Theory and Algorithms   140

8.1	Mixed Integer Programming	140
8.2	Numerical Mixed Integer Programming Solvers	143
8.3	Relaxations and Duality	145
8.4	Algorithms for Solving Mixed Integer Programs	150
8.5	Exercises	157

9   **Mixed Integer Programming Models: Portfolios with Combinatorial Constraints**   161
              9.1   Combinatorial Auctions   161
              9.2   The Lockbox Problem   163
              9.3   Constructing an Index Fund   165
              9.4   Cardinality Constraints   167
              9.5   Minimum Position Constraints   168
              9.6   Risk-Parity Portfolios and Clustering   169
              9.7   Exercises   169
              9.8   Case Study   171

10   **Stochastic Programming: Theory and Algorithms**   173
              10.1   Examples of Stochastic Optimization Models   173
              10.2   Two-Stage Stochastic Optimization   174
              10.3   Linear Two-Stage Stochastic Programming   175
              10.4   Scenario Optimization   176
              10.5   *The L-Shaped Method   177
              10.6   Exercises   179

11   **Stochastic Programming Models: Risk Measures**   181
              11.1   Risk Measures   181
              11.2   A Key Property of CVaR   185
              11.3   Portfolio Optimization with CVaR   186
              11.4   Notes   190
              11.5   Exercises   190

Part III   Multi-Period Models   195

12   **Multi-Period Models: Simple Examples**   197
              12.1   The Kelly Criterion   197
              12.2   Dynamic Portfolio Optimization   198
              12.3   Execution Costs   201
              12.4   Exercises   209

13   **Dynamic Programming: Theory and Algorithms**   212
              13.1   Some Examples   212
              13.2   Model of a Sequential System (Deterministic Case)   214
              13.3   Bellman's Principle of Optimality   215
              13.4   Linear–Quadratic Regulator   216
              13.5   Sequential Decision Problem with Infinite Horizon   218
              13.6   Linear–Quadratic Regulator with Infinite Horizon   219
              13.7   Model of Sequential System (Stochastic Case)   221
              13.8   Notes   222
              13.9   Exercises   222

14   **Dynamic Programming Models: Multi-Period Portfolio Optimization**   225
     14.1   Utility of Terminal Wealth   225
     14.2   Optimal Consumption and Investment   227
     14.3   Dynamic Trading with Predictable Returns and Transaction Costs   228
     14.4   Dynamic Portfolio Optimization with Taxes   230
     14.5   Exercises   234

15   **Dynamic Programming Models: the Binomial Pricing Model**   238
     15.1   Binomial Lattice Model   238
     15.2   Option Pricing   238
     15.3   Option Pricing in Continuous Time   244
     15.4   Specifying the Model Parameters   245
     15.5   Exercises   246

16   **Multi-Stage Stochastic Programming**   248
     16.1   Multi-Stage Stochastic Programming   248
     16.2   Scenario Optimization   250
     16.3   Scenario Generation   255
     16.4   Exercises   259

17   **Stochastic Programming Models: Asset–Liability Management**   262
     17.1   Asset–Liability Management   262
     17.2   The Case of an Insurance Company   263
     17.3   Option Pricing via Stochastic Programming   265
     17.4   Synthetic Options   270
     17.5   Exercises   273

Part IV   **Other Optimization Techniques**   275

18   **Conic Programming: Theory and Algorithms**   277
     18.1   Conic Programming   277
     18.2   Numerical Conic Programming Solvers   282
     18.3   Duality and Optimality Conditions   282
     18.4   Algorithms   284
     18.5   Notes   287
     18.6   Exercises   287

19   **Robust Optimization**   289
     19.1   Uncertainty Sets   289
     19.2   Different Flavors of Robustness   290
     19.3   Techniques for Solving Robust Optimization Models   294
     19.4   Some Robust Optimization Models in Finance   297
     19.5   Notes   302
     19.6   Exercises   302

20   **Nonlinear Programming: Theory and Algorithms**   305
        20.1   Nonlinear Programming   305
        20.2   Numerical Nonlinear Programming Solvers   306
        20.3   Optimality Conditions   306
        20.4   Algorithms   308
        20.5   Estimating a Volatility Surface   315
        20.6   Exercises   319

Appendices   321

Appendix   Basic Mathematical Facts   323
        A.1   Matrices and Vectors   323
        A.2   Convex Sets and Convex Functions   324
        A.3   Calculus of Variations: the Euler Equation   325

References   327

Index   334

Preface

The use of sophisticated mathematical tools in modern finance is now common-place. Researchers and practitioners routinely run simulations or solve differential equations to price securities, estimate risks, or determine hedging strategies. Some of the most important tools employed in these computations are opti-mization algorithms. Many computational finance problems ranging from asset allocation to risk management, from option pricing to model calibration, can be solved by optimization techniques. This book is devoted to explaining how to solve such problems efficiently and accurately using the state of the art in optimization models, methods, and software.

Optimization is a mature branch of applied mathematics. Typical optimization problems have the goal of allocating limited resources to alternative activities in order to maximize the total benefit obtained from these activities. Through decades of intensive and innovative research, fast and reliable algorithms and software have become available for many classes of optimization prob-lems. Consequently, optimization is now being used as an effective manage-ment and decision-support tool in many industries, including the financial industry.

This book discusses several classes of optimization problems encountered in financial models, including linear, quadratic, integer, dynamic, stochastic, conic, and nonlinear programming. For each problem class, after introducing the rele-vant theory (optimality conditions, duality, etc.) and efficient solution methods, we discuss several problems of mathematical finance that can be modeled within this problem class.

The second edition includes a more detailed discussion of mean–variance opti-mization, multi-period models, and additional material to highlight the relevance to finance.

The book's structure has also been clarified for the second edition; it is now organized in four main parts, each comprising several chapters. Part I guides the reader through the solution of asset liability cash flow matching using lin-ear programming techniques, which are also used to explain asset pricing and arbitrage. Part II is devoted to single-period models. It provides a thorough treatment of mean–variance portfolio optimization models, including derivations of the one-fund and two-fund theorems and their connection to the capital asset pricing model, a discussion of linear factor models that are used extensively in risk and portfolio management, and techniques to deal with the sensitivity of mean–variance models to parameter estimation. We discuss integer programming formulations for portfolio construction problems with cardinality constraints, and we explain how this is relevant to constructing an index fund. The final chapters of Part II present a stochastic programming approach to modeling measures of risk other than the variance, including the popular value at risk and conditional value at risk.

Part III of the book discusses multi-period models such as the iconic Kelly criterion and binomial lattice models for asset pricing as well as more elaborate and modern models for optimal trade execution, dynamic portfolio optimization with transaction costs and taxes, and asset–liability management. These applications showcase techniques from dynamic and stochastic programming.

Part IV is devoted to more advanced optimization techniques. We introduce conic programming and discuss applications such as the approximation of covariance matrices and robust portfolio optimization. The final chapter of Part IV covers one of the most general classes of optimization models, namely nonlinear programming, and applies it to volatility estimation.

This book is intended as a textbook for Master's programs in financial engineering, finance, or computational finance. In addition, the structure of chapters, alternating between optimization methods and financial models that employ these methods, allows the book to be used as a primary or secondary text in upper-level undergraduate or introductory graduate courses in operations research, management science, and applied mathematics. A few sections are marked with a '*' to indicate that the material they contain is more technical and can be safely skipped without loss of continuity.

Optimization algorithms are sophisticated tools and the relationship between their inputs and outputs is sometimes opaque. To maximize the value from using these tools and to understand how they work, users often need a significant amount of guidance and practical experience with them. This book aims to provide this guidance and serve as a reference tool for the finance practitioners who use or want to use optimization techniques.

This book has benefited from the input provided by instructors and students in courses at various institutions. We thank them for their valuable feedback and for many stimulating discussions. We would also like to thank the colleagues who provided the initial impetus for this book and colleagues who collaborated with us on various research projects that are reflected in the book. We especially thank Kathie Cameron, the late Rick Green, Raphael Hauser, John Hooker, Miroslav Karamanov, Mark Koenig, Masakazu Kojima, Vijay Krishnamurthy, Miguel Lejeune, Yanjun Li, François Margot, Ana Margarida Monteiro, Mustafa Pınar, Sebastian Pokutta, Sanjay Srivastava, Michael Trick, and Luís Vicente.

Part I

Introduction

1 Overview of Optimization Models

Optimization is the process of finding the *best* way of making decisions that satisfy a set of constraints. In mathematical terms, an optimization model is a problem of the form

$$\min_{\mathbf{x}} \quad f(\mathbf{x})$$
$$\text{s.t.} \quad \mathbf{x} \in \mathcal{X}, \tag{1.1}$$

where $f : \mathbb{R}^n \to \mathbb{R}$ and $\mathcal{X} \subseteq \mathbb{R}^n$.

Model (1.1) has three main components, namely the vector of *decision variables* $\mathbf{x} := \begin{bmatrix} x_1 & \cdots & x_n \end{bmatrix}^{\mathsf{T}} \in \mathbb{R}^n$; the *objective function* $f(\mathbf{x})$; and the *constraint set* or *feasible region* $\mathcal{X}$. The constraint set is often expressed in terms of equalities and inequalities involving additional functions. More precisely, the constraint set $\mathcal{X}$ is often of the form

$$\mathcal{X} = \{\mathbf{x} \in \mathbb{R}^n : g_i(\mathbf{x}) = b_i, \text{ for } i = 1, \dots, m, \text{ and } h_j(\mathbf{x}) \leq d_j, \text{ for } j = 1, \dots, p\}, \tag{1.2}$$

for some $g_i, h_j : \mathbb{R}^n \to \mathbb{R}$, $i = 1, \dots, m$, $j = 1, \dots, p$. When this is the case, the optimization problem (1.1) is usually written in the form

$$\min_{\mathbf{x}} \quad f(\mathbf{x})$$
$$\text{s.t.} \quad g_i(\mathbf{x}) = b_i, \text{ for } i = 1, \dots, m$$
$$\qquad h_j(\mathbf{x}) \leq d_j, \text{ for } j = 1, \dots, p,$$

or in the more concise form

$$\min_{\mathbf{x}} \quad f(\mathbf{x})$$
$$\text{s.t.} \quad \mathbf{g}(\mathbf{x}) = \mathbf{b}$$
$$\qquad \mathbf{h}(\mathbf{x}) \leq \mathbf{d}.$$

We will use the following terminology. A *feasible point* or *feasible solution* to (1.1) is a point in the constraint set $\mathcal{X}$. An *optimal solution* to (1.1) is a feasible point that attains the best possible objective value; that is, a point $\mathbf{x}^* \in \mathcal{X}$ such that $f(\mathbf{x}^*) \leq f(\mathbf{x})$ for all $\mathbf{x} \in \mathcal{X}$. The *optimal value* of (1.1) is the value of the objective function at an optimal solution; that is, $f(\mathbf{x}^*)$ where $\mathbf{x}^*$ is an optimal solution to (1.1). If the feasible region $\mathcal{X}$ is of the form (1.2) and $\mathbf{x} \in \mathcal{X}$, the *binding constraints* at $\mathbf{x}$ are the equality constraints and those inequality constraints that hold with equality at $\mathbf{x}$. The term *active constraint* is also often used in lieu of "binding constraint". The problem (1.1) is *infeasible* if $\mathcal{X} = \emptyset$. On the other hand, (1.1) is *unbounded* if there exist $\mathbf{x}_k \in \mathcal{X}$, $k = 1, 2, \ldots$, such that $f(\mathbf{x}_k) \to -\infty$.

1.1   Types of Optimization Models

For optimization models to be of practical interest, their computational tractability, that is, the ability to find the optimal solution efficiently, is a critical issue. Particular structural assumptions on the objective and constraints of the problem give rise to different classes of optimization models with various degrees of computational difficulty. We should note that the following is only a partial classification based on the current generic tractability of various types of optimization models. However, what is "tractable" in some specific context may be more nuanced. Furthermore, tractability evolves as new algorithms and technologies are developed.

Convex optimization: These are problems where the objective $f(\mathbf{x})$ is a convex function and the constraint set $\mathcal{X}$ is a convex set. This class of optimization models is tractable most of the time. By this we mean that a user can expect any of these models to be amenable to an efficient algorithm. We will emphasize this class of optimization models throughout the book.

Mixed integer optimization: These are problems where some of the variables are restricted to take integer values. This restriction makes the constraint set $\mathcal{X}$ non-convex. This class of optimization models is somewhat tractable a fair portion of the time. By this we mean that a model of this class may be solvable provided the user does some judicious modeling and has access to high computational power.

Stochastic and dynamic optimization: These are problems involving random and time-dependent features. This class of optimization models is tractable only in some special cases. By this we mean that, unless some specific structure and assumptions hold, a model of this class would typically be insoluble with any realistic amount of computational power at our disposal. Current research is expected to enrich the class of tractable models in this area.

The modeling of time and uncertainty is pervasive in almost every financial problem. The various types of optimization problems that we will discuss are based on how they deal with these two issues. Generally speaking, *static models* are associated with simple single-period models where the future is modeled as a single stage. By contrast, in *multi-period* models the future is modeled as a sequence, or possibly as a continuum, of stages. With regard to uncertainty, *deterministic models* are those where all the defining data are assumed to be known with certainty. By contrast, *stochastic models* are ones that incorporate probabilistic or other types of uncertainty in the data.

A good portion of the models that we will present in this book will be convex optimization models due to their favorable mathematical and computational properties. There are two special types of convex optimization problems that we will use particularly often: *linear* and *quadratic* programming, the latter being an extension of the former. These two types of optimization models will be discussed in more detail in Chapters 2 and 5. We now present a high-level description of four major classes of optimization models: linear programming, quadratic programming, mixed integer programming, and stochastic optimization.

Linear Programming

A linear programming model is an optimization problem where the objective is a linear function and the constraint set is defined by finitely many linear equalities and linear inequalities. In other words, a linear program is a problem of the form

$$\min_{\mathbf{x}} \quad \mathbf{c}^\mathsf{T}\mathbf{x}$$
$$\text{s.t.} \quad \mathbf{A}\mathbf{x} = \mathbf{b}$$
$$\mathbf{D}\mathbf{x} \geq \mathbf{d}$$

for some vectors $\mathbf{c} \in \mathbb{R}^n, \mathbf{b} \in \mathbb{R}^m, \mathbf{d} \in \mathbb{R}^p$ and matrices $\mathbf{A} \in \mathbb{R}^{m \times n}, \mathbf{D} \in \mathbb{R}^{p \times n}$.

The term *linear optimization* is sometimes used in place of linear programming. The wide popularity of linear programming is due in good part to the availability of very efficient algorithms. The two best known and most successful methods for solving linear programs are the *simplex method* and *interior-point methods*. We briefly discuss these algorithms in Chapter 2.

Quadratic Programming

Quadratic programming, also known as quadratic optimization, is an extension of linear programming where the objective function includes a quadratic term. In other words, a quadratic program is a problem of the form

$$\min_{\mathbf{x}} \quad \tfrac{1}{2}\mathbf{x}^\mathsf{T}\mathbf{Q}\mathbf{x} + \mathbf{c}^\mathsf{T}\mathbf{x}$$
$$\text{s.t.} \quad \mathbf{A}\mathbf{x} = \mathbf{b}$$
$$\mathbf{D}\mathbf{x} \geq \mathbf{d}$$

for some vectors and matrices $\mathbf{Q} \in \mathbb{R}^{n \times n}$, $\mathbf{c} \in \mathbb{R}^n$, $\mathbf{b} \in \mathbb{R}^m$, $\mathbf{d} \in \mathbb{R}^p$, $\mathbf{A} \in \mathbb{R}^{m \times n}$, $\mathbf{D} \in \mathbb{R}^{p \times n}$. It is customary to assume that the matrix $\mathbf{Q}$ is symmetric. This assumption can be made without loss of generality since

$$\mathbf{x}^\mathsf{T}\mathbf{Q}\mathbf{x} = \mathbf{x}^\mathsf{T}\tilde{\mathbf{Q}}\mathbf{x}$$

where $\tilde{\mathbf{Q}} = \tfrac{1}{2}(\mathbf{Q} + \mathbf{Q}^\mathsf{T})$, which is clearly a symmetric matrix.

We note that a quadratic function $\tfrac{1}{2}\mathbf{x}^\mathsf{T}\mathbf{Q}\mathbf{x} + \mathbf{c}^\mathsf{T}\mathbf{x}$ is convex if and only if the matrix $\mathbf{Q}$ is positive semidefinite ($\mathbf{x}^\mathsf{T}\mathbf{Q}\mathbf{x} \geq 0$ for all $x \in \mathbb{R}^n$). In this case the above quadratic program is a convex optimization problem and can be solved efficiently. The two best known methods for solving convex quadratic programs are *active-set methods* and *interior-point methods*. We briefly discuss these algorithms in Chapter 5.

Mixed Integer Programming

A mixed-integer program is an optimization problem that restricts some or all of the decision variables to take integer values. In particular, a mixed integer linear programming model is a problem of the form

$$\min_{\mathbf{x}} \quad \mathbf{c}^\mathsf{T}\mathbf{x}$$
$$\text{s.t.} \quad \mathbf{Ax} = \mathbf{b}$$
$$\mathbf{Dx} \geq \mathbf{d}$$
$$x_j \in \mathbb{Z},\ j \in J$$

for some vectors and matrices $\mathbf{c} \in \mathbb{R}^n$, $\mathbf{b} \in \mathbb{R}^m$, $\mathbf{d} \in \mathbb{R}^p$, $\mathbf{A} \in \mathbb{R}^{m \times n}$, $\mathbf{D} \in \mathbb{R}^{p \times n}$ and some $J \subseteq \{1, \ldots, n\}$.

An important case occurs when the model includes *binary* variables, that is, variables that are restricted to take values 0 or 1. As we will see, the inclusion of this type of constraint increases the modeling power but comes at a cost in terms of computational tractability. It is noteworthy that the computational and algorithmic machinery for solving mixed integer programs has vastly improved during the last couple of decades. The main classes of methods for solving mixed integer programs are *branch and bound, cutting planes*, and a combination of these two approaches known as *branch and cut*. We briefly discuss these algorithms in Chapter 8.

Stochastic Optimization

Stochastic optimization models are optimization problems that account for randomness in their objective or constraints. The following formulation illustrates a generic type of stochastic optimization problem

$$\min_{\mathbf{x}} \quad \mathbb{E}(F(\mathbf{x}, \omega))$$
$$\mathbf{x} \in \mathcal{X}.$$

In this problem the set of decisions $\mathbf{x}$ must be made before a random outcome ω occurs. The goal is to optimize the expectation of some function that depends on both the decision vector $\mathbf{x}$ and the random outcome ω. A variation of this formulation, that has led to important developments, is to replace the expectation by some kind of *risk measure* ϱ in the objective:

$$\min_{\mathbf{x}} \quad \varrho(F(\mathbf{x}, \omega))$$
$$\mathbf{x} \in \mathcal{X}.$$

There are numerous refinements and variants of the above two formulations. In particular, the class of *two-stage stochastic optimization with recourse* has been widely studied in the stochastic programming community. In this setting a set of decisions $\mathbf{x}$ must be made in stage one. Between stage one and stage two a random outcome ω occurs. At stage two we have the opportunity to make some second-stage *recourse* decisions $\mathbf{y}(\omega)$ that may depend on the random outcome ω.

The two-stage stochastic optimization problem with recourse can be formally stated as

$$\min_{\mathbf{x}} \quad f(\mathbf{x}) + \mathbb{E}[Q(\mathbf{x}, \omega)]$$
$$\mathbf{x} \in \mathcal{X}.$$

The *recourse* term $Q(\mathbf{x}, \omega)$ depends on the first-stage decisions $\mathbf{x}$ and the random outcome ω. It is of the form

$$Q(\mathbf{x}, \omega) := \min_{\mathbf{y}(\omega)} \quad g(\mathbf{y}(\omega), \omega)$$
$$\mathbf{y}(\omega) \in \mathcal{Y}(\mathbf{x}, \omega).$$

The second-stage decisions $\mathbf{y}(\omega)$ are *adaptive* to the random outcome ω because they are made after ω is revealed. The objective function in a two-stage stochastic optimization problem contains a term for the stage-one decisions and a term for the stage-two decisions where the latter term involves an expectation over the random outcomes. The intuition of this objective function is that the stage-one decisions should be made considering what is to be expected in stage two.

The above two-stage setting generalizes to a multi-stage context where the random outcome is revealed over time and decisions are made dynamically at multiple stages and can adapt to the information revealed up to their stage.

1.2 Solution to Optimization Problems

The solution to an optimization problem can often be characterized in terms of a set of *optimality conditions*. Optimality conditions are derived from the mathematical relationship between the objective and constraints in the problem. Subsequent chapters discuss optimality conditions for various types of optimization problems. In special cases, these optimality conditions can be solved analytically and used to infer properties about the optimal solution. However, in many cases we rely on numerical solvers to obtain the solution to the optimization models.

There are numerous software vendors that provide solvers for optimization problems. Throughout this book we will illustrate examples with two popular solvers, namely Excel **Solver** and the MATLAB®-based optimization modeling framework CVX. Excel and MATLAB files for the examples and exercises in the book are available at:

www.andrew.cmu.edu/user/jfp/OIFbook/

Both Excel **Solver** and CVX enable us to solve small to medium-sized problems and are fairly easy to use. There are far more sophisticated solvers such as the commercial solvers IBM®-ILOG® CPLEX®, Gurobi, FICO® Xpress, and the ones available via the open-source projects COIN-OR or SCIP.

Optimization problems can be formulated using modeling languages such as AMPL, GAMS, MOSEL, or OPL. The need for these modeling languages arises when the size of the formulation is large. A modeling language lets people use common notation and familiar concepts to formulate optimization models and examine solutions. Most importantly, large problems can be formulated in a compact way. Once the problem has been formulated using a modeling language, it can be solved using any number of solvers. A user can switch between solvers with a single command and select options that may improve solver performance.

1.3   Financial Optimization Models

In this book we will focus on the use of optimization models for financial problems such as portfolio management, risk management, asset and liability management, trade execution, and dynamic asset management. Optimization models are also widely used in other areas of business, science, and engineering, but this will not be the subject of our discussion.

Portfolio Management

One of the best known optimization models in finance is the portfolio selection model of Markowitz (1952). Markowitz's mean–variance approach led to major developments in financial economics including Tobin's mutual fund theorem (Tobin, 1958) and the capital asset pricing model of Treynor[1], Sharpe (1964), Lintner (1965), and Mossin (1966). Markowitz was awarded the Nobel Prize in Economics in 1990 for the enormous influence of his work in financial theory and practice. The gist of this model is to formalize the principle of diversification when selecting a portfolio in a universe of risky assets. As we discuss in detail in Chapter 6, Markowitz's mean–variance model and a wide range of its variations can be stated as a quadratic programming problem of the form

$$\min_{\mathbf{x}} \quad \tfrac{1}{2}\gamma \cdot \mathbf{x}^\mathsf{T}\mathbf{V}\mathbf{x} - \boldsymbol{\mu}^\mathsf{T}\mathbf{x}$$
$$\mathbf{A}\mathbf{x} = \mathbf{b} \tag{1.3}$$
$$\mathbf{D}\mathbf{x} \geq \mathbf{d}.$$

The vector of decision variables $\mathbf{x}$ in model (1.3) represents the portfolio holdings. These holdings typically represent the percentages invested in each asset and thus are often subject to the full investment constraint $\mathbf{1}^\mathsf{T}\mathbf{x} = 1$. Other common constraints include the long-only constraint $\mathbf{x} \geq \mathbf{0}$, as well as restrictions related to sector or industry composition, turnover, etc. The terms $\mathbf{x}^\mathsf{T}\mathbf{V}\mathbf{x}$ and $\boldsymbol{\mu}^\mathsf{T}\mathbf{x}$ in the objective function are respectively the variance, which is a measure of risk,

---

[1] "Toward a theory of market value of risky assets". Unpublished manuscript, 1961.

and the expected return of the portfolio defined by $\mathbf{x}$. The risk-aversion constant $\gamma > 0$ in the objective determines the tradeoff between risk and return of the portfolio.

Risk Management

Risk is inherent in most economic activities. This is especially true of financial activities where results of decisions made today may have many possible different outcomes depending on future events. Since companies cannot usually insure themselves completely against risk, they have to manage it. This is a hard task even with the support of advanced mathematical techniques. Poor risk management led to several spectacular failures in the financial industry in the 1990s (e.g., Barings Bank, Long Term Capital Management, Orange County). It was also responsible for failures and bailouts of a number of institutions (e.g., Lehman Brothers, Bear Stearns, AIG) during the far more severe global financial crisis of 2007–2008. Regulations, such as those prescribed by the Basel Accord (see Basel Committee on Banking Supervision, 2011), mandate that financial institutions control their risk via a variety of measurable requirements. The modeling of regulatory constraints as well as other risk-related constraints that the firm wishes to impose to prevent vulnerabilities can often be stated as a set of constraints

$$\mathbf{RM}(\mathbf{x}) \leq \mathbf{b}. \tag{1.4}$$

The vector $\mathbf{x}$ in (1.4) represents the holdings in a set of risky securities. The entries of the vector-valued function $\mathbf{RM}(\mathbf{x})$ represent one or more measures of risk and the vector $\mathbf{b}$ represents the acceptable upper limits on these measures. The set of risk management constraints (1.4) may be embedded in a more elaborate model that aims to optimize some kind of performance measure such as expected investment return.

In Chapter 2 we discuss a linear programming model for optimal bank planning under Basel III regulations. In this case the components of the function $\mathbf{RM}(\mathbf{x})$ are linear functions of $\mathbf{x}$. In Chapter 11 we discuss more sophisticated risk measures such as value at risk and conditional value at risk that typically make $\mathbf{RM}(\mathbf{x})$ a nonlinear function of $\mathbf{x}$.

Asset and Liability Management

How should a financial institution manage its assets and liabilities? A static model, such as the Markowitz mean–variance portfolio selection model, fails to incorporate the multi-period nature of typical liabilities faced by financial institutions. Furthermore, it penalizes returns both above and below the mean. A multi-period model that emphasizes the need to meet liabilities in each period for a finite (or possibly infinite) horizon is often more appropriate. Since liabilities and asset returns usually have random components, their optimal management requires techniques to optimize under uncertainty such as stochastic optimization.

We discuss several asset and liability management models in Chapters 3, 16, and 17. A generic asset and liability management model can often be formulated as a stochastic programming problem of the form

$$\max_{\mathbf{x}} \quad \mathbb{E}(U(\mathbf{x}))$$
$$\mathbf{Fx} = \mathbf{L} \qquad\qquad (1.5)$$
$$\mathbf{Dx} \geq \mathbf{0}.$$

The vector $\mathbf{x}$ in (1.5) represents the investment decisions for the available assets at the dates in the planning horizon. The vector $\mathbf{L}$ in (1.5) represents the liabilities that the institution faces at the dates in the planning horizon. The constraints $\mathbf{Fx} = \mathbf{L}, \mathbf{Dx} \geq \mathbf{0}$ represent the cash flow rules and restrictions applicable to the assets during the planning horizon. The term $U(\mathbf{x})$ in the objective function is some appropriate measure of utility. For instance, it could be the value of terminal wealth at the end of the planning horizon. In general, the components $\mathbf{F}, \mathbf{L}, \mathbf{D}$ are discrete-time random processes and thus (1.5) is a multi-stage stochastic programming model with recourse. In Chapter 3 we discuss some special cases of (1.5) with no randomness.

1.4 Notes

George Dantzig was the inventor of linear programming and author of many related articles as well as a classical reference on the subject (Dantzig, 1963). A particularly colorful and entertaining description of the diet problem, a classical linear programming model, can be found in Dantzig (1990).

Boyd and Vandenberghe (2004) give an excellent exposition of convex optimization appropriate for senior or first-year graduate students in engineering. This book is freely available at:

www.stanford.edu/~boyd/cvxbook/

Ragsdale (2007) gives a practical exposition of optimization and related spreadsheet models that circumvent most technical issues. It is appropriate for senior or Master's students in business.

2 Linear Programming: Theory and Algorithms

Linear programming is one of the most significant contributions to computational mathematics made in the twentieth century. This chapter introduces the main ideas behind linear programming theory and algorithms. It also introduces two easy-to-use solvers.

2.1 Linear Programming

A *linear program* is an optimization problem whose objective is to minimize or maximize a linear function subject to a finite set of linear equality and linear inequality constraints. By flipping signs if necessary, a linear program can always be written in the generic form:

$$\min_{\mathbf{x}} \quad \mathbf{c}^{\mathsf{T}}\mathbf{x}$$
$$\text{s.t.} \quad \mathbf{A}\mathbf{x} = \mathbf{b}$$
$$\mathbf{D}\mathbf{x} \geq \mathbf{d}$$

for some vectors and matrices $\mathbf{c} \in \mathbb{R}^n, \mathbf{b} \in \mathbb{R}^m, \mathbf{d} \in \mathbb{R}^p, \mathbf{A} \in \mathbb{R}^{m \times n}, \mathbf{D} \in \mathbb{R}^{p \times n}$. The terms *linear programming model* or *linear optimization model* are also used to refer to a linear program. We will use these terms interchangeably throughout the book.

The following two simplified portfolio construction examples illustrate the use of linear programming as a modeling tool.

Example 2.1 (Fund allocation) *You would like to allocate $80,000 among four mutual funds that have different expected returns as well as different weights in large-, medium- and small-capitalization stocks.*

Capitalization	Fund 1	Fund 2	Fund 3	Fund 4
Large	50%	30%	25%	60%
Medium	30%	10%	40%	20%
Small	20%	60%	35%	20%
Exp. return	10%	15%	16%	8%

The allocation must contain at least 35% large-cap, 30% mid-cap, and 15% small-cap stocks. Find an acceptable allocation with the highest expected return assuming you are only allowed to hold long positions in the funds.

This problem can be formulated as the following linear programming model.

Linear programming model for fund allocation
Variables:

$$x_i: \text{ amount (in \$1000s) invested in fund } i \text{ for } i = 1, \ldots, 4.$$

Objective:

$$\max \quad 0.10x_1 + 0.15x_2 + 0.16x_3 + 0.08x_4.$$

Constraints:

$$
\begin{array}{lll}
0.50x_1 + 0.30x_2 + 0.25x_3 + 0.60x_4 & \geq & 0.35 * 80 \quad \text{(large-cap)} \\
0.30x_1 + 0.10x_2 + 0.40x_3 + 0.20x_4 & \geq & 0.30 * 80 \quad \text{(mid-cap)} \\
0.20x_1 + 0.60x_2 + 0.35x_3 + 0.20x_4 & \geq & 0.15 * 80 \quad \text{(small-cap)} \\
x_1 + x_2 + x_3 + x_4 & = & 80 \qquad\quad \text{(money to allocate)} \\
x_1, \ldots, x_4 & \geq & 0 \qquad\qquad \text{(long-only positions)}.
\end{array}
$$

Example 2.2 (Bond allocation)   A bond portfolio manager has \$100,000 to allocate to two different bonds: a corporate bond and a government bond. These bonds have the following yield, risk level, and maturity:

Bond	Yield	Risk level	Maturity
Corporate	4%	2	3 years
Government	3%	1	4 years

The portfolio manager would like to allocate the funds so that the average risk level of the portfolio is at most 1.5 and the average maturity is at most 3.6 years. Any amount not invested in the bonds will be kept in a cash account that is assumed to generate no interest and does not contribute to the average risk level or maturity. In other words, assume cash has zero yield, zero risk level, and zero maturity.

How should the manager allocate funds to the two bonds to maximize yield? Assume the portfolio can only include long positions.

This problem can be formulated as the following linear programming model.

Linear programming model for bond allocation
Variables:

$$x_1, x_2: \text{ amounts (in \$1000s) invested in the corporate and government}$$
$$\text{bonds respectively.}$$

Objective:

$$\max \quad 4x_1 + 3x_2.$$

Constraints:

$$
\begin{aligned}
x_1 + x_2 &\leq 100 \quad \text{(total funds)} \\
\frac{2x_1 + x_2}{100} &\leq 1.5 \quad \text{(risk level)} \\
\frac{3x_1 + 4x_2}{100} &\leq 3.6 \quad \text{(maturity)} \\
x_1, x_2 &\geq 0 \quad\;\; \text{(long-only positions)}
\end{aligned}
$$

or equivalently

$$
\begin{aligned}
\max\quad & 4x_1 + 3x_2 \\
\text{s.t.}\quad &
\end{aligned}
$$

$$
\begin{aligned}
x_1 + x_2 &\leq 100 \quad \text{(total funds)} \\
2x_1 + x_2 &\leq 150 \quad \text{(risk level)} \\
3x_1 + 4x_2 &\leq 360 \quad \text{(maturity)} \\
x_1, x_2 &\geq 0 \quad\;\; \text{(long-only positions).}
\end{aligned}
$$

The linear programming model in Example 2.1 can be written more concisely using matrix–vector notation as follows:

$$
\begin{aligned}
\max\quad & \mathbf{r}^\mathsf{T}\mathbf{x} \\
\text{s.t.}\quad & \mathbf{A}\mathbf{x} = \mathbf{b} \\
& \mathbf{D}\mathbf{x} \geq \mathbf{d} \\
& \mathbf{x} \geq \mathbf{0},
\end{aligned}
$$

where $\mathbf{r} = \begin{bmatrix} 0.10 \\ 0.15 \\ 0.16 \\ 0.08 \end{bmatrix}$, $\mathbf{A} = \begin{bmatrix} 1 & 1 & 1 & 1 \end{bmatrix}$, $\mathbf{b} = 80$, $\mathbf{D} = \begin{bmatrix} 0.5 & 0.3 & 0.25 & 0.6 \\ 0.3 & 0.1 & 0.4 & 0.2 \\ 0.2 & 0.6 & 0.35 & 0.2 \end{bmatrix}$, and

$\mathbf{d} = \begin{bmatrix} 28 \\ 24 \\ 12 \end{bmatrix}$.

Likewise, the linear programming model in Example 2.2 can be written as

$$
\begin{aligned}
\max\quad & \mathbf{r}^\mathsf{T}\mathbf{x} \\
\text{s.t.}\quad & \mathbf{A}\mathbf{x} \leq \mathbf{b} \\
& \mathbf{x} \geq \mathbf{0},
\end{aligned}
$$

for $\mathbf{r} = \begin{bmatrix} 4 \\ 3 \end{bmatrix}$, $\mathbf{A} = \begin{bmatrix} 1 & 1 \\ 2 & 1 \\ 3 & 4 \end{bmatrix}$, and $\mathbf{b} = \begin{bmatrix} 100 \\ 150 \\ 360 \end{bmatrix}$.

A linear programming model is in *standard form* if it is written as follows:

$$
\begin{aligned}
\min\quad & \mathbf{c}^\mathsf{T}\mathbf{x} \\
\text{s.t.}\quad & \mathbf{A}\mathbf{x} = \mathbf{b} \\
& \mathbf{x} \geq \mathbf{0}.
\end{aligned}
$$

The standard form is a kind of formatting convention that is used by some solvers. It is also particularly convenient to describe the most popular algorithms for solving linear programming, namely the simplex and interior-point methods.

The standard form is not restrictive. Any linear program can be rewritten in standard form. In particular, inequality constraints (other than non-negativity) can be rewritten as equality constraints after the introduction of a so-called *slack* or *surplus* variable. For instance, the linear program from Example 2.2 can be written as

$$
\begin{array}{lll}
\max & 4x_1 + 3x_2 & \\
\text{s.t.} & & \\
& x_1 + x_2 + x_3 & = 100 \\
& 2x_1 + x_2 + x_4 & = 150 \\
& 3x_1 + 4x_2 + x_5 & = 360 \\
& x_1, x_2, x_3, x_4, x_5 & \geq 0.
\end{array}
$$

More generally, a linear program of the form

$$
\begin{array}{ll}
\min & \mathbf{c}^\mathsf{T}\mathbf{x} \\
\text{s.t.} & \mathbf{A}\mathbf{x} \leq \mathbf{b} \\
& \mathbf{x} \geq \mathbf{0}
\end{array}
$$

can be rewritten as

$$
\begin{array}{ll}
\min & \mathbf{c}^\mathsf{T}\mathbf{x} \\
\text{s.t.} & \mathbf{A}\mathbf{x} + \mathbf{s} = \mathbf{b} \\
& \mathbf{x}, \mathbf{s} \geq \mathbf{0}.
\end{array}
$$

It can then be rewritten, using matrix notation, in the following standard form:

$$
\begin{array}{ll}
\min & \begin{bmatrix} \mathbf{c} \\ \mathbf{0} \end{bmatrix}^\mathsf{T} \begin{bmatrix} \mathbf{x} \\ \mathbf{s} \end{bmatrix} \\
\text{s.t.} & \begin{bmatrix} \mathbf{A} & \mathbf{I} \end{bmatrix} \begin{bmatrix} \mathbf{x} \\ \mathbf{s} \end{bmatrix} = \mathbf{b} \\
& \begin{bmatrix} \mathbf{x} \\ \mathbf{s} \end{bmatrix} \geq \mathbf{0}.
\end{array}
$$

Unrestricted variables can be expressed as the difference of two new non-negative variables. For example, consider the linear program

$$
\begin{array}{ll}
\min & \mathbf{c}^\mathsf{T}\mathbf{x} \\
\text{s.t.} & \mathbf{A}\mathbf{x} \leq \mathbf{b}.
\end{array}
$$

The unrestricted variable $\mathbf{x}$ can be replaced by $\mathbf{u} - \mathbf{v}$ where $\mathbf{u}, \mathbf{v} \geq \mathbf{0}$. Hence the above linear program can be rewritten as

$$
\begin{array}{ll}
\min & \mathbf{c}^\mathsf{T}(\mathbf{u} - \mathbf{v}) \\
\text{s.t.} & \mathbf{A}(\mathbf{u} - \mathbf{v}) \leq \mathbf{b} \\
& \mathbf{u}, \mathbf{v} \geq \mathbf{0}.
\end{array}
$$

It can also be rewritten, after adding slack variables and using matrix notation, in the following standard form:

$$\min \quad \begin{bmatrix} \mathbf{c} \\ -\mathbf{c} \\ \mathbf{0} \end{bmatrix}^{\mathsf{T}} \begin{bmatrix} \mathbf{u} \\ \mathbf{v} \\ \mathbf{s} \end{bmatrix}$$

$$\text{s.t.} \quad \begin{bmatrix} \mathbf{A} & -\mathbf{A} & \mathbf{I} \end{bmatrix} \begin{bmatrix} \mathbf{u} \\ \mathbf{v} \\ \mathbf{s} \end{bmatrix} = \mathbf{b}$$

$$\begin{bmatrix} \mathbf{u} \\ \mathbf{v} \\ \mathbf{s} \end{bmatrix} \geq \mathbf{0}.$$

2.2 Graphical Interpretation of a Two-Variable Example

Banks need to consider regulations when determining their business strategy. In this section, we consider the Basel III regulations (Basel Committee on Banking Supervision, 2011). We present a simplified example following the paper of Pokutta and Schmaltz (2012). Consider a bank with total deposits D and loans L. The loans may default and the deposits are exposed to early withdrawal. The bank holds capital C in order to buffer against possible default losses on the loans, and it holds a liquidity reserve R to buffer against early withdrawals on the deposits. The balance sheet of the bank satisfies $L + R = D + C$. Normalizing the total assets to 1, we have $R = 1 - L$ and $C = 1 - D$. Basel III regulations require banks to satisfy four minimum ratio constraints in order to buffer against different types of risk:

Capital ratio: $\dfrac{C}{L} \geq r_1$

Leverage ratio: $C \geq r_2$

Liquidity coverage ratio: $\dfrac{R}{D} \geq r_3$

Net stable funding ratio: $\dfrac{\alpha D + C}{L} \geq r_4,$

where the ratios $r_1, r_2, r_3, r_4, \alpha$ are computed for each bank based on the riskiness of its loans and the likelihood of early withdrawals on deposits. To illustrate, consider a bank with $r_1 = 0.3$, $r_2 = 0.1$, $r_3 = 0.25$, $r_4 = 0.7$, $\alpha = 0.3$. Expressing the four ratio constraints in terms of the variables D and L, we get

$$D + 0.3L \leq 1$$
$$D \leq 0.9$$
$$0.25D + L \leq 1$$
$$0.7D + 0.7L \leq 1.$$

Figure 2.1 displays a plot of the feasible region of this system of inequalities in the plane (D, L).

Given this feasible region, the objective of the bank is to maximize the margin income $m_D D + m_L L$ that it makes on its products; where m_D is the margin that

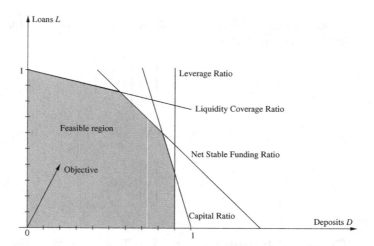

Figure 2.1 Basel III regulations

the bank makes on its deposits and m_L is the margin charged on its loans. For example, if $m_D = 0.02$ and $m_L = 0.03$, the best solution that satisfies all the constraints corresponds to the vertex $D = 0.571, L = 0.857$ on the boundary of the feasible region, at the intersection of the lines $0.25D + L = 1$ and $0.7D + 0.7L = 1$. This means that the bank should have 57.1% of its liabilities in deposits and 42.9% in capital, and it should have 85.7% of its assets in loans and the remaining 14.3% in liquidity reserve. The fact that an optimal solution occurs at a vertex of the feasible region is a property of linear programs that extends to higher dimensions than 2: To find an optimal solution of a linear program, it suffices to restrict the search to vertices of the feasible region. This geometric insight is the basis of the simplex method, which goes from one vertex of the feasible region to an adjacent one with a better objective value until it reaches an optimum. An algebraic description of the simplex method that can be coded on a computer is presented in Section 2.7.1.

2.3   Numerical Linear Programming Solvers

There are a variety of both commercial and open-source software packages for linear programming. Most of these packages implement the algorithms described in Section 2.7 below. Next we illustrate two of these solvers by applying them to Example 2.1.

Excel Solver

Figure 2.2 displays a printout of an Excel spreadsheet implementation of the linear programming model for Example 2.1 as well as the dialog box obtained when we run the Excel add-in `Solver`. The spreadsheet model contains the three components of the linear program. The decision variables are in the range B4:E4. The objective is in cell F3. The left- and right-hand sides of the equality constraint are in the cells F4 and H4 respectively. Likewise, the left- and right-hand sides of the three inequality constraints are in the ranges F8:F10 and H8:H10 respectively. These components are specified in the Solver dialog box. In addition, the Solver options are used to indicate that this is a linear model and that the variables are non-negative.

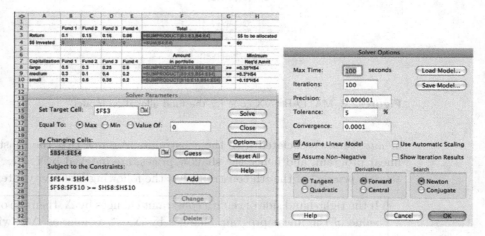

Figure 2.2 Spreadsheet implementation and the Solver dialog box for the fund allocation model

MATLAB CVX

Figure 2.3 displays a CVX script for the same problem. The script can be run provided the freely available CVX toolbox is installed.

Either Excel Solver or MATLAB CVX find the following optimal solution to the problem in Example 2.1:

$$\mathbf{x}^* = \begin{bmatrix} 0.0000 \\ 12.6316 \\ 46.3158 \\ 21.0526 \end{bmatrix},$$

and the corresponding optimal objective value 10.9895 (recall that the units are in $1000s).

2.4 Sensitivity Analysis

In addition to the optimal solution, the process of solving a linear program also generates some interesting *sensitivity information* via the so-called *shadow prices*

```
File  Edit  Text  Go  Cell  Tools  Debug  Desktop  Window

 1       % Matlab CVX code for the fund allocation problem
 2 -       n = 4 ;
 3 -       r = [0.10;0.15;0.16;0.08] ;
 4 -       A = [1 1 1 1] ;
 5 -       b = 80 ;
 6 -       D = [0.5 0.3 0.25 0.6;
 7            0.3 0.1 0.4  0.2;
 8            0.2 0.6 0.35 0.2] ;
 9 -       d = [28;24;12] ;
10
11 -      cvx_begin
12 -           variables x(n) ;
13 -           maximize (r'*x)
14 -           A*x == b ;
15 -           D*x >= d ;
16 -           x >= zeros(n,1) ;
17 -      cvx_end
18
```

Figure 2.3 MATLAB CVX code for the fund allocation model

or *dual values* associated with the constraints. Assume that the constraints of a linear program, and hence the shadow prices, are indexed by $i = 1, \ldots, m$. The *shadow price* y_i^* of the ith constraint has the following sensitivity interpretation:

> If the right-hand side of the ith constraint changes by Δ, then the optimal value of the linear program changes by $\Delta \cdot y_i^*$ as long as Δ is within a certain range.

Both Excel **Solver** and MATLAB **CVX** compute the shadow prices implicitly. To make this information explicit in Excel **Solver** we request a sensitivity report after running it as shown in Figure 2.4.

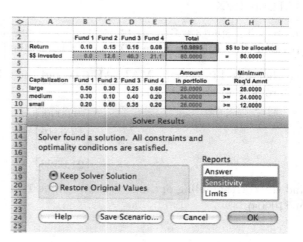

Figure 2.4 Requesting sensitivity report in Solver

Figure 2.5 displays the sensitivity report for Example 2.1.

Adjustable Cells

Cell	Name	Final Value	Reduced Cost	Objective Coefficient	Allowable Increase	Allowable Decrease
B4	$$ invested Fund 1	0	-0.00263158	0.1	0.00263158	1.00E+30
C4	$$ invested Fund 2	12.63157895	0	0.15	0.0166667	0.00142857
D4	$$ invested Fund 3	46.31578947	0	0.16	0.00166667	0.00625
E4	$$ invested Fund 4	21.05263158	0	0.08	0.01	0.00357143

Constraints

Cell	Name	Final Value	Shadow Price	Constraint R.H. Side	Allowable Increase	Allowable Decrease
F4	$$ invested Total	80.0000	0.22	80	21.0526	6.31579
F8	large in portfolio	28.0000	-0.231579	28	6	6.66667
F9	medium in portfolio	24.0000	-0.00526316	24	3.42857	14.6667
F10	small in portfolio	28.0000	0	12	16	1.00E+30

Figure 2.5 Sensitivity report

The values y_i^* can be found in the column labeled "Shadow Price". In addition, the "Allowable Increase" and "Allowable Decrease" columns indicate the range of change for each right-hand side of a constraint where the sensitivity analysis holds. For example, if the right-hand side of the large-capitalization constraint

$$0.5x_1 + 0.3x_2 + 0.25x_3 + 0.6x_4 \geq 28$$

changes from 28 to $28 + \Delta$, then the optimal value changes by $-0.231579 \cdot \Delta$. This holds provided Δ is within the allowable range $[-6.6666, 6]$. If the requirement on large-cap stocks is reduced from 35% to 30%, the change in right-hand side is $\Delta = -0.05 * 80 = -4$, which is within the allowable range. Therefore the optimal objective value increases by $-0.231579 \cdot (-4) = 0.926316$. Because our units are in $1000, this means that the expected return on an optimal portfolio would increase by $926.32 if we relaxed the constraint on large-cap stocks by 5%, from 35% to 30%.

The shadow prices of the non-negativity constraints are the "Reduced Cost" displayed in the initial part of the sensitivity report. This is also the convention for more general lower and upper bounds on the decision variables. Observe that in Example 2.1 the reduced costs of the non-zero variables are zero. The reduced costs also have a deeper meaning in the context of the simplex algorithm for linear programming as described in Section 2.7.1 below.

A linear programming model is *non-degenerate* if all of the allowable increase and allowable decrease limits are positive. The above linear programming model is non-degenerate.

In CVX this information can also be obtained by including a few additional pieces of code to save the dual information in the dual variables y,z as shown in Figure 2.6.

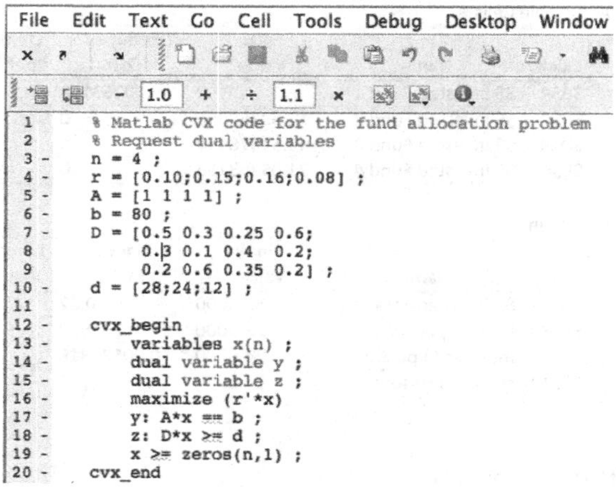

Figure 2.6 MATLAB CVX code with dual variables

Both solvers yield the following dual values: $\mathbf{y}^* = 0.22$, $\mathbf{z}^* = \begin{bmatrix} -0.231579 \\ -0.005263 \\ 0 \end{bmatrix}$.

We note that some solvers may flip the sign of the dual values. In particular, the output of the above CVX code yields the values -0.22 and $\begin{bmatrix} 0.231579 \\ 0.005263 \\ 0 \end{bmatrix}$. It is important to be mindful of this subtlety when interpreting the dual information. The ambiguity can be easily resolved by thinking in terms of sensitivity analysis. In this particular example, it is clear that the shadow price of the first constraint should be non-negative as more capital should lead to a higher return. Likewise, it is clear that the shadow prices of the other constraints should be non-positive as more stringent diversification constraints, e.g., higher percentage in large cap, reduces the set of feasible portfolios and hence can only lead to portfolios with return less than or equal to the optimal return of the original problem.

2.5   *Duality

Every linear program has an associated *dual* linear programming problem. The properties of these two linear programs and how they are related to each other have deep implications. In particular, duality enables us to answer the following kinds of questions:

- Can we recognize an optimal solution?
- Can we construct an algorithm to find an optimal solution?
- Can we assess how suboptimal a current feasible solution is?

The attentive reader may have noticed that dual variables were already mentioned in Section 2.4 when discussing sensitivity analysis with CVX. This is not a coincidence. There is a close connection between duality and sensitivity analysis. The vector of shadow prices of the constraints of a linear program corresponds precisely to the optimal solution of its dual.

Consider the following linear program in standard form, which we shall refer to as the *primal* problem:

$$\begin{aligned} \min \quad & \mathbf{c}^\mathsf{T}\mathbf{x} \\ \text{s.t.} \quad & \mathbf{Ax} = \mathbf{b} \\ & \mathbf{x} \ge \mathbf{0}. \end{aligned} \tag{2.1}$$

The following linear program is called the *dual* problem:

$$\begin{aligned} \max \quad & \mathbf{b}^\mathsf{T}\mathbf{y} \\ \text{s.t.} \quad & \mathbf{A}^\mathsf{T}\mathbf{y} \le \mathbf{c}. \end{aligned} \tag{2.2}$$

Sometimes it is convenient to rewrite the constraints in the dual problem as equality constraints by means of slack variables. That is, problem (2.2) can also be written as

$$\begin{aligned} \max \quad & \mathbf{b}^\mathsf{T}\mathbf{y} \\ \text{s.t.} \quad & \mathbf{A}^\mathsf{T}\mathbf{y} + \mathbf{s} = \mathbf{c} \\ & \mathbf{s} \ge \mathbf{0}. \end{aligned} \tag{2.3}$$

There is a deep connection between the primal and dual problems. The next result follows by construction.

Theorem 2.3 (Weak duality) *Assume* $\mathbf{x}$ *is a feasible point for* (2.1) *and* $\mathbf{y}$ *is a feasible point for* (2.2). *Then*

$$\mathbf{b}^\mathsf{T}\mathbf{y} \le \mathbf{c}^\mathsf{T}\mathbf{x}.$$

Proof Under the assumptions on $\mathbf{x}$ and $\mathbf{y}$ it follows that

$$\mathbf{b}^\mathsf{T}\mathbf{y} = (\mathbf{Ax})^\mathsf{T}\mathbf{y} = (\mathbf{A}^\mathsf{T}\mathbf{y})^\mathsf{T}\mathbf{x} \le \mathbf{c}^\mathsf{T}\mathbf{x}. \qquad \square$$

The following (not so straightforward) result also holds.

Theorem 2.4 (Strong duality) *Assume one of the problems* (2.1) *or* (2.2) *is feasible. Then this problem is bounded if and only if the other one is feasible. In that case both problems have optimal solutions and their optimal values are the same.*

We refer the reader to Bertsimas and Tsitsiklis (1997) or Chvátal (1983) for a proof of Theorem 2.4. This result is closely related to the following classical properties of linear inequality systems.

Theorem 2.5 *Assume* $\mathbf{A} \in \mathbb{R}^{m\times n}$ *and* $\mathbf{b} \in \mathbb{R}^m$. *In each of the following cases exactly one of the systems* (I) *or* (II) *has a solution but not both.*

(a) *Farkas's lemma*

$$\mathbf{Ax} = \mathbf{b}, \ \mathbf{x} \ge \mathbf{0}, \tag{I}$$

$$\mathbf{A}^\mathsf{T}\mathbf{y} \le \mathbf{0}, \ \mathbf{b}^\mathsf{T}\mathbf{y} < 0. \tag{II}$$

(b) *Gordan's theorem*

$$\mathbf{Ax} = \mathbf{0}, \ \mathbf{x} \gneq \mathbf{0}, \tag{I}$$

$$\mathbf{A}^\mathsf{T}\mathbf{y} > \mathbf{0}. \tag{II}$$

(c) *Stiemke's theorem*

$$\mathbf{Ax} = \mathbf{0}, \ \mathbf{x} > \mathbf{0}, \tag{I}$$

$$\mathbf{A}^\mathsf{T}\mathbf{y} \gneq \mathbf{0}. \tag{II}$$

The equivalence between Theorems 2.4 and 2.5 is explored in Exercises 2.11 and 2.12.

We next present a derivation of the dual problem via the so-called Lagrangian function. This derivation has the advantage of introducing an important concept that we will encounter again in later chapters. Associated with the optimization problem (2.1) consider the *Lagrangian* function defined by

$$L(\mathbf{x}, \mathbf{y}, \mathbf{s}) := \mathbf{c}^\mathsf{T}\mathbf{x} + \mathbf{y}^\mathsf{T}(\mathbf{b} - \mathbf{Ax}) - \mathbf{s}^\mathsf{T}\mathbf{x}.$$

The constraints of (2.1) can be encoded using the Lagrangian function via the following observation: For a given vector $\mathbf{x}$

$$\max_{\substack{\mathbf{y}, \mathbf{s} \\ \mathbf{s} \geq 0}} L(\mathbf{x}, \mathbf{y}, \mathbf{s}) = \begin{cases} \mathbf{c}^\mathsf{T}\mathbf{x} & \text{if } \mathbf{Ax} = \mathbf{b} \text{ and } \mathbf{x} \geq 0 \\ +\infty & \text{otherwise.} \end{cases}$$

Therefore the primal problem (2.1) can be written as

$$\min_{\mathbf{x}} \max_{\substack{\mathbf{y}, \mathbf{s} \\ \mathbf{s} \geq 0}} L(\mathbf{x}, \mathbf{y}, \mathbf{s}). \tag{2.4}$$

On the other hand, observe that $L(\mathbf{x}, \mathbf{y}, \mathbf{s}) = \mathbf{b}^\mathsf{T}\mathbf{y} + (\mathbf{c} - \mathbf{A}^\mathsf{T}\mathbf{y} - \mathbf{s})^\mathsf{T}\mathbf{x}$. Hence for a given pair of vectors $(\mathbf{y}, \mathbf{s})$

$$\min_{\mathbf{x}} L(\mathbf{x}, \mathbf{y}, \mathbf{s}) = \begin{cases} \mathbf{b}^\mathsf{T}\mathbf{y} & \text{if } \mathbf{A}^\mathsf{T}\mathbf{y} + \mathbf{s} = \mathbf{c} \\ -\infty & \text{otherwise.} \end{cases}$$

The dual problem is obtained by flipping the order of the min and max operations in (2.4). Indeed, observe that the dual problem (2.3) can be written as

$$\max_{\substack{\mathbf{y}, \mathbf{s} \\ \mathbf{s} \geq 0}} \min_{\mathbf{x}} L(\mathbf{x}, \mathbf{y}, \mathbf{s}).$$

A similar procedure can be applied to obtain the dual of a linear program that is not necessarily in standard form. For example, the primal problem

$$\begin{aligned} \min \quad & \mathbf{c}^\mathsf{T}\mathbf{x} \\ \text{s.t.} \quad & \mathbf{Ax} \geq \mathbf{b} \\ & \mathbf{x} \geq \mathbf{0} \end{aligned} \tag{2.5}$$

can be written as

$$\min_{\mathbf{x}} \max_{\mathbf{y} \geq 0, \, \mathbf{s} \geq 0} L(\mathbf{x}, \mathbf{y}, \mathbf{s}),$$

for $L(\mathbf{x}, \mathbf{y}, \mathbf{s}) = \mathbf{c}^T\mathbf{x} + \mathbf{y}^T(\mathbf{b} - \mathbf{A}\mathbf{x}) - \mathbf{s}^T\mathbf{x}$. In this case the dual problem is

$$\max_{\mathbf{y}\geq 0,\, \mathbf{s}\geq 0}\ \min_{\mathbf{x}} L(\mathbf{x}, \mathbf{y}, \mathbf{s}),$$

and can be rewritten as

$$\begin{aligned}
\max_{\mathbf{y}}\quad & \mathbf{b}^T\mathbf{y} \\
\text{s.t.}\quad & \mathbf{A}^T\mathbf{y} \leq \mathbf{c} \\
& \mathbf{y} \geq \mathbf{0}.
\end{aligned} \tag{2.6}$$

Again the weak and strong duality properties hold for the pair of problems (2.5) and (2.6).

Consider the linear programming model of Example 2.1, namely

$$\begin{aligned}
\max_{\mathbf{x}}\quad & \mathbf{r}^T\mathbf{x} \\
\text{s.t.}\quad & \mathbf{A}\mathbf{x} = \mathbf{b} \\
& \mathbf{D}\mathbf{x} \geq \mathbf{d} \\
& \mathbf{x} \geq \mathbf{0}.
\end{aligned} \tag{2.7}$$

We give a derivation for its dual. Observe that (2.7) can be recast as

$$\max_{\mathbf{x}}\ \min_{\substack{\mathbf{y}, \mathbf{w}, \mathbf{s} \\ \mathbf{w}\geq 0,\, \mathbf{s}\geq 0}}\ L(\mathbf{x}, \mathbf{y}, \mathbf{w}, \mathbf{s})$$

for

$$\begin{aligned}
L(\mathbf{x}, \mathbf{y}, \mathbf{w}, \mathbf{s}) &= \mathbf{r}^T\mathbf{x} + \mathbf{y}^T(\mathbf{b} - \mathbf{A}\mathbf{x}) + \mathbf{w}^T(\mathbf{D}\mathbf{x} - \mathbf{d}) + \mathbf{s}^T\mathbf{x} \\
&= \mathbf{b}^T\mathbf{y} - \mathbf{d}^T\mathbf{w} + \mathbf{x}^T(\mathbf{r} - \mathbf{A}^T\mathbf{y} + \mathbf{D}^T\mathbf{w} + \mathbf{s}).
\end{aligned}$$

It follows that its dual $\displaystyle \min_{\substack{\mathbf{y}, \mathbf{w}, \mathbf{s} \\ \mathbf{w}\geq 0,\, \mathbf{s}\geq 0}}\ \max_{\mathbf{x}} L(\mathbf{x}, \mathbf{y}, \mathbf{s}, \mathbf{z})$ can be rewritten as

$$\begin{aligned}
\min_{\mathbf{y}, \mathbf{z}}\quad & \mathbf{b}^T\mathbf{y} - \mathbf{d}^T\mathbf{w} \\
\text{s.t.}\quad & \mathbf{A}^T\mathbf{y} - \mathbf{D}^T\mathbf{w} \geq \mathbf{r} \\
& \mathbf{w} \geq \mathbf{0}.
\end{aligned} \tag{2.8}$$

An alternative way to obtain the dual (2.8) is to rewrite (2.7) in standard form and derive its standard dual. The latter turns out to be equivalent to (2.8). (See Exercise 2.6.)

2.6   *Optimality Conditions

Consider again the linear programming problem (2.1). A powerful consequence of Theorem 2.4 is a set of *optimality conditions* that completely characterize the solutions to both (2.1) and (2.2).

Theorem 2.6 (Optimality conditions)   *The vectors* $\mathbf{x} \in \mathbb{R}^n$ *and* $(\mathbf{y}, \mathbf{s}) \in \mathbb{R}^m \times \mathbb{R}^n$ *are respectively optimal solutions to* (2.1) *and* (2.3) *if and only if they satisfy the following system of equations and inequalities:*

$$
\begin{aligned}
\mathbf{A}^\mathsf{T}\mathbf{y} + \mathbf{s} &= \mathbf{c} \\
\mathbf{A}\mathbf{x} &= \mathbf{b} \\
\mathbf{x}, \mathbf{s} &\geq \mathbf{0} \\
x_i s_i &= 0, \quad i = 1, \ldots, n.
\end{aligned}
\tag{2.9}
$$

The equations $x_i s_i = 0$ are known as the complementary slackness conditions. They imply that, if a dual constraint $(\mathbf{A}^\mathsf{T}\mathbf{y})_i \leq c_i$ holds strictly (that is, $(\mathbf{A}^\mathsf{T}\mathbf{y})_i < c_i$), then the corresponding primal variable x_i must be 0. And conversely, if $x_i > 0$, the corresponding dual constraint is tight, that is, $(\mathbf{A}^\mathsf{T}\mathbf{y})_i = c_i$.

The optimality conditions (2.9) provide an avenue for constructing algorithms to solve the linear programming problems (2.1) and (2.3). To lay the groundwork for discussing them, we next state two interesting results concerning the optimal solutions of a linear programming problem and its dual.

Theorem 2.7 (Strictly complementary solutions)   *Assume* $\mathbf{A} \in \mathbb{R}^{m \times n}$ *is full row rank,* $\mathbf{b} \in \mathbb{R}^m$, *and* $\mathbf{c} \in \mathbb{R}^n$ *are such that both* (2.1) *and* (2.2) *are feasible. Then there exist optimal solutions* $\mathbf{x}^*$ *to* (2.1) *and* $(\mathbf{y}^*, \mathbf{s}^*)$ *to* (2.3) *such that*

$$
\mathbf{x}^* + \mathbf{s}^* > \mathbf{0}.
$$

For a matrix $\mathbf{A}$ and a subset B of its columns, let $\mathbf{A}_B$ denote the submatrix of $\mathbf{A}$ containing the columns in B. For a square non-singular matrix $\mathbf{D}$, the notation $\mathbf{D}^{-\mathsf{T}}$ stands for $(\mathbf{D}^{-1})^\mathsf{T}$.

Theorem 2.8 (Optimal basic feasible solutions)   *Assume* $\mathbf{A} \in \mathbb{R}^{m \times n}$ *is full row rank,* $\mathbf{b} \in \mathbb{R}^m$, *and* $\mathbf{c} \in \mathbb{R}^n$ *are such that both* (2.1) *and* (2.2) *are feasible. Then there exists a partition* $B \cup N = \{1, \ldots, n\}$ *with* $|B| = m$ *and* $\mathbf{A}_B$ *non-singular, such that*

$$
\mathbf{x}_B^* = \mathbf{A}_B^{-1}\mathbf{b}, \quad \mathbf{x}_N^* = \mathbf{0}, \quad \mathbf{y}^* = \mathbf{A}_B^{-\mathsf{T}}\mathbf{c}_B
$$

are optimal solutions to (2.1) *and* (2.2) *respectively.*

2.7   *Algorithms for Linear Programming

We next sketch the two main algorithmic schemes for solving linear programs, namely the *simplex method* and *interior-point methods*. Our discussion of these two important topics is only intended to give the reader a basic understanding of the main solution techniques for linear programming. For a more detailed and thorough discussion of these two classes of algorithms, see Bertsimas and Tsitsiklis (1997), Boyd and Vandenberghe (2004), Chvátal (1983), Renegar (2001), and Ye (1997).

We follow the usual convention of assuming that the problem of interest is in standard form as in (2.1) and (2.3) and A has full row rank.

2.7.1   The Simplex Method

One of the most popular algorithms for linear programming is the *simplex method*. It generates a sequence of iterates that satisfy $\mathbf{A}\mathbf{x} = \mathbf{b}$, $\mathbf{x} \geq \mathbf{0}$, $\mathbf{A}^{\mathsf{T}}\mathbf{y} + \mathbf{s} = \mathbf{c}$ and $x_i s_i = 0$, with $i = 1,\ldots,n$. Each iteration of the algorithm aims to make progress towards satisfying $\mathbf{s} \geq \mathbf{0}$. Theorem 2.6 guarantees that the algorithm terminates with an optimal solution when this goal is attained. The *dual* simplex method is a variant that generates a sequence of iterates satisfying $\mathbf{A}\mathbf{x} = \mathbf{b}$, $\mathbf{A}^{\mathsf{T}}\mathbf{y} + \mathbf{s} = \mathbf{c}$, $\mathbf{s} \geq \mathbf{0}$, and $x_i s_i = 0$, for $i = 1,\ldots,n$. Each iteration of the algorithm aims to make progress towards satisfying $\mathbf{x} \geq \mathbf{0}$.

The simplex method relies on the property stated in Theorem 2.8. The gist of the method is to search for an *optimal basis*; that is, a subset $B \subseteq \{1,\ldots,n\}$ as in Theorem 2.8. To motivate and describe the algorithm we next introduce some terminology and key observations.

A *basis* is a subset $B \subseteq \{1,\ldots,n\}$ such that $|B| = m$ and $\mathbf{A}_B$ is a non-singular matrix. A basis B defines the *basic solution* $\bar{\mathbf{x}} = (\bar{\mathbf{x}}_B, \bar{\mathbf{x}}_N)$ where $\bar{\mathbf{x}}_B = \mathbf{A}_B^{-1}\mathbf{b}$, $\bar{\mathbf{x}}_N = \mathbf{0}$. Observe that $\bar{\mathbf{x}}$ solves the system of equations $\mathbf{A}\mathbf{x} = \mathbf{b}$. The vector $\bar{\mathbf{x}}$ is a *basic feasible solution* if in addition $\bar{\mathbf{x}} \geq 0$. A basis B also defines the *reduced cost* $\bar{\mathbf{c}} = \mathbf{c} - \mathbf{A}^{\mathsf{T}}\mathbf{A}_B^{-\mathsf{T}}\mathbf{c}_B$. The following fact suggests the main idea for the simplex method.

Proposition 2.9   *Assume $B \subseteq \{1,\ldots,n\}$ is a basis. Let $\bar{\mathbf{x}}$ and $\bar{\mathbf{c}}$ be respectively the corresponding basic solution and reduced cost vector. If $\bar{\mathbf{x}} \geq 0$ and $\bar{\mathbf{c}} \geq 0$ then $\bar{\mathbf{x}}$ is an optimal solution to (2.1). Furthermore, in this case $\bar{\mathbf{y}} = \mathbf{A}_B^{-\mathsf{T}}\mathbf{c}_B$ is an optimal solution to (2.2).*

An *optimal basis* is a basis that satisfies the conditions $\bar{\mathbf{x}} \geq 0$ and $\bar{\mathbf{c}} \geq 0$ stated above. Given a basis B that is not optimal, the main idea of the simplex method is to generate a better basis by replacing an index from B. To that end, a possible avenue is as follows. Suppose B is a basis with a basic feasible solution $\bar{\mathbf{x}}$. If B is not an optimal basis, then $\bar{c}_j < 0$ for some $j \notin B$. Thus for $\alpha > 0$ the point $\mathbf{x}(\alpha)$ defined by

$$\mathbf{x}_B(\alpha) = \bar{\mathbf{x}}_B - \alpha\mathbf{A}_B^{-1}\mathbf{A}_j,$$

$$x_j(\alpha) = \alpha, \quad x_i(\alpha) = 0 \ \text{ for all other indices } i \notin B \cup \{j\}$$

satisfies

$$\mathbf{c}^{\mathsf{T}}\mathbf{x}(\alpha) = \mathbf{c}^{\mathsf{T}}\bar{\mathbf{x}} + \alpha\bar{c}_j < \mathbf{c}^{\mathsf{T}}\bar{\mathbf{x}}.$$

Hence we can get a point with better (lower) objective value than the current basic feasible solution $\bar{x}$. We would like this new point to remain feasible. Unless the problem is unbounded, there is a length $\alpha^* \geq 0$ that makes one of the current basic components ℓ of $\mathbf{x}$ drop to zero while keeping all of them non-negative. When $\alpha^* > 0$, a basis with a better basic feasible solution can be obtained by replacing ℓ with j. The simplex method modifies the basis in this way, even in the degenerate case when $\alpha^* = 0$, which may occur in some iterations. Algorithm 2.1 gives a formal description of the simplex method.

Algorithm 2.1 The simplex method

1: start with a basis $B \subseteq \{1, \ldots, n\}$ such that $\bar{\mathbf{x}}$ is a basic feasible solution
2: **while** $\bar{\mathbf{c}} \not\geq \mathbf{0}$ **do**
3:     choose an index j such that $\bar{c}_j < 0$
4:     compute $\mathbf{u} = \mathbf{A}_B^{-1}\mathbf{A}_j$
5:     **if** $\mathbf{u} \leq \mathbf{0}$ **then** HALT; the problem is unbounded **end if**
6:     let $\alpha^* := \min\limits_{i:u_i>0} \dfrac{\bar{x}_i}{u_i} = \dfrac{\bar{x}_\ell}{u_\ell}$
7:     form a new basis by replacing ℓ with j
8:     update the basic feasible solution by replacing $\bar{\mathbf{x}}$ with $\mathbf{x}(\alpha^*)$
9: **end while**

Observe that the basic feasible solution $\bar{\mathbf{x}}$ and the reduced cost $\bar{\mathbf{c}}$ corresponding to a basis B satisfy $\bar{\mathbf{x}}_N = \mathbf{0}$ and $\bar{\mathbf{c}}_B = \mathbf{0}$ where $N = \{1, \ldots, n\} \setminus B$. Hence the simplex method only needs to keep track of $\bar{\mathbf{x}}_B$ and $\bar{\mathbf{c}}_N$. We next illustrate the simplex method in the linear programming model from Example 2.2. If we start with the initial basis $B = \{3, 4, 5\}$ the algorithm proceeds as follows.

Iteration 1: $B = \{3, 4, 5\}$, $\bar{\mathbf{x}}_B = \begin{bmatrix} \bar{x}_3 & \bar{x}_4 & \bar{x}_5 \end{bmatrix}^\mathsf{T} = \begin{bmatrix} 100 & 150 & 360 \end{bmatrix}^\mathsf{T}$, $\bar{\mathbf{c}}_N = \begin{bmatrix} \bar{c}_1 & \bar{c}_2 \end{bmatrix}^\mathsf{T} = \begin{bmatrix} -4 & -3 \end{bmatrix} \not\geq \mathbf{0}$. Choose $j = 1$ as the new index to enter the basis. Compute $\mathbf{u} = \begin{bmatrix} u_3 & u_4 & u_5 \end{bmatrix} = \mathbf{A}_B^{-1}\mathbf{A}_j = \begin{bmatrix} 1 & 2 & 3 \end{bmatrix}^\mathsf{T}$ and $\alpha^* := \min\limits_{i:u_i>0} \dfrac{\bar{x}_i}{u_i} = \dfrac{150}{2} = \dfrac{\bar{x}_4}{u_4}$. Hence $\ell = 4$ is the index leaving the basis. Update the basis and basic feasible solution to $B = \{3, 1, 5\}$ and $\bar{\mathbf{x}}_B = \begin{bmatrix} \bar{x}_3 & \bar{x}_1 & \bar{x}_5 \end{bmatrix}^\mathsf{T} = \begin{bmatrix} 25 & 75 & 135 \end{bmatrix}^\mathsf{T}$.

Iteration 2: $B = \{3, 1, 5\}$, $\bar{\mathbf{x}}_B = \begin{bmatrix} \bar{x}_3 & \bar{x}_1 & \bar{x}_5 \end{bmatrix}^\mathsf{T} = \begin{bmatrix} 25 & 75 & 135 \end{bmatrix}^\mathsf{T}$, $\bar{\mathbf{c}}_N = \begin{bmatrix} \bar{c}_2 & \bar{c}_4 \end{bmatrix} = \begin{bmatrix} -1 & 0 \end{bmatrix}^\mathsf{T} \not\geq \mathbf{0}$. Choose $j = 2$ as the new index to enter the basis. Compute $\mathbf{u} = \begin{bmatrix} u_3 & u_1 & u_5 \end{bmatrix}^\mathsf{T} = \mathbf{A}_B^{-1}\mathbf{A}_j = \begin{bmatrix} 1/2 & 1/2 & 5/2 \end{bmatrix}^\mathsf{T}$ and $\alpha^* := \min\limits_{i:u_i>0} \dfrac{\bar{x}_i}{u_i} = \dfrac{25}{1/2} = \dfrac{\bar{x}_3}{u_3}$. Hence $\ell = 3$ is the index leaving the basis. Update the basis and basic feasible solution to $B = \{2, 1, 5\}$ and $\bar{\mathbf{x}}_B = \begin{bmatrix} \bar{x}_2 & \bar{x}_1 & \bar{x}_5 \end{bmatrix}^\mathsf{T} = \begin{bmatrix} 50 & 50 & 10 \end{bmatrix}^\mathsf{T}$.

Iteration 3: $B = \{2, 1, 5\}$, $\bar{\mathbf{x}}_B = \begin{bmatrix} \bar{x}_2 & \bar{x}_1 & \bar{x}_5 \end{bmatrix}^\mathsf{T} = \begin{bmatrix} 50 & 50 & 10 \end{bmatrix}^\mathsf{T}$, $\bar{\mathbf{c}}_N = \begin{bmatrix} c_3 & c_4 \end{bmatrix}^\mathsf{T} = \begin{bmatrix} 2 & 1 \end{bmatrix}^\mathsf{T} \geq \mathbf{0}$. Hence B is an optimal basis and

$$\bar{\mathbf{x}} = \begin{bmatrix} 50 & 50 & 0 & 0 & 10 \end{bmatrix}^\mathsf{T}$$

is an optimal solution.

Notice how, geometrically, the simplex iterations move from one vertex of the feasible region to an adjacent vertex until an optimum solution is identified. See Figure 2.7.

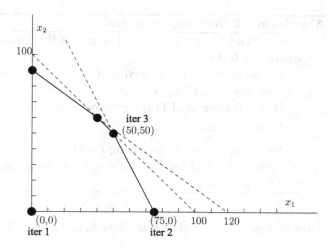

Figure 2.7 Simplex iterations

2.7.2   Dual Simplex Method

The above version of the simplex method is a *primal version* that generates primal feasible iterates and aims for dual feasibility. The *dual simplex method* generates dual feasible iterates and aims for primal feasibility. The logic behind the algorithm is similar. Suppose B is a basis with reduced cost $\bar{\mathbf{c}} \geq \mathbf{0}$. This means that $\bar{\mathbf{y}} = \mathbf{A}_B^{-\mathsf{T}}\mathbf{c}_B$ is a dual feasible solution with slack $\bar{\mathbf{c}} = \mathbf{c} - \mathbf{A}^{\mathsf{T}}\mathbf{y} \geq \mathbf{0}$. If B is not an optimal basis, then $\bar{x}_\ell < 0$ for some $\ell \in B$. Let $\mathbf{e}_\ell \in \mathbb{R}^n$ denote the vector with ℓth component equal to 1 and all others equal to 0. For $\alpha > 0$ the vector $\mathbf{y}(\alpha)$ defined by

$$\mathbf{y}(\alpha) = \bar{\mathbf{y}} - \alpha\mathbf{A}_B^{-\mathsf{T}}\mathbf{e}_\ell$$

satisfies

$$\mathbf{b}^{\mathsf{T}}\mathbf{y}(\alpha) = \mathbf{b}^{\mathsf{T}}\bar{\mathbf{y}} - \alpha\bar{x}_\ell > \mathbf{b}^{\mathsf{T}}\bar{\mathbf{y}}.$$

Observe that the slack of $\mathbf{y}(\alpha)$ is

$$\mathbf{c}(\alpha) = \bar{\mathbf{c}} + \alpha\mathbf{A}^{\mathsf{T}}\mathbf{A}_B^{-\mathsf{T}}\mathbf{e}_\ell.$$

Hence we can get a point with better (lower) objective value than the current basic feasible solution $\bar{\mathbf{x}}$. Unless the problem is unbounded, there is a length $\alpha^* \geq 0$ that makes one of the non-basic components j of $\mathbf{c}(\alpha)$ drop to zero while keeping all of them non-negative. In this case a basis with a better dual solution can be obtained by replacing ℓ with j. Algorithm 2.2 gives a formal description of the dual simplex method.

We illustrate the dual simplex in the following variation of Example 2.2. Suppose we add the constraint

$$6x_1 + 5x_2 \leq 500.$$

Algorithm 2.2 Dual simplex method

1: start with a basis $B \subseteq \{1, \ldots, n\}$ such that the reduced cost $\bar{c}$ is non-negative
2: **while** $\bar{x} \not\geq 0$ **do**
3:     choose an index $\ell \in B$ such that $\bar{x}_\ell < 0$
4:     compute $\mathbf{v} = \mathbf{A}^T \mathbf{A}_B^{-T} e_\ell$
5:     **if** $\mathbf{v} \geq 0$ **then** HALT; the problem is unbounded **end if**
6:     let $\alpha^* := \min\limits_{i:v_i < 0} \dfrac{\bar{c}_i}{|v_i|} = \dfrac{\bar{c}_j}{|v_j|}$
7:     form a new basis by replacing ℓ with j
8:     update the dual feasible solution by replacing $\bar{\mathbf{y}}$ with $\mathbf{y}(\alpha^*)$
9: **end while**

After adding the relevant new slack variable the new linear program is

$$
\begin{aligned}
\min \quad & -4x_1 - 3x_2 \\
\text{s.t.} \quad &
\end{aligned}
$$

$$
\begin{aligned}
x_1 + x_2 + x_3 \qquad\qquad\qquad &= 100 \\
2x_1 + x_2 \qquad + x_4 \qquad\qquad &= 150 \\
3x_1 + 4x_2 \qquad\qquad + x_5 \qquad &= 360 \\
6x_1 + 5x_2 \qquad\qquad\qquad + x_6 &= 500 \\
x_1, x_2, x_3, x_4, x_5, x_6 \quad &\geq 0.
\end{aligned}
$$

If we start with the initial basis $B = \{1, 2, 5, 6\}$ the algorithm proceeds as follows.

Iteration 1: $B = \{1, 2, 5, 6\}$, $\bar{c}_N = \begin{bmatrix} \bar{c}_3 & \bar{c}_4 \end{bmatrix}^T = \begin{bmatrix} 2 & 1 \end{bmatrix}^T \geq 0$, and

$$
\bar{x}_B = \begin{bmatrix} \bar{x}_1 & \bar{x}_2 & \bar{x}_5 & \bar{x}_6 \end{bmatrix}^T = \begin{bmatrix} 50 & 50 & 10 & -50 \end{bmatrix}^T \not\geq 0.
$$

Choose $\ell = 6$ as the index to leave the basis. Compute $\mathbf{v} = \mathbf{A}^T \mathbf{A}_B^{-T} e_\ell = \begin{bmatrix} 0 & 0 & -4 & -1 & 0 & 1 \end{bmatrix}^T$ and $\alpha^* := \min\limits_{i:v_i < 0} \dfrac{\bar{c}_i}{|v_i|} = \dfrac{2}{4} = \dfrac{\bar{c}_3}{|v_3|}$. Hence $j = 3$ is the index entering the basis. Update the basis, reduced cost, and basic solution respectively to $B = \{1, 2, 5, 3\}$, $\bar{c}_N = \begin{bmatrix} \bar{c}_4 & \bar{c}_6 \end{bmatrix} = \begin{bmatrix} 0.5 & 0.5 \end{bmatrix}^T$, and $\bar{x}_B = \begin{bmatrix} \bar{x}_1 & \bar{x}_2 & \bar{x}_5 & \bar{x}_3 \end{bmatrix}^T = \begin{bmatrix} 62.5 & 25 & 72.5 & 12.5 \end{bmatrix}^T$. The new basic solution is non-negative and hence it is an optimal solution.

2.7.3   Interior-Point Methods

In contrast to the simplex method, interior-point methods generate a sequence of iterates that satisfy $\mathbf{x}, \mathbf{s} > 0$. Each iteration of the algorithm aims to make progress towards satisfying $\mathbf{A}\mathbf{x} = \mathbf{b}$, $\mathbf{A}^T \mathbf{y} + \mathbf{s} = \mathbf{c}$, and $x_i s_i = 0$, $i = 1, \ldots, n$.

Throughout this section we use the following notational convention: Given a vector $\mathbf{x} \in \mathbb{R}^n$, let $\mathbf{X} \in \mathbb{R}^{n \times n}$ denote the diagonal matrix defined by $X_{ii} = x_i$, $i = 1, \ldots, n$, and let $\mathbf{1} \in \mathbb{R}^n$ denote the vector whose components are all ones.

The optimality conditions (2.9) can be restated as

$$
\begin{bmatrix} \mathbf{A}^\mathsf{T}\mathbf{y} + \mathbf{s} - \mathbf{c} \\ \mathbf{A}\mathbf{x} - \mathbf{b} \\ \mathbf{XS1} \end{bmatrix} = \begin{bmatrix} 0 \\ 0 \\ 0 \end{bmatrix}, \quad \mathbf{x}, \mathbf{s} \geq 0.
$$

Given $\mu > 0$, let $(\mathbf{x}(\mu), \mathbf{y}(\mu), \mathbf{s}(\mu))$ be the solution to the following perturbed version of the above optimality conditions:

$$
\begin{bmatrix} \mathbf{A}^\mathsf{T}\mathbf{y} + \mathbf{s} - \mathbf{c} \\ \mathbf{A}\mathbf{x} - \mathbf{b} \\ \mathbf{XS1} \end{bmatrix} = \begin{bmatrix} 0 \\ 0 \\ \mu\mathbf{1} \end{bmatrix}, \quad \mathbf{x}, \mathbf{s} > 0.
$$

The first condition above can be written as $\mathbf{r}_\mu(\mathbf{x}, \mathbf{y}, \mathbf{s}) = 0$ for the *residual vector*

$$
\mathbf{r}_\mu(\mathbf{x}, \mathbf{y}, \mathbf{s}) := \begin{bmatrix} \mathbf{A}^\mathsf{T}\mathbf{y} + \mathbf{s} - \mathbf{c} \\ \mathbf{A}\mathbf{x} - \mathbf{b} \\ \mathbf{XS1} - \mu\mathbf{1} \end{bmatrix}.
$$

The *central path* is the set $\{(\mathbf{x}(\mu), \mathbf{y}(\mu), \mathbf{s}(\mu)) : \mu > 0\}$. It is intuitively clear that $(\mathbf{x}(\mu), \mathbf{y}(\mu), \mathbf{s}(\mu))$ converges to an optimal solution to both (2.1) and (2.3) as μ goes to 0. This suggests the following algorithmic strategy: Suppose $(\mathbf{x}, \mathbf{y}, \mathbf{s})$ is "near" $(\mathbf{x}(\mu), \mathbf{y}(\mu), \mathbf{s}(\mu))$ for some $\mu > 0$. Use $(\mathbf{x}, \mathbf{y}, \mathbf{s})$ to move to a better point $(\mathbf{x}^+, \mathbf{y}^+, \mathbf{s}^+)$ "near" $(\mathbf{x}(\mu^+), \mathbf{y}(\mu^+), \mathbf{s}(\mu^+))$ for some $\mu^+ < \mu$.

It can be shown that if a point $(\mathbf{x}, \mathbf{y}, \mathbf{s})$ is on the central path, then the corresponding value of μ satisfies $\mathbf{x}^\mathsf{T}\mathbf{s} = n\mu$. Likewise, given $\mathbf{x}, \mathbf{s} > 0$, define

$$
\mu(\mathbf{x}, \mathbf{s}) := \frac{\mathbf{x}^\mathsf{T}\mathbf{s}}{n}.
$$

To move from a current point $(\mathbf{x}, \mathbf{y}, \mathbf{s})$ to a new point, we use the so-called *Newton step*; that is, the solution to the following system of equations:

$$
\begin{bmatrix} 0 & \mathbf{A}^\mathsf{T} & \mathbf{I} \\ \mathbf{A} & 0 & 0 \\ \mathbf{S} & 0 & \mathbf{X} \end{bmatrix} \begin{bmatrix} \Delta\mathbf{x} \\ \Delta\mathbf{y} \\ \Delta\mathbf{s} \end{bmatrix} = \begin{bmatrix} \mathbf{c} - \mathbf{A}^\mathsf{T}\mathbf{y} - \mathbf{s} \\ \mathbf{b} - \mathbf{A}\mathbf{x} \\ \mu\mathbf{1} - \mathbf{XS1} \end{bmatrix}. \tag{2.10}
$$

Algorithm 2.3 presents a template for an interior-point method.

Algorithm 2.3 Interior-point method

1: choose $\mathbf{x}^0, \mathbf{s}^0 > 0$
2: **for** $k = 0, 1, \ldots$ **do**
3:      solve the Newton system (2.10) for $(\mathbf{x}, \mathbf{y}, \mathbf{s}) = (\mathbf{x}^k, \mathbf{y}^k, \mathbf{s}^k)$ and $\mu :=$ $0.1\mu(\mathbf{x}^k, \mathbf{s}^k)$
4:      choose a step length $\alpha \in (0, 1]$ and set $(\mathbf{x}^{k+1}, \mathbf{y}^{k+1}, \mathbf{s}^{k+1}) = (\mathbf{x}^k, \mathbf{y}^k, \mathbf{s}^k) +$ $\alpha(\Delta\mathbf{x}, \Delta\mathbf{y}, \Delta\mathbf{s})$
5: **end for**

The step length α in step 4 should be chosen so that $\mathbf{x}^{k+1}, \mathbf{s}^{k+1} > 0$ and the size of $\mathbf{r}_\mu(\mathbf{x}^{k+1}, \mathbf{y}^{k+1}, \mathbf{s}^{k+1})$ is sufficiently smaller than $\mathbf{r}_\mu(\mathbf{x}^k, \mathbf{y}^k, \mathbf{s}^k)$. A *line-search procedure* as the one described in Algorithm 2.4 is a popular strategy for choosing the step length α.

Algorithm 2.4 Line search to select the step length α

1: let $\alpha_{\max} := \max\{\alpha : (\mathbf{x}^k, \mathbf{y}^k, \mathbf{s}^k) + \alpha(\Delta\mathbf{x}, \Delta\mathbf{y}, \Delta\mathbf{s}) \geq 0\}$
2: start with $\alpha := 0.99\alpha_{\max}$
3: **while** $\|\mathbf{r}_\mu((\mathbf{x}^k, \mathbf{y}^k, \mathbf{s}^k) + \alpha(\Delta\mathbf{x}, \Delta\mathbf{y}, \Delta\mathbf{s}))\| \geq (1 - 0.01\alpha)\|\mathbf{r}_\mu(\mathbf{x}^k, \mathbf{y}^k, \mathbf{s}^k)\|$ **do**
4:     $\alpha := \alpha/2$
5: **end while**

In contrast to the primal and dual simplex methods, which in principle generate either primal or dual feasible iterates and terminate after finitely many iterations, interior-point methods typically generate infeasible iterates and converge to the optimal solution in the limit. In practice, the convergence is so fast that in a few iterations the algorithm yields iterates that are within machine precision of an exact optimal solution. The algorithm can also be enhanced to detect infeasibility. It relies on the fact that when the primal or dual problem is infeasible, the norm of the residual $\mathbf{r}_\mu(\mathbf{x}, \mathbf{y}, \mathbf{s})$ cannot be driven to zero.

When applied to Example 2.2 starting from $\mathbf{x}^0 = \mathbf{s}^0 = \begin{bmatrix} 100 & \cdots & 100 \end{bmatrix}^\mathsf{T}$, $\mathbf{y}^0 = \mathbf{0}$, the above interior-point algorithm generates the following sequence of iterates. (For ease of notation we only display the first two entries of each iterate.)

Iteration	0	1	2	$\cdots$	8	9
x_1	100	19.7084	17.2976	$\cdots$	49.9383	49.9962
x_2	100	57.3930	41.6795	$\cdots$	50.0436	50.0011

2.8   Notes

The simplex method was developed by George Dantzig (1963). Clever implementations of the simplex method, such as the *revised simplex* and *simplex tableau*, perform iterations far more efficiently than what a naive recalculation of the basic solution and reduced cost from scratch at each iteration would involve. Detailed discussions on these implementations and other related issues can be found in the books by Bertsimas and Tsitsiklis (1997) and Chvátal (1983).

Interior-point methods for linear programming were introduced in a landmark paper by Karmarkar (1984) and subsequently triggered a massive burst of research in optimization during the 1990s and early 2000s. The books by Renegar (2001), Ye (1997), and Roos et al. (2005) present the main developments on this topic.

State-of-the-art linear programming solvers such as CPLEX, MOSEK, Gurobi, and others use implementations of both the simplex and interior-point methods. These solvers can easily solve linear programs with millions of variables and constraints.

2.9 Exercises

Exercise 2.1 Draw the feasible region of the following two-variable linear program:

$$\begin{aligned}
\max \quad & 2x_1 - x_2 \\
& x_1 + x_2 \geq 1 \\
& x_1 - x_2 \leq 0 \\
& 3x_1 + x_2 \leq 6 \\
& x_1, x_2 \geq 0.
\end{aligned}$$

Determine the optimal solution to this problem by inspection.

Exercise 2.2 Consider the following two-variable linear program:

$$\begin{aligned}
\min \quad & 2x_1 + 3x_2 \\
& x_1 + x_2 \geq 5 \\
& x_1 \geq 1 \\
& x_2 \geq 2.
\end{aligned}$$

Prove that $\mathbf{x}^* = \begin{bmatrix} 3 \\ 2 \end{bmatrix}$ is an optimal solution by showing that the objective value of any feasible solution is at least 12. Hint: Use an appropriate combination of the constraints.

Exercise 2.3 Consider the linear programming problem

$$\begin{aligned}
\max \quad & \mathbf{c}^\mathsf{T}\mathbf{x} \\
& \mathbf{A}\mathbf{x} \leq \mathbf{b} \\
& \mathbf{x} \geq \mathbf{0},
\end{aligned}$$

where

$$\mathbf{A} = \begin{bmatrix} 1 & 1 & 1 & 1 & 1 \\ 5 & 4 & 3 & 2 & 1 \end{bmatrix}, \quad \mathbf{b} = \begin{bmatrix} 3 \\ 14 \end{bmatrix}, \quad \mathbf{c}^\mathsf{T} = \begin{bmatrix} 6 & 5 & 4 & 3 & 5 & 4 \end{bmatrix}.$$

Solve this problem using the following strategy:

(a) Find the dual of the above primal linear program. The dual has only two variables. Solve the dual by inspection after drawing a graph of its feasible set.

(b) Using the optimal solution to the dual problem and the optimality conditions, determine what primal constraints are binding and what primal variables must be zero at an optimal solution. Using this information, determine the optimal solution to the primal linear program.

Exercise 2.4

(a) Give an example of a two-variable linear program that is infeasible.
(b) Give an example of a two-variable linear program that is unbounded.

Exercise 2.5   Consider the linear programming problem

$$
\begin{aligned}
\max \quad & c_1 x_1 + \cdots + c_n x_n \\
\text{s.t.} \quad & a_1 x_1 + \cdots + a_n x_n = b \\
& x_1, \ldots, x_n \geq 0,
\end{aligned}
$$

where $b > 0$ and $c_j, a_j > 0$, $j = 1, \ldots, n$. Characterize the optimal solution(s) to this problem. Could there be more than one?

Exercise 2.6

(a) Write the linear programming model in Example 2.1 (fund allocation) in standard form. More precisely, show that for suitable $\tilde{\mathbf{A}}$, $\tilde{\mathbf{b}}$, $\tilde{\mathbf{c}}$, $\tilde{\mathbf{x}}$ the linear programming model in Example 2.1 can be rewritten as

$$
\begin{aligned}
\min_{\tilde{\mathbf{x}}} \quad & \tilde{\mathbf{c}}^\mathsf{T} \tilde{\mathbf{x}} \\
& \tilde{\mathbf{A}} \tilde{\mathbf{x}} = \tilde{\mathbf{b}} \\
& \tilde{\mathbf{x}} \geq \mathbf{0}.
\end{aligned}
$$

Hint: Introduce additional variables.

(b) Show that the standard dual of the model in part (a), namely

$$
\begin{aligned}
\max_{\tilde{\mathbf{y}}} \quad & \tilde{\mathbf{b}}^\mathsf{T} \tilde{\mathbf{y}} \\
& \tilde{\mathbf{A}}^\mathsf{T} \tilde{\mathbf{y}} \leq \tilde{\mathbf{c}},
\end{aligned}
$$

is equivalent to (2.8).

Exercise 2.7   Consider the linear programming problem (2.5). In principle the dual of this problem can be obtained as follows: First, rewrite it in standard form (2.1) by using slack variables. Then obtain the "standard" dual, as in (2.2), for the problem rewritten in this standard form. Prove that the dual problem obtained in this fashion is equivalent to (2.6).

Exercise 2.8   Consider the following investment problem over T years, where the objective is to maximize the value of the investments in year T. We assume a perfect capital market with the same annual lending and borrowing rate $r > 0$ each year. We also assume that exogenous investment funds b_t are available in year t, for $t = 1, \ldots, T$. Let n be the number of possible investments. We assume that each investment can be undertaken fractionally (between 0 and 1). Let a_{tj} denote the cash flow associated with investment j in year t. Let c_j be the value of investment j in year T (including all cash flows subsequent to year T discounted at the interest rate r).

   The linear program that maximizes the value of the investments in year T is the following. Denote by x_j the fraction of investment j undertaken, and let y_t be the amount borrowed (if negative) or lent (if positive) in year t:

$$\max \quad \sum_{j=1}^{n} c_j x_j + y_T$$

$$\text{s.t.} \quad -\sum_{j=1}^{n} a_{1j} x_j + y_1 \leq b_1$$

$$-\sum_{j=1}^{n} a_{tj} x_j - (1+r)y_{t-1} + y_t \leq b_t \quad \text{for } t = 2, \ldots, T$$

$$0 \leq x_j \leq 1 \quad \text{for } j = 1, \ldots, n.$$

(a) Write the dual of the above linear program.

(b) Solve the dual linear program found in part (a).

    Hint: Note that some of the dual variables can be computed by backward substitution.

(c) Write the complementary slackness conditions.

(d) Deduce that the first T constraints in the primal linear program hold as equalities.

(e) Use the complementary slackness conditions to show that the solution obtained by setting $x_j = 1$ if $c_j + \sum_{t=1}^{T}(1+r)^{T-t} a_{tj} > 0$, and $x_j = 0$ otherwise, is an optimal solution.

(f) Conclude that the above investment problem always has an optimal solution where each investment is either undertaken completely or not at all.

Exercise 2.9   Consider the following variation of Exercise 2.5 where there are upper bounds u_i on each of the variables:

$$\max \quad c_1 x_1 + \cdots + c_n x_n$$

$$\text{s.t.} \quad a_1 x_1 + \cdots + a_n x_n \leq b$$

$$0 \leq x_i \leq u_i \quad \text{for } i = 1, \ldots, n.$$

Assume that $b > 0$ and all a_i, c_i, u_i are also strictly positive for $i = 1, \ldots, n$. Furthermore, assume

$$\frac{c_1}{a_1} \geq \frac{c_2}{a_2} \geq \cdots \geq \frac{c_n}{a_n}.$$

Write the problem in standard form and apply the simplex method to it. What steps will the simplex method take? In other words, in what order will the variables enter and leave the basis?

Exercise 2.10   Install and get acquainted with CVX. This package is freely available and extremely easy to install and use. It can be downloaded from http://cvxr.com/cvx/download/.

    Write a MATLAB script that takes as inputs an $m \times n$ matrix $\mathbf{A}$, an m-dimensional vector $\mathbf{b}$, and an n-dimensional vector $\mathbf{c}$ and solves the optimization problem

$$\min \quad \mathbf{c}^T \mathbf{x}$$

$$\text{s.t.} \quad \mathbf{A}\mathbf{x} = \mathbf{b}$$

$$\mathbf{x} \geq \mathbf{0}.$$

(a) Test your script on instances generated as follows:

```
>> m=1, n=5, b=1, c=ones(n,1), A=rand(m,n);
```

and

```
>> m=1, n=5, b=1, c=rand(n,1), A=ones(m,n);
```

Are the results consistent with your answer to Exercise 2.5?

(b) Test your script on instances generated as follows:

```
>> m=2, n=6,  b=rand(m,1), c=rand(n,1), A=rand(m,n);
>> m=2, n=10, b=rand(m,1), c=rand(n,1), A=rand(m,n);
>> m=4, n=10, b=rand(m,1), c=rand(n,1), A=rand(m,n);
>> m=4, n=20, b=rand(m,1), c=rand(n,1), A=rand(m,n);
```

Do you notice anything peculiar about the number of non-zero entries in the optimal solution $\mathbf{x}$ in each case?

Exercise 2.11   Use Theorem 2.4 to prove Theorem 2.5. To that end, proceed as follows.

(a) (Farkas's lemma) Consider the linear programming problem

$$
\begin{array}{ll}
\min & \mathbf{b}^\mathsf{T}\mathbf{y} \\
\text{s.t.} & \mathbf{A}^\mathsf{T}\mathbf{y} \leq \mathbf{0}.
\end{array}
$$

Show that the dual of this problem is

$$
\begin{array}{ll}
\max & 0 \\
\text{s.t.} & \mathbf{A}\mathbf{x} = \mathbf{b} \\
& \mathbf{x} \geq \mathbf{0}.
\end{array}
$$

Now apply Theorem 2.4.

(b) (Gordan's theorem) Proceed as in (a) but this time start with the linear programming problem

$$
\begin{array}{ll}
\max & t \\
\text{s.t.} & \mathbf{A}^\mathsf{T}\mathbf{y} - \mathbf{1}t \geq \mathbf{0}.
\end{array}
$$

(c) (Stiemke's theorem) Proceed as in (a) and (b) but this time start with the linear programming problem

$$
\begin{array}{ll}
\max & t \\
\text{s.t.} & \mathbf{A}\mathbf{x} = \mathbf{0} \\
& \mathbf{x} - \mathbf{1}t \geq \mathbf{0}.
\end{array}
$$

Exercise 2.12   Use Theorem 2.5 to prove Theorem 2.4.

Exercise 2.13   To break the circular argument in the above two exercises, prove Theorem 2.5 using the following *hyperplane separation theorem*: If $S \subseteq \mathbb{R}^n$ is closed and convex and $\mathbf{x} \notin S$ then there exists a hyperplane separating $\mathbf{x}$ and S. That is, there exists $\mathbf{a} \in \mathbb{R}^n \setminus \{\mathbf{0}\}$ and $b \in \mathbb{R}$ such that $\mathbf{a}^\mathsf{T}\mathbf{x} < b \leq \mathbf{a}^\mathsf{T}\mathbf{y}$ for all $\mathbf{y} \in S$.

3 Linear Programming Models: Asset–Liability Management

This chapter presents a classical application of linear programming to covering known liabilities by constructing a dedicated fixed-income portfolio. When the liabilities span multiple years, the model assumes that the only sources of risk are changes in the term structure of interest rates. We also discuss a short-term financing problem.

3.1   Dedication

Consider the problem of funding a stream of liabilities that extends over the future. Assume the forecast of liabilities is accurate. This problem arises in certain practical situations such as the liabilities of a pension fund. It also arises in non-financial institutions planning acquisitions, expansion, or product development. A *dedicated* bond portfolio is a portfolio of bonds constructed today and whose cash flows offset the liabilities.

Example 3.1 (Bond dedication)   Suppose a pension fund needs to cover some liabilities in the next six years. Cash requirements (in million \$) are:

Year	1	2	3	4	5	6
Required	100	200	800	100	800	1200

Suppose the pension fund can invest in ten government bonds with the cash flows and current prices in Table 3.1.

Find the least expensive portfolio of bonds whose cash flows will be sufficient to cover the cash requirements. Assume surplus cash can be carried from one year to the next but earn no interest.

We can formulate this problem as the following linear programming model.

Linear programming model for bond dedication
Variables:

x_j: amount of bonds j in the portfolio, for $j = 1, \ldots, 10$;
s_t: surplus cash in year t, for $t = 1, \ldots, 6$.

Table 3.1

| | Year | | | | | | |
	1	2	3	4	5	6	Price
Bond 1	10	10	10	10	10	110	109
Bond 2	7	7	7	7	7	107	94.8
Bond 3	8	8	8	8	8	108	99.5
Bond 4	6	6	6	6	106		93.1
Bond 5	7	7	7	7	107		97.2
Bond 6	5	5	5	105			92.9
Bond 7	10	10	110				110
Bond 8	8	8	108				104
Bond 9	7	107					102
Bond 10	100						95.2

Objective:

$$\min \quad 109x_1 + 94.8x_2 + \cdots + 102x_9 + 95.2x_{10}.$$

Constraints:

$$
\begin{aligned}
10x_1 + 7x_2 + \cdots + 7x_9 + 100x_{10} \quad\quad\quad &= \quad 100 + s_1 \\
10x_1 + 7x_2 + \cdots + 107x_9 \quad\quad\quad +s_1 &= \quad 200 + s_2 \\
&\;\;\vdots \\
110x_1 + 107x_2 + 108x_3 \quad\quad\quad +s_5 &= \quad 1200 + s_6 \\
\end{aligned}
$$
$$x_j \geq 0, \; j = 1,\ldots,10$$
$$s_t \geq 0, \; t = 1,\ldots,6.$$

Notice that we can write the equality constraints also as

$$
\begin{aligned}
10x_1 + 7x_2 + \cdots + 7x_9 + 100x_{10} \quad\quad -s_1 &= \quad 100 \\
10x_1 + 7x_2 + \cdots + 107x_9 \quad\quad\quad +s_1 - s_2 &= \quad 200 \\
&\;\;\vdots \\
110x_1 + 107x_2 + 108x_3 \quad\quad\quad +s_5 - s_6 &= \quad 1200 \\
\end{aligned}
$$
$$x_j \geq 0, \; j = 1,\ldots,10$$
$$s_t \geq 0, \; t = 1,\ldots,6,$$

or as

$$
\begin{aligned}
10x_1 + 7x_2 + \cdots + 7x_9 + 100x_{10} \quad\quad -100 &= \quad s_1 \\
10x_1 + 7x_2 + \cdots + 107x_9 \quad\quad\quad +s_1 - 200 &= \quad s_2 \\
&\;\;\vdots \\
110x_1 + 107x_2 + 108x_3 \quad\quad\quad +s_5 - 1200 &= \quad s_6 \\
\end{aligned}
$$
$$x_j \geq 0, \; j = 1,\ldots,10$$
$$s_t \geq 0, \; t = 1,\ldots,6.$$

In general, for a given problem with liabilities projected over m points in time over the future, the stream of liabilities is a vector:

Date	1	2	$\cdots$	m
Required	L_1	L_2	$\cdots$	L_m

Suppose we can use n bonds with the following cash flows and prices:

Date	1	2	$\cdots$	m	Prices
Bond 1	F_{11}	F_{21}	$\cdots$	F_{m1}	p_1
$\vdots$					$\vdots$
Bond j	F_{1j}	F_{2j}	$\cdots$	F_{mj}	p_j
$\vdots$					$\vdots$
Bond n	F_{1n}	F_{2n}	$\cdots$	F_{mn}	p_n

The linear programming formulation of the cash matching problem is as follows.

Linear programming model for bond dedication (general version)
Variables:

x_j: amount of bonds j in the portfolio, for $j = 1, \ldots, n$;
s_t: surplus cash in year t, for $t = 1, \ldots, m$.

Linear programming model:

$$\min \quad \sum_{j=1}^{n} p_j x_j$$

$$\text{s.t.} \quad \sum_{j=1}^{n} F_{1j} x_j - s_1 \quad = \quad L_1$$

$$\sum_{j=1}^{n} F_{tj} x_j + s_{t-1} - s_t \quad = \quad L_t, \ t = 2, \ldots, m$$

$$x_j \geq 0, \ j = 1, \ldots, n$$

$$s_t \geq 0, \ t = 1, \ldots, m.$$

The problem can be written more concisely as follows:

$$\min \quad \mathbf{p}^{\mathsf{T}} \mathbf{x}$$
$$\text{s.t.} \quad \mathbf{F}\mathbf{x} + \mathbf{R}\mathbf{s} = \mathbf{L}$$
$$\mathbf{x} \geq \mathbf{0}$$
$$\mathbf{s} \geq \mathbf{0},$$

where

$$\mathbf{F} = \begin{bmatrix} F_{11} & \cdots & F_{1n} \\ \vdots & \ddots & \vdots \\ F_{m1} & \cdots & F_{mn} \end{bmatrix}, \quad \mathbf{R} = \begin{bmatrix} -1 & 0 & 0 & 0 & \cdots & 0 \\ 1 & -1 & 0 & 0 & \cdots & 0 \\ 0 & 1 & -1 & 0 & \cdots & 0 \\ \vdots & \ddots & \ddots & \ddots & \ddots & \vdots \\ 0 & \cdots & 0 & 1 & -1 & 0 \\ 0 & 0 & \cdots & 0 & 1 & -1 \end{bmatrix},$$

$$
\mathbf{p} = \begin{bmatrix} p_1 \\ \vdots \\ p_n \end{bmatrix}, \quad \mathbf{L} = \begin{bmatrix} L_1 \\ \vdots \\ L_m \end{bmatrix}.
$$

3.2   Sensitivity Analysis

As noted in Section 2.4, when a linear programming model is solved, the dual solution yields a great deal of *sensitivity information,* or information about what happens when data values are changed. Recall the sensitivity interpretation associated with the *shadow price.* Assume λ is the shadow price of a constraint:

> If the right-hand side of a constraint changes by Δ, then the optimal objective value changes by $\lambda \cdot \Delta$, as long as the change of the right-hand side is within the allowable increase or decrease.

This concept is particularly insightful in the bond dedication problem. In a nutshell, the sensitivity information of the linear optimization model leads to an *implied term structure* as we next explain. Recall our linear programming model for portfolio dedication:

$$
\begin{aligned}
\min \quad & \sum_{j=1}^{n} p_j x_j \\
\text{s.t.} \quad & \sum_{j=1}^{n} F_{1j} x_j - s_1 && = L_1 \\
& \sum_{j=1}^{n} F_{tj} x_j + s_{t-1} - s_t && = L_t, \ t = 2, \ldots, m \\
& x_j \geq 0, \ j = 1, \ldots, n \\
& s_t \geq 0, \ t = 1, \ldots, m.
\end{aligned}
$$

The shadow price of constraint at time t is the extra amount of money needed today to cover an extra unit of liability at time t. In other words, the shadow price λ_t gives the discount factor for time t. The current portfolio therefore *implies* the following term structure of interest rates:

$$
r_t = \frac{1}{(\lambda_t)^{1/t}} - 1.
$$

3.3   Immunization

Consider again the problem of covering a stream of liabilities $(L_1, \ldots, L_m)$ due at m different dates in the future. In principle, the stream of liabilities is equivalent to a lump sum of cash today equal to its present value, obtained by discounting the future liabilities. Setting aside an amount equal to this present value seems simpler than constructing a dedicated portfolio. A problem with this approach is that it is fully exposed to interest-rate risk. By contrast, a dedicated portfolio is not subject to interest-rate risk since it matches the liabilities at the time they occur. *Immunization* is an approach that reduces interest-rate risk as compared to the simple-minded present value approach, but it does not completely protect against it as dedication would. The advantage is that immunized portfolios are typically cheaper than dedicated portfolios. The idea is simple: construct a portfolio with the same present value as the stream of liabilities, and further require that this present value has the same sensitivity to changes in interest rates as the stream of liabilities.

More precisely, suppose $r_1, \ldots, r_m$ is the *term structure* of risk-free interest rates. This means that the value r_t is the yield on a risk-free zero-coupon bond with maturity t. In other words, r_t is the interest rate that applies to money invested between now and time t. By discounting each of the cash flows with the appropriate discount rate, it follows that the present value (PV) of a stream of cash flows $(F_1, \ldots, F_m)$, where F_t occurs at time t, is

$$PV = \frac{F_1}{1 + r_1} + \frac{F_2}{(1 + r_2)^2} + \cdots + \frac{F_m}{(1 + r_m)^m}.$$

If interest rates shift by δ, we get

$$PV(\delta) = \frac{F_1}{1 + r_1 + \delta} + \frac{F_2}{(1 + r_2 + \delta)^2} + \cdots + \frac{F_m}{(1 + r_m + \delta)^m}.$$

Notice that

$$PV(\delta) - PV \approx -\delta \left(\frac{F_1}{(1 + r_1)^2} + \frac{2F_2}{(1 + r_2)^3} + \cdots + \frac{mF_m}{(1 + r_m)^{m+1}} \right).$$

This motivates the following concept.

Definition 3.2 The Fisher–Weil dollar duration (DD) of the stream of cash flows $(F_1, \ldots, F_m)$ is

$$DD := \sum_{t=1}^{m} \frac{tF_t}{(1 + r_t)^{t+1}}.$$

An *immunized portfolio* is a portfolio of bonds whose present value and duration match those of the stream of liabilities. In optimization terms, this corresponds to a portfolio that satisfies the following constraints:

$$\sum_{j=1}^{n} PV_j x_j = PV_L$$

$$\sum_{j=1}^{n} DD_j x_j = DD_L.$$

A closer look at the difference between PV and PV(δ) suggests that we can get an even better matching of sensitivity to changes in the term structure by looking at second-order terms:

$$\mathrm{PV}(\delta) - \mathrm{PV} \approx -\delta \sum_{t=1}^{m} \frac{tF_t}{(1+r_t)^{t+1}} + \frac{1}{2}\delta^2 \sum_{t=1}^{m} \frac{t(t+1)F_t}{(1+r_t)^{t+2}}.$$

This leads to the so-called Fisher–Weil *dollar convexity* (DC) of $(F_1, \ldots, F_m)$:

$$\mathrm{DC} := \sum_{t=1}^{m} \frac{t(t+1)F_t}{(1+r_t)^{t+2}},$$

as well as the Fisher–Weil *convexity* (C):

$$\mathrm{C} := \frac{1}{\mathrm{PV}} \sum_{t=1}^{m} \frac{t(t+1)F_t}{(1+r_t)^{t+2}}.$$

A portfolio can therefore be further immunized by matching present value, dollar duration, and dollar convexity:

$$\sum_{j=1}^{n} \mathrm{PV}_j x_j = \mathrm{PV}_L$$

$$\sum_{j=1}^{n} \mathrm{DD}_j x_j = \mathrm{DD}_L \tag{3.1}$$

$$\sum_{j=1}^{n} \mathrm{DC}_j x_j \geq \mathrm{DC}_L.$$

Here the subindices $j = 1, \ldots, n$ and L refer to the bonds and liabilities respectively. Note that since having net positive convexity is favorable, the last constraint is an inequality constraint.

The immunization constraints (3.1) are generally less stringent than the bond dedication constraints, namely

$$\sum_{j=1}^{n} F_{1j} x_j - s_1 \qquad = \quad L_1$$

$$\sum_{j=1}^{n} F_{tj} x_j + s_{t-1} - s_t \quad = \quad L_t, \ t = 2, \ldots, m \tag{3.2}$$

$$x_j \geq 0, \ j = 1, \ldots, n$$

$$s_t \geq 0, \ t = 1, \ldots, m.$$

Indeed, if the surplus variables s_t, $t = 1, \ldots, m$, are all zero in (3.2), then some straightforward algebra shows that any $x_1, \ldots, x_n$ satisfying (3.2) also satisfies (3.1).

The previous discussion assumes that interest is compounded at discrete time intervals, e.g., annually or semiannually. In some practical circumstances cash flows may occur at irregular times. In those cases it could be more convenient to assume that interest is continuously compounded. Suppose r_t is the *continuously compounded* spot rate for a risk-free zero-coupon bond with maturity t. Then the present value of a stream of cash flows $(F_1, \ldots, F_m)$ is

$$PV = \sum_{t=1}^{m} F_t e^{-t \cdot r_t}.$$

Consequently its dollar duration is

$$DD = \sum_{t=1}^{m} t F_t e^{-t \cdot r_t},$$

and its dollar convexity is

$$DC = \sum_{t=1}^{m} t^2 F_t e^{-t \cdot r_t}.$$

A nice feature of continuous compounding is that the formulas for an irregular stream of cash flows $(F_{t_1}, \ldots, F_{t_m})$ are very similar:

$$PV = \sum_{i=1}^{m} F_{t_i} e^{-t_i \cdot r_{t_i}},$$

$$DD = \sum_{i=1}^{m} t_i F_{t_i} e^{-t_i \cdot r_{t_i}},$$

$$DC = \sum_{i=1}^{m} t_i^2 F_{t_i} e^{-t_i \cdot r_{t_i}}.$$

The kind of immunization via duration and convexity enforced by the constraints (3.1) provides hedging against parallel shocks in the term structure. This implicitly assumes a *one-factor* interest risk model. There are enhancements based on a *multi-factor* interest risk model. Two popular ones are the *key-rate model* and the *shift–twist–butterfly* model as discussed in Tuckman (2002). The logic of immunization naturally extends to a multi-factor interest risk model. In such a context an immunized portfolio should be hedged against changes in each of the risk factors.

3.4 Some Practical Details about Bonds

There are certain details about the way bonds are quoted and traded in actual exchanges. The discussion below applies only to plain vanilla treasury bonds. For a more detailed discussion, see Fabozzi (2004) or Tuckman (2002).

Principal Value, Coupon Payments, Clean and Dirty Prices

The *principal value*, or *par*, or *principal* of a bond is the amount that the issuer agrees to repay the bondholder. The *term to maturity* of a bond is the time remaining until principal payment. The *maturity date* of a bond is that date when the issuer will pay the principal.

The *coupon rate* or *nominal rate* of a bond is the annual interest that the issuer pays the bondholder. Treasury bonds pay their coupons semiannually. For example, a bond with an 8% coupon rate and a principal of $1,000 will pay a $40 installment to the holder every six months. At the maturity date, it will pay the $40 installment plus the $1,000 principal.

When an investor purchases a bond between coupon payments, the investor must compensate the seller of the bond for the coupon interest earned since the last coupon payment. This is called the *accrued interest* and is computed based on the proportion of time since the last coupon payment. The convention for United States treasuries is not to include the accrued interest in the price quote. This price is called the *clean price* or simply the *price*. It is customary to present the price quote as a percentage of the par value of the bond. The clean price plus the accrued interest is called the *dirty price* or *full price.*

For example, suppose that on February 15, 2051 investor B buys a treasury bond with $10,000 face value, 5.5% coupon rate that matures on January 31, 2053. In this case the coupon payment is $275 = 2.75\%$ of $10,000 and the accrued interest is

$$\frac{15}{181} \cdot 275 = 22.79.$$

Suppose the price quote on February 15, 2051 is 101.145. Then the full price of the bond, that is, the price paid by the buyer to the seller, is $10,114.5+$22.79 = $10,137.29.

Yield Curve and Term Structure

Recall that the *yield* of a bond is the interest rate that makes the discounted value of the cash flows match the current price of the bond. By convention, the yield is quoted on an annual basis. The *treasury yield curve* is the curve of yields for *on-the-run* (most recently auctioned) treasuries. It should be noted that the yield curve is not the same as the term structure of interest rates. This is the case because treasuries with maturity greater than one year are not zero-coupon bonds. Indeed, the term structure of interest rates is actually a theoretical construct that must be estimated from actual bonds.

There are various ways of estimating the term structure. A quick and dirty (perhaps too dirty) approach is to ignore the difference described above and to use the yield curve as a proxy for the term structure of interest rates.

A second approach is to use the following *bootstrapping* approach: Use several coupon-bearing bonds with various maturities. Determine the spot rate implied by the bond with the shortest maturity. Use that knowledge to compute the spot rate implied by the bond with the next shortest maturity and so on. For example, suppose we have a 0.5-year 5.25% bill, a 1-year 5.75% note, and a 1.5-year 6% note. For simplicity assume they all are trading at par. Let z_1, z_2, z_3 denote the one-half annualized 0.5-year, 1-year, and 1.5-year spot rates.

Using the 0.5-year bond, we readily get

$$z_1 = 0.0525 \cdot 0.5 = 0.02625.$$

Using $100 as par, the cash flows for the 1-year bond are

0.5-year: $0.0575 \cdot 100 \cdot 0.5 = 2.875$
1-year: $0.0575 \cdot 100 \cdot 0.5 + 100 = 102.875.$

Now compute its present value using the spot rates z_1, z_2 and equate that to its current price:

$$100 = \frac{2.875}{1 + z_1} + \frac{102.875}{(1 + z_2)^2}.$$

Because we already know z_1, we can solve for z_2 and obtain

$$z_2 = 0.028786.$$

Repeat with the 1.5-year bond: Using $100 as par, the cash flows for the 1.5-year bond are

0.5-year: $0.06 \cdot 100 \cdot 0.5 = 3$
1-year: $0.06 \cdot 100 \cdot 0.5 = 3$
1.5-year: $0.06 \cdot 100 \cdot 0.5 + 100 = 103.$

Now compute its present value using the spot rates z_1, z_2, z_3 and equate that to its current price:

$$100 = \frac{3}{1 + z_1} + \frac{3}{(1 + z_2)^2} + \frac{103}{(1 + z_3)^3}.$$

Because we already know z_1, z_2, we can solve for z_3 and obtain

$$z_3 = 0.030063097.$$

Thus the annualized spot rates are

$$r_{0.5} = 0.0525, \ r_1 = 0.057572, \ r_2 = 0.06012.$$

Yet a third, and much more elaborate, approach to estimating the term structure is to take into consideration all bonds with similar characteristics available in the market and perform an elaborate regression model. This approach requires advanced statistical techniques and is beyond the scope of this book. For a related discussion see Campbell et al. (1997) and Heath et al. (1992)

It should also be noted that the previous two estimation approaches only give spot rates at specific points in time. The spot rates at other times can be obtained by interpolation. The simplest type of interpolation is piecewise linear.

3.5   Other Cash Flow Problems

The dedication model discussed in Section 3.1 belongs to the broader class of *cash flow problems*. A firm faces a stream of both positive (inflows) and negative (outflows) flows of cash. The negative flows are considered liabilities that must be met when they occur. To meet the liabilities, the firm can purchase a variety of instruments each with a different cash flow pattern.

The following *short-term financing problem* is of this kind. Corporations routinely face the problem of financing short-term cash commitments. Linear programming can help in figuring out an optimal combination of financial instruments to meet these commitments. For illustration, consider the following problem. For the sake of exposition, we keep the example small.

Example 3.3 (Short-term financing)   A company has the following short-term financing problem (net cash flow requirements are given in $1000s).

Month	J	F	M	A	M	J
Net cash flow	−150	−100	200	−200	50	300

The company has the following sources of funds:

- A line of credit of up to $100,000 at an interest rate of 1% per month.
- It can issue 90-day commercial paper bearing a total interest of 2% for the 3-month period.
- Each month excess funds can be invested at an interest rate of 0.3% per month.

There are many questions that the company might want to answer. Is it economical to use the line of credit in some of the months? If so, when? How much? What interest payments will the company need to make between January and June? Linear programming gives us a mechanism for answering these questions quickly and easily.

Linear programming model for short-term financing problem
Variables:

x_j: amount drawn from the line of credit in month j, for $j = 1, \ldots, 5$

y_j: amount of commercial paper issued in month j, for $j = 1, \ldots, 3$

z_j: excess funds in month j, for $j = 1, \ldots, 6$.

Objective:

$$\max \ z_6.$$

Constraints: Cash balance constraints in each month and bounds on x_j, y_j and z_j:

$$
\begin{aligned}
x_1 + y_1 && && - z_1 &= 150 \\
x_2 + y_2 &&- 1.01x_1 + 1.003z_1 &- z_2 &= 100 \\
x_3 + y_3 &&- 1.01x_2 + 1.003z_2 &- z_3 &= -200 \\
x_4 - 1.02y_1 &&- 1.01x_3 + 1.003z_3 &- z_4 &= 200 \\
x_5 - 1.02y_2 &&- 1.01x_4 + 1.003z_4 &- z_5 &= -50 \\
- 1.02y_3 &&- 1.01x_5 + 1.003z_5 &- z_6 &= -300 \\
&& x_j &\leq 100 && \text{for } j = 1, \ldots, 5 \\
&& x_j &\geq 0 && \text{for } j = 1, \ldots, 5 \\
&& y_j &\geq 0 && \text{for } j = 1, \ldots, 3 \\
&& z_j &\geq 0 && \text{for } j = 1, \ldots, 5.
\end{aligned}
$$

Solving this linear program using either Excel `Solver` or MATLAB CVX, we obtain the following optimal solution:

$$
\mathbf{x}^* = \begin{bmatrix} 0 \\ 0 \\ 0 \\ 0 \\ 52 \end{bmatrix}, \quad
\mathbf{y}^* = \begin{bmatrix} 150 \\ 100 \\ 151.944 \end{bmatrix}, \quad
\mathbf{z}^* = \begin{bmatrix} 0 \\ 0 \\ 351.944 \\ 0 \\ 0 \\ 92.497 \end{bmatrix}.
$$

Thus the company can attain an optimal wealth of \$92,497 in June. To achieve this, the company will issue \$150,000 in commercial paper in January, \$100,000 in February and \$151,944 in March. In addition, it will draw \$52,000 from its line of credit in May. Excess cash of \$351,944 in March will be invested for one month.

Figure 3.1 displays the Excel `Solver` sensitivity report for this model.

The key columns for sensitivity analysis are the "Reduced Cost" and "Shadow Price" columns. Recall that the *shadow price* u of a constraint C has the following interpretation:

> If the right-hand side of the constraint C changes by an amount Δ, the optimal objective value changes by $u \cdot \Delta$ as long as Δ is within a certain range.

The above sensitivity information allows us to perform various kinds of "what if" analysis.

- For example, assume that net cash flow in January were -200 (instead of -150). By how much would the wealth of the company decrease at the end of June? The answer is in the shadow price of the January constraint, $u = -1.0373$. The right-hand side of the January constraint would go from 150 to 200, an increase of $\Delta = 50$, which is within the allowable

Variable Cells

Cell	Name	Final Value	Reduced Cost	Objective Coefficient	Allowable Increase	Allowable Decrease
I8	credit J Amount	0	-0.003214	0	0.00321386	1E+30
I9	credit F Amount	0	-3.89E-16	0	3.8858E-16	1E+30
I10	credit M Amount	0	-0.007119	0	0.00711864	1E+30
I11	credit A Amount	0	-0.003151	0	0.00315085	1E+30
I12	credit M Amount	52	0	0	0.00311965	3.8096E-16
I13	commercial J Amount	150	0	0	0.00399754	0.00321386
I14	commercial F Amount	100	0	0	0.00318204	3.8858E-16
I15	commercial M Amount	151.9441675	0	0	3.8473E-16	0.0031603
B18	Surplus January	0	-0.003998	0	0.00399754	1E+30
C18	Surplus February	0	-0.00714	0	0.00714	1E+30
D18	Surplus March	351.9441675	0	0	0.00393091	0.0031603
E18	Surplus April	0	-0.003919	0	0.00391915	1E+30
F18	Surplus May	0	-0.007	0	0.007	1E+30
G18	Surplus June	92.49694915	0	1	1E+30	1

Constraints

Cell	Name	Final Value	Shadow Price	Constraint R.H. Side	Allowable Increase	Allowable Decrease
B19	Total January	150	-1.037288	150	89.1718954	149.411765
C19	Total February	100	-1.0302	100	47.0588235	50.9803922
D19	Total March	-200	-1.02	-200	90.6832835	151.944167
E19	Total April	200	-1.016949	200	90.9553333	152.4
F19	Total May	-50	-1.01	-50	48	52
G19	Total June	-300	-1	-300	92.4969492	1E+30

Figure 3.1 Sensitivity report for short-term financing model

increase (89.17). So the wealth of the company in June would decrease by $1.0373 \cdot 50{,}000 = \$51{,}865$.

- Now assume that net cash flow in March were 250 (instead of 200). By how much would the wealth of the company increase at the end of June? Again, the change $\Delta = -50$ is within the allowable decrease (151.944), so we can use the shadow price $u = -1.02$ to calculate the change in objective value. The increase is $(-1.02) \cdot (-50) = \$51{,}000$.

- Assume that the negative net cash flow in January is due in part to the purchase of a machine worth \$100,000. The vendor allows the payment to be made in June at an interest rate of 3% for the 5-month period. Would the wealth of the company increase or decrease by using this option? What if the interest rate for the 5-month period were 4%? The shadow price of the January constraint is -1.0373. This means that reducing cash requirements in January by \$1 increases the wealth in June by \$1.0373. In other words, the break-even interest rate for the 5-month period is 3.73%. So, if the vendor charges 3%, we should accept, but if he charges 4% we should not. Note that the analysis is valid since the amount $\Delta = -100$ is within the allowable decrease.

Next, let us consider the reduced costs. Recall that these are the shadow prices of the upper and lower bounds placed directly on the variables. The reduced cost of a variable is non-zero only when the variable is equal to one of its bounds. Assume x is equal to its lower bound b and its reduced cost is c. There are two useful interpretations of the reduced cost c.

- First, if the value of x is set to a value $b + \Delta$ for $\Delta > 0$ instead of its optimal value b then the objective value is changed by $c \cdot \Delta$. For example, what would be the effect of financing part of the January cash needs through the line of credit? The answer is in the reduced cost of the first variable. Because this reduced cost -0.0032 is strictly negative, the objective function would decrease. Specifically, each dollar financed through the line of credit in January would result in a decrease of \$3.2 in the wealth of the company in June.

- The second interpretation of c is that its magnitude $|c|$ is the minimum amount by which the objective coefficient of x must be changed in order for the variable x to move away from its bound in an optimal solution. For example, consider the first variable again. Its value is zero in the current optimal solution, with objective function z_6. However, if we changed the objective to $z_6 + 0.0032 x_1$, it would now be optimal to use the line of credit in January. In other words, the reduced cost on x_1 can be viewed as the minimum rebate that the bank would have to offer (payable in June) to make it attractive to use the line of credit in January.

3.6 Exercises

Exercise 3.1 You need to create a portfolio to cover the following stream of liabilities for the next six future dates:

Date	1	2	3	4	5	6
Required	500	200	800	200	800	1200

You may purchase the bonds in Table 3.2.
The term structure of risk-free interest rates is:

Date	1	2	3	4	5	6
Rate	5.04%	5.94%	6.36%	7.18%	7.89%	8.39%

(a) Formulate a linear programming model to find the lowest-cost long-only dedicated portfolio that covers the stream of liabilities with the bonds above. Assume surplus balances can be carried from one date to the next but earn no interest. What is the cost of your portfolio? What is the composition of your portfolio?

Table 3.2

Bond	1	2	3	4	5	6	Price
			Year				
1	10	10	10	10	10	110	109
2	7	7	7	7	7	107	94.8
3	8	8	8	8	8	108	99.5
4	6	6	6	6	106		93.1
5	7	7	7	7	107		97.2
6	6	6	6	106			96.3
7	5	5	5	105			92.9
8	10	10	110				110
9	8	8	108				104
10	6	6	106				101
11	10	110					107
12	7	107					102
13	100						95.2

(b) Formulate a linear programming model to find the lowest-cost portfolio that matches the present value and dollar duration of the stream of liabilities. What is the cost of your portfolio? How do the two present values change if interest rates decrease by one percentage point? How do they change if interest rates increase by one percentage point? How do they change if the interest rates in dates 1 and 2 decrease by one percentage point, the rates in dates 3, 4, and 5 remain the same, and the rate in date 6 increases by one percentage point?

(c) Use the linear programming sensitivity information from part (a) to determine the implied term structure of interest rates.

(d) Suppose that the stream of liabilities changes to:

Date	1	2	3	4	5	6
Required	100	200	800	500	800	1200

Find the new optimal dedicated portfolio and determine the new implied term structure. Is it different from the one you obtained in part (c)? Can you provide an intuitive explanation for the difference or lack thereof?

(e) Assume the liabilities occur at irregular time intervals:

Date	1.25	2.5	3.5	4.5	5.75	6.5
Required	500	200	800	200	800	1200

(i) Repeat part (b) for this irregular stream of liabilities.

You will need to do some kind of interpolation to estimate the term structure at the relevant times. You will also need to make an assumption about how to discount at irregular time intervals.

(ii) Repeat part (a) for this irregular stream of liabilities.

(f) Formulate a linear programming model to find the lowest-cost long-only dedicated portfolio that covers the stream of liabilities:

Date	1	2	3	4	5	6
Required	500	200	800	400	700	900

with the following new set of bonds:

Bond	1	2	3	4	5	6	Price	Rating
1	10	10	10	10	10	110	108	B
2	7	7	7	7	7	107	94	B
3	8	8	8	8	8	108	99	B
4	6	6	6	6	106		92.7	B
5	7	7	7	7	107		96.6	B
6	6	6	6	106			95.9	B
7	5	5	5	105			92.9	A
8	10	10	110				110	A
9	8	8	108				104	A
10	6	6	106				101	A
11	10	110					107	A
12	7	107					102	A
13	100						95.2	A

This time assume that at most 50% of your portfolio's value can be in bonds rated B. Again, assume surplus balances can be carried from one date to the next but earn no interest. What is the cost of your portfolio? What is the composition of your portfolio?

Exercise 3.2 Suppose today is November 30, 2052. A pension fund will need to cover the following stream of liabilities over the subsequent four years (in million dollars):

5/31/53	11/30/53	5/31/54	11/30/54	5/31/55	11/30/55	5/31/56	11/31/56
12	10	10	10	9	9	9	15

To cover these liabilities, the pension fund intends to use a portfolio comprised of the following 14 US treasury notes:

Description	Coupon	Maturity date	Clean price
US TREAS NTS 3.500% 05/31/2053	3.5	5/31/53	101.563
US TREAS NTS 0.500% 05/31/2053	0.5	5/31/53	100.188
US TREAS NTS 2.000% 11/30/2053	2	11/30/53	101.746
US TREAS NTS 0.250% 11/30/2053	0.25	11/30/53	100.078
US TREAS NTS 2.250% 05/31/2054	2.25	5/31/54	102.941
US TREAS NTS 0.250% 05/31/2054	0.25	5/31/54	100.023
US TREAS NTS 2.125% 11/30/2054	2.125	11/30/54	103.656
US TREAS NTS 0.250% 11/30/2054	0.25	11/30/54	100.016
US TREAS NTS 2.125% 05/31/2055	2.125	5/31/55	104.461
US TREAS NTS 1.375% 11/30/2055	1.375	11/30/55	103.031
US TREAS NTS 3.250% 05/31/2056	3.25	5/31/56	109.738
US TREAS NTS 1.750% 05/31/2056	1.75	5/31/56	104.570
US TREAS NTS 2.750% 11/30/2056	2.75	11/30/56	108.879
US TREAS NTS 0.875% 11/30/2056	0.875	11/30/56	101.516

(a) Compute the dirty (full) price of each of the above 14 bonds. For consistency, assume today is November 30, 2052.

(b) Formulate a linear programming model to find the lowest-cost dedicated portfolio that covers the stream of liabilities. To eliminate the possibility of any interest risk, assume a 0% reinvestment rate on cash balances carried from one date to the next. Assume no short sales are allowed. What is the cost of your portfolio? What is the composition of your portfolio?

(c) Use the linear programming sensitivity information from part (b) to determine the term structure of interest rates implied by the portfolio.

Exercise 3.3   Prove that, if $s_1 = s_2 = \cdots = s_m = 0$ in (3.2), each of the immunization constraints in (3.1) is implied by the dedication constraints (3.2).

Exercise 3.4   A company will face the following cash requirements in the next eight quarters (positive entries represent cash needs while negative entries represent cash surpluses):

Q1	Q2	Q3	Q4	Q5	Q6	Q7	Q8
100	500	100	−600	−500	200	600	−900

The company has three borrowing possibilities.

• A 2-year loan available at the beginning of Q1, with a 1% interest per quarter.

• The other two borrowing opportunities are available at the beginning of every quarter: a 6-month loan with a 1.8% interest per quarter, and a quarterly loan with a 2.5% interest for the quarter.

• Any surplus can be invested at a 0.5% interest per quarter.

Formulate a linear program that maximizes the wealth of the company at the beginning of Q9.

Exercise 3.5 Generate the sensitivity report for Exercise 3.4 with your favorite LP solver.

(a) Suppose the cash requirement in Q2 is 300 (instead of 500). How would this affect the wealth in Q9?

(b) Suppose the cash requirement in Q2 is 100 (instead of 500). Can the sensitivity report be used to determine the wealth in Q9?

(c) One of the company's suppliers may allow deferred payments of $50 from Q3 to Q4. What would be the value of this?

Exercise 3.6 A home buyer can combine several mortgage loans to finance the purchase of a house. Regulations impose limits on the amount that can be borrowed from certain sources as well as a limit on the total reimbursement each month. Let B be the borrowing needs and T the number of months over which the loans will be paid back. There are n different loan opportunities available. Loan i has a fixed interest rate r_i, a length $T_i \leq T$ and a maximum amount borrowed b_i. The monthly payment on loan i is not required to be the same every month, but a minimum payment m_i is required each month. Furthermore, we would like the total monthly payment p over all loans to be constant. Formulate a linear program that finds a combination of loans that minimizes the home buyer's cost of borrowing.

Hint: In addition to variables x_{ti} for the payment on loan i in month t, it may be useful to introduce a variable for the amount of outstanding principal on loan i in month t.

3.7   Case Study

Let y denote the current year. A municipality sends you the following liability stream (in million dollars):

12/15/y	6/15/y+1	12/15/y+1	6/15/y+2	12/15/y+2	6/15/y+3	12/15/y+3	6/15/y+4
11	9	8	7	9	10	9	12

12/15/y+4	6/15/y+5	12/15/y+5	6/15/y+6	12/15/y+6	6/15/y+7	12/15/y+7	6/15/y+8
9	6	5	7	9	7	8	7

Questions
1. Determine the current term structure of treasury rates, and find the present value, dollar duration, and dollar convexity of the stream of liabilities. Please explain the main steps (interest rates, discount factors, compounding, etc.) followed in your calculations. You can find current data on numerous websites such as

    http://finance.yahoo.com/bonds
    http://fixedincome.fidelity.com/fi/FILanding

2. Identify at least 30 fixed-income assets that are suitable for a dedicated portfolio. Use assets that are considered risk-free, e.g., US government non-callable treasury bonds, treasury bills, or treasury notes. Display a succinct summary of the main characteristics of the bonds you chose (prices, coupon rates, maturity dates).

3. Formulate a linear programming model to find the lowest-cost dedicated portfolio that covers the stream of liabilities. To eliminate the possibility of any interest risk, assume a 0% reinvestment rate on cash balances carried from one date to the next. Assume no short sales are allowed. What is the cost of your portfolio? What is the composition of your portfolio?

4. Use the linear programming sensitivity information to determine the term structure of interest rates implied by the portfolio. Use a plot to compare it with the current term structure of treasury rates.

5. Formulate a linear programming model to find the lowest-cost portfolio that matches the present value, dollar duration, and dollar convexity of the stream of liabilities. Assume no short sales are allowed. What is the cost of your portfolio? How much would you save by using this immunization strategy instead of dedication? Is your portfolio immunized against non-parallel shifts in the term structure? Explain why or why not.

6. Combine a cash matching strategy for the liabilities during the first three years and an immunization strategy based on present value, duration and convexity for the liabilities during the last five years. Compare the cost of this portfolio with the cost of the two previous portfolios.

7. The municipality would like you to make a second bid: What is your best dedicated portfolio of risk-free bonds you can create _if short sales are allowed?_ Did you find arbitrage opportunities? Did you take into consideration the bid–ask spread? Did you set limits on the transaction amounts? Discuss the practical feasibility of your solution.

4 Linear Programming Models: Arbitrage and Asset Pricing

In this chapter, we prove the fundamental theorem of asset pricing and we give several applications, from arbitrage detection in the foreign exchange market, to pricing of options, and clientele effects in bond portfolio management.

4.1 Arbitrage Detection in the Foreign Exchange Market

The foreign exchange market includes the trading of currencies. It is one of the markets with largest trading volume. Given two currencies at any particular time, say the US dollar and the euro, there are two exchange rates between them: one dollar will buy r_1 euros, and one euro will buy r_2 dollars. It is evident that an arbitrage opportunity would arise if $r_1 r_2 > 1$ since one could simultaneously convert 1 dollar into r_1 euros and the r_1 euros into $r_1 r_2 > 1$ dollars. These two transactions would net $r_1 r_2 - 1$ dollars without any risk.

An interesting related question is: Can one detect a similar type of arbitrage opportunity involving more than two currencies? In particular, consider the following hypothetical exchange rates among the currencies USD (US Dollars), EUR (Euros), GBP (British Pounds), AUD (Australian Dollars), and JPY (Japanese Yen).

	USD	EUR	GBP	AUD	JPY
USD	1	0.639	0.537	1.0835	98.89
EUR	1.564	1	0.843	1.6958	154.773
GBP	1.856	1.186	1	2.014	184.122
AUD	0.9223	0.589	0.496	1	91.263
JPY	0.01011	0.00645	0.00543	0.01095	1

A simple verification shows that there are no arbitrage opportunities involving only two currencies. However, could there be one involving more than two currencies? Could you simply eyeball such an opportunity? If you cannot, can you prove that such an opportunity does not exist?

We next show how to answer these questions using linear programming. For convenience, use $i = 1, \ldots, 5$ to index the above five currencies USD, EUR, GBP, AUD, and JPY in that order. We let a_{ij} denote the exchange rate from currency i to currency j. For instance $a_{34} = 2.014$ and $a_{25} = 154.773$. To model a set of transactions with potential for arbitrage, consider the following decision variables:

- x_{ij}: amount of currency i converted to currency j.
- y_k: net amount of currency k after all transactions.

These variables are related via the following constraints:

$$y_k = \sum_{i=1}^{5} a_{ik} x_{ik} - \sum_{j=1}^{5} x_{kj}, \ k = 1, \ldots, 5.$$

An arbitrage would exist if there is a set of transactions so that after all transactions the net amount for each currency is non-negative and at least one of them is strictly positive. To find such a set of transactions we could solve the following linear programming problem:

$$
\begin{aligned}
\max \quad & y_1 \\
\text{s.t.} \quad & y_k = \sum_{i=1}^{5} a_{ik} x_{ik} - \sum_{j=1}^{5} x_{kj}, \ k = 1, \ldots, 5 \\
& x_{ij} \geq 0 \\
& y_k \geq 0.
\end{aligned}
$$

However, if there is indeed an arbitrage opportunity, then the above problem would be unbounded. We can easily amend the above model so that the arbitrage can be revealed by introducing a bound on the objective function:

$$
\begin{aligned}
\max \quad & y_1 \\
\text{s.t.} \quad & y_k = \sum_{i=1}^{5} a_{ik} x_{ik} - \sum_{j=1}^{5} x_{kj}, \ k = 1, \ldots, 5 \\
& y_1 \leq 1 \\
& x_{ij} \geq 0 \\
& y_k \geq 0.
\end{aligned}
$$

Solving this linear programming model, we find that indeed there are arbitrage opportunities. However, to obtain \$1 in arbitrage, we have to exchange about 1669.172 US dollars into 1066.601 euros, then convert these euros into 899.1446 pounds, then convert these pounds into 1810.877 Australian dollars, and finally change these Australian dollars into 1670.172 US dollars. The arbitrage opportunity is so tight that, depending on the numerical precision used, a linear programming solver may not find it. Furthermore, even if a solver does find it, the tightness of the arbitrage may render it impractical when accounting for market frictions such as transaction costs.

4.2 The Fundamental Theorem of Asset Pricing

One of the most widely studied problems in financial mathematics is the pricing of *contingent claims*. These are securities whose price depends on the value of another *underlying security*. Under the assumption of no arbitrage, the price of such a contingent claim should match the price of a portfolio that replicates the payoff of the contingent claim. This basic principle underlies the powerful option pricing machinery dating back to the pioneering work of Merton (1973) and Black and Scholes (1973). The absence of arbitrage and the replication argument can be cleverly stated in terms of a so-called *risk-neutral probability measure*. The latter concept can be equivalently stated in terms of a *stochastic discount factor* or a *positive linear pricing rule*.

We next use linear programming duality to give a formal derivation of the equivalence between the absence of arbitrage and the existence of a risk-neutral probability measure for the special case of a simple economy in a single-period framework. Assume the economy contains m assets. Let $\mathbf{S}_0 := \begin{bmatrix} S_0^1 & \cdots & S_0^m \end{bmatrix}^\mathsf{T}$ denote the vector of prices per share of the m assets at time 0 (beginning of the period). Assume there are n possible states $\Omega = \{\omega_1, \ldots, \omega_n\}$ at time 1 (end of the period). Let $\mathbf{S}_1(\omega_j) = \begin{bmatrix} S_1^1(\omega_j) & \cdots & S_1^m(\omega_j) \end{bmatrix}^\mathsf{T}$ denote the vector of prices per share of the m assets at time 1 in state ω_j.

An *arbitrage opportunity* in this economy is an opportunity to make money without any cost and without any risk. Mathematically, an arbitrage opportunity is a portfolio of the m assets that has non-positive cost, yields non-negative payoffs in all future states, and in addition either has strictly negative cost or generates a strictly positive payoff in some future state. In other words, an arbitrage portfolio is a set of holdings $y_1, \ldots, y_m$ in the m assets such that

$$\mathbf{S}_0^\mathsf{T}\mathbf{y} \leq 0, \quad \mathbf{S}_1(\omega_j)^\mathsf{T}\mathbf{y} \geq 0, \quad j = 1, \ldots, n$$

and such that at least one of these inequalities is strict.

A *positive linear pricing rule* is a set of positive numbers $x_1, \ldots, x_n$ such that

$$\mathbf{S}_0 = \sum_{j=1}^n \mathbf{S}_1(\omega_j)x_j, \quad i = 1, \ldots, m.$$

Proposition 4.1 *In the above single-period economy with m assets and n future states there is no arbitrage if and only if there exists a positive linear pricing rule.*

Proof Let $\mathbf{S} := \begin{bmatrix} \mathbf{S}_1(\omega_1) & \cdots & \mathbf{S}_1(\omega_n) \end{bmatrix}$. An arbitrage portfolio is precisely a solution to the following system of inequalities:

$$\begin{bmatrix} \mathbf{S} & -\mathbf{S}_0 \end{bmatrix}^\mathsf{T}\mathbf{y} \gneq \mathbf{0}. \tag{4.1}$$

Similarly, a positive linear pricing rule is precisely a solution to the following system of inequalities:

$$\mathbf{S}\mathbf{x} = \mathbf{S}_0 \\ \mathbf{x} > \mathbf{0}. \tag{4.2}$$

Observe that (4.2) has a solution if and only if the following system of inequalities has a solution:

$$\begin{bmatrix} \mathbf{S} & -\mathbf{S}_0 \end{bmatrix} \mathbf{u} = \mathbf{0}$$
$$\mathbf{u} > \mathbf{0}. \tag{4.3}$$

Hence it suffices to show that (4.1) does not have a solution if and only if (4.3) has a solution. This readily follows from Theorem 2.5(c). $\square$

The existence of a positive linear pricing rule can be equivalently stated in terms of a stochastic discount factor or in terms of a risk-neutral measure. In both of these interpretations the set of future states $\Omega = \{\omega_1, \ldots, \omega_n\}$ is seen as a probability space. Assume Ω is endowed with a probability measure $\mathbb{P}$. Then the future payoff of each asset i can be seen as a random variable $S_i : \Omega \to \mathbb{R}$. A *stochastic discount factor* is a random variable $D : \Omega \to \mathbb{R}$ such that

$$\mathbf{S}_0 = \mathbb{E}(D\mathbf{S}_1) = \sum_{j=1}^{n} D(\omega_j)\mathbf{S}_1(\omega_j)\mathbb{P}(\omega_j), \quad i = 1, \ldots, m.$$

For convenience, assume there is a risk-free asset in the above economy; that is, an asset i such that $S_0^i = 1$ and $S_1^i(\omega_j) = 1 + r$ for $j = 1, \ldots, n$. A *risk-neutral probability measure* is a probability measure $\mathbb{Q}$ in the space $\Omega = \{\omega_1, \ldots, \omega_n\}$ such that

$$\mathbf{S}_0 = \frac{1}{1+r}\widetilde{\mathbb{E}}(\mathbf{S}_1) = \frac{1}{1+r}\sum_{j=1}^{n} \mathbf{S}_1(\omega_j)\mathbb{Q}(\omega_j).$$

Here $\widetilde{\mathbb{E}}$ indicates that the expectation is taken with respect to the risk-neutral probability measure $\mathbb{Q}$, as opposed to the original probability measure $\mathbb{P}$.

We can now formally state the fundamental theorem of asset pricing.

Theorem 4.2 (Fundamental theorem of asset pricing)　*Consider the above single-period economy with n future states and m assets, one of which is risk-free. The following conditions are equivalent:*

(i) *There are no arbitrage opportunities.*
(ii) *There exists a positive linear pricing rule.*
(iii) *There exists a positive stochastic discount factor.*
(iv) *There exists a risk-neutral probability measure.*

Proposition 4.1, which gives the equivalence between (i) and (ii), provides the crux of the proof of Theorem 4.2. The proofs of the other equivalences are a straightforward exercise.

4.3　One-Period Binomial Pricing Model

This section illustrates the *pricing* of a contingent claim on an underlying risky security in a simple one-period binomial model. This model provides the building block for the powerful and widely used multi-period *binomial pricing model* that we will discuss in Chapter 15.

Consider a single-period economy with a risk-free asset and a risky asset. Let r denote the risk-free rate and S_0 denote the price per share of the risky asset at time 0. Assume there are two possible future states $\Omega = \{H, T\}$ at time 1. Assume the price per share of the risky asset at time 1 is $S_1(H) = u \cdot S_0$ in state H and $S_1(T) = d \cdot S_0$ in state T for some "up" and "down" factors $u > d > 0$:

$$S_0 \diagup \begin{array}{l} S_1(H) = uS_0 \\ \\ S_1(T) = dS_0 \end{array}$$

In this economy there is no arbitrage if and only if $u > 1 + r > d$ and in this case the risk-neutral probability measure should satisfy

$$S_0 = \frac{1}{1+r}(\mathbb{Q}(H)S_1(H) + \mathbb{Q}(T)S_2(T)) = \frac{S_0}{1+r}(\mathbb{Q}(H)u + \mathbb{Q}(T)d)$$

$$1 = \mathbb{Q}(H) + \mathbb{Q}(T).$$

Therefore,

$$\mathbb{Q}(H) = \frac{1+r-d}{u-d}, \qquad \mathbb{Q}(T) = \frac{u-1-r}{u-d}. \tag{4.4}$$

It is customary to write $\tilde{p} := \mathbb{Q}(H)$ and $\tilde{q} := 1 - \tilde{p} = \mathbb{Q}(T)$ as shorthand for the risk-neutral probabilities and $p = \mathbb{P}(H)$ and $q = 1 - p = \mathbb{P}(T)$ for the actual probabilities:

Consider the problem of pricing a *contingent claim* on the risky asset with the following payoff structure:

$$V_0 = ? \diagup \begin{array}{l} V_1(H) \\ \\ V_1(T) \end{array}$$

For example, the contingent claim could be a *European call option* – that is, a contract with the following conditions. At time 1, the *holder* of the option has the right, but not the obligation, to purchase a share of the risky asset, known as the *underlying security*, for a prescribed amount, known as the *strike price*. Thus the payoff of a European call option with strike K is $V_1 = (S_1 - K)^+ := \max\{S_1 - K, 0\}$. The payoff structure of this option in our one-period binomial model is as follows:

$$V_0 = ? \diagup \begin{array}{l} V_1(H) = (uS_0 - K)^+ \\ \\ V_1(T) = (dS_0 - K)^+ \end{array}$$

A *European put option* is a similar contract, except that it confers the right to sell the underlying security for a prescribed strike price.

The fundamental theorem of asset pricing implies that the fair price V_0 of a general contingent claim with payoffs $V_1(H)$ and $V_1(T)$ is

$$V_0 = \frac{1}{1+r}(\tilde{p}V_1(H) + \tilde{q}V_1(T)).$$

Furthermore, the binomial pricing model yields the following *delta-hedging* formula to construct a portfolio of the underlying risky asset and the risk-free asset that replicates the payoff of the contingent claim. At time 0 construct a portfolio with Δ shares of the underlying risky asset and B shares of the risk-free asset where

$$\Delta := \frac{V_1(H) - V_1(T)}{S_1(H) - S_1(T)} = \frac{V_1(H) - V_1(T)}{S_0(u-d)}, \quad B := \frac{uV_1(T) - dV_1(H)}{(1+r)(u-d)}.$$

A straightforward verification shows that this portfolio replicates the payoff of the contingent claim. That is, the payoff of the portfolio (Δ, B) is as follows:

$$\Delta S_0 + B \begin{cases} \Delta S_1(H) + (1+r)B = V_1(H) \\ \\ \Delta S_1(T) + (1+r)B = V_1(T) \end{cases}$$

Thus the value of this replicating portfolio at time 0 must be V_0 to rule out arbitrage. Indeed, the value of the replicating portfolio at time 0 is

$$\Delta S_0 + B = \frac{(1+r)(V_1(H) - V_1(T)) + uV_1(T) - dV_1(H)}{(1+r)(u-d)}$$

$$= \frac{1}{1+r}(\tilde{p}V_1(H) + \tilde{q}V_1(T)) = V_0.$$

Example 4.3   Suppose stock XYZ has share price $S_0 = 40$ today. Suppose the share price of stock XYZ a month from today will either double or halve with equal probabilities:

$$S_0 = 40 \begin{cases} S_1(H) = 80 \\ \\ S_1(T) = 20 \end{cases}$$

Assume also that the one-month risk-free rate is zero. Consider a European call option to buy one share of XYZ stock for $50 a month from today. What is the fair price of this option?

In Example 4.3 we have $u = 2, d = \frac{1}{2}$ and $r = 0$. Thus the risk-neutral probabilities are $\tilde{p} = \frac{1}{3}$ and $\tilde{q} = \frac{2}{3}$. Next, observe that a month from now the call option with strike price $50 will be worth $30 = \$80 - \$50$ in the H state and it will be worthless in the T state. Thus the fair price of the option is the price of the following contract:

$$? \begin{cases} (80-50)^+ = 30 \\ \\ (20-50)^+ = 0 \end{cases}$$

The fundamental theorem of asset pricing implies that the fair price of this contract is

$$30 \cdot \tilde{p} + 0 \cdot \tilde{q} = 30 \cdot \frac{1}{3} = 10.$$

Furthermore, from the delta-hedging formula it follows that a replicating portfolio can be constructed by buying $\frac{1}{2}$ share of stock XYZ and borrowing 10 shares of the risk-free asset. Observe that the value of this replicating portfolio at time 0 is

$$\frac{1}{2} \cdot 40 - 10 = 10.$$

Using the risk-neutral probability measure we can also price other derivative securities on the XYZ stock. For example, consider a European put option on the XYZ stock with strike price \$60 and with the same expiration date:

$$? \begin{cases} (60-80)^+ = 0 \\ (60-20)^+ = 40 \end{cases}$$

It readily follows that the fair price of this option is

$$0 \cdot \tilde{p} + 40 \cdot \tilde{q} = 40 \cdot \frac{2}{3} = \frac{80}{3}.$$

Observe that in the one-period binomial pricing model the risk-neutral probability is unique and the payoff of any contingent claim can be replicated via delta-hedging. In general, uniqueness of the risk-neutral probability corresponds to *completeness* of the market. The latter concept means that the payoff of any contract can be replicated with a portfolio of the existing underlying assets in the economy as detailed in Exercise 4.6.

4.4 Static Arbitrage Bounds

The no-arbitrage approach discussed in Section 4.2 has the drawback that it assumes only a finite number of possible future states. In this section, we do not make this assumption. Instead, we assume that there is a finite set of derivative securities written on the same underlying asset and with the same maturity. We show how the no-arbitrage approach can be used to obtain so-called *static arbitrage bounds* on the price of a new derivative security implied by the prices of the other derivative securities. As in Section 4.2, the gist of this approach is to use linear programming to detect arbitrage opportunities in a single-period economy. This discussion is based on Herzel (2005).

Consider an underlying security with a (random) price S_T at a future time T. Consider n derivative securities written on this security that mature at time T, and have *piecewise linear* payoff functions $\Psi_i(S_T)$, each with a single breakpoint K_i, for $i = 1, \ldots, n$. The obvious motivation is the collection of calls and puts written on the underlying security with strike prices K_i, $i = 1, \ldots, n$. More precisely, if the ith derivative security were a European call with strike price K_i, we would have $\Psi_i(S_T) = (S_T - K_i)^+$. If it were a European put with strike price K_i, we would have $\Psi_i(S_T) = (K_i - S_T)^+$.

We shall assume without loss of generality that the K_is are in increasing order. Also, we let p_i denote the current price of the ith derivative security. Consider a portfolio $\mathbf{x} = \begin{bmatrix} x_1 & \cdots & x_n \end{bmatrix}^\mathsf{T}$ of the derivative securities 1 to n and let $\Psi^\times(S_T)$ denote the payoff function of the portfolio:

$$\Psi^\times(S_T) = \sum_{i=1}^{n} \Psi_i(S_T) x_i.$$

The cost of the portfolio $\mathbf{x}$ is given by

$$\sum_{i=1}^{n} p_i x_i. \tag{4.5}$$

To determine whether there exists an arbitrage opportunity in the above set of n derivative securities, we consider the following question: Is it possible to construct a portfolio of the derivative securities $1, \ldots, n$ with negative cost and whose payoff function $\Psi^\times(S_T)$ at time T is non-negative for all $S_T \in [0, \infty)$? Since non-negativity of $\Psi^\times(S_T)$ corresponds to "no future obligations" such a portfolio would be an arbitrage opportunity.

Since all $\Psi_i(S_T)$s are piecewise linear, so is $\Psi^\times(S_T)$ with breakpoints in $K_1, \ldots, K_n$. Note that a piecewise linear function is non-negative over $[0, \infty)$ if and only if it is non-negative at 0 and all the breakpoints, and if the slope of the function is non-negative to the right of the largest breakpoint. In other words, $\Psi^\times(S_T)$ is non-negative for all $S_T \geq 0$ if and only if the following three conditions hold:

(i) $\Psi^\times(0) \geq 0$,
(ii) $\Psi^\times(K_j) \geq 0$, $j = 1, \ldots, n$,
(iii) $[(\Psi^\times)'_+(K_n)] \geq 0$.

These three conditions can be written as the following system of linear inequalities:

$$\sum_{i=1}^{n} \Psi_i(0) x_i \quad \geq \quad 0$$

$$\sum_{i=1}^{n} \Psi_i(K_j) x_i \quad \geq \quad 0, \quad j = 1, \ldots, n \tag{4.6}$$

$$\sum_{i=1}^{n} (\Psi_i(K_n + 1) - \Psi_i(K_n)) x_i \quad \geq \quad 0.$$

Since all $\Psi_i(S_T)$s are piecewise linear, the quantity $\Psi_i(K_n + 1) - \Psi_i(K_n)$ gives the right derivative of $\Psi_i(S_T)$ at K_n and the expression in the last constraint is the right derivative of $\Psi^\times(S_T)$ at K_n. The system of linear inequalities (4.6) can be more succinctly written as

$$\mathbf{Kx} \geq \mathbf{0}$$

for

$$
\mathbf{K} := \begin{bmatrix}
\Psi_1(0) & \cdots & \Psi_n(0) \\
\Psi_1(K_1) & \cdots & \Psi_n(K_1) \\
\vdots & & \vdots \\
\Psi_1(K_n) & \cdots & \Psi_n(K_n) \\
\Psi_1(K_n+1) - \Psi_1(K_n) & \cdots & \Psi_n(K_n+1) - \Psi(K_n)
\end{bmatrix}.
$$

It thus follows that the above type of arbitrage opportunity exists if and only if the following problem has a solution:

$$
\mathbf{Kx} \geq 0, \quad \mathbf{p}^{\mathsf{T}}\mathbf{x} < 0.
$$

Next, we focus on the special case where the derivative securities under consideration are European call options with strikes K_i for $i = 1, \ldots, n$. In this case $\Psi_i(S_T) = (S_T - K_i)^+$ and hence

$$
\Psi_i(K_j) = (K_j - K_i)^+.
$$

In this case, (4.6) can be written as

$$
\mathbf{Ax} \geq 0 \tag{4.7}
$$

for

$$
\mathbf{A} = \begin{bmatrix}
K_2 - K_1 & 0 & 0 & \cdots & 0 \\
K_3 - K_1 & K_3 - K_2 & 0 & \cdots & 0 \\
\vdots & \vdots & \vdots & \ddots & \vdots \\
K_n - K_1 & K_n - K_2 & K_n - K_3 & \cdots & 0 \\
1 & 1 & 1 & \cdots & 1
\end{bmatrix}.
$$

This formulation is obtained by removing the first two constraints of (4.6) which are redundant in this particular case. Using this formulation, we obtain the following theorem giving necessary and sufficient conditions for a set of call option prices to contain no arbitrage opportunities.

Theorem 4.4 *Let $K_1 \leq K_2 \leq \cdots \leq K_n$ denote the strike prices of European call options written on the same underlying security with the same maturity. For $i = 1, \ldots, n$ let p_i denote the price of the ith call option. There are no arbitrage opportunities if and only if the prices p_i, $i = 1, \ldots, n$, satisfy the following conditions:*

(i) *$0 \leq p_n \leq p_{n-1} \leq \cdots \leq p_1$.*
(ii) *The piecewise linear function $C : [K_1, K_n] \rightarrow \mathbb{R}$ with breakpoints $K_1, \ldots, K_n$ defined by $C(K_i) := p_i$, $i = 1, \ldots, n$, is convex.*

The previous approach can be further extended to infer both lower and upper bounds on the current price p_{new} of a new derivative with maturity T and payoff $\Psi_{\text{new}}(S_T)$ given prices of other derivatives on the same underlying security and with the same maturity. As before, assume $\Psi_{\text{new}}(S_T)$ and $\Psi_i(S_T)$ are piecewise linear functions each with a single breakpoint K and K_i, $i = 1, \ldots, n$, respectively. Assume $K_1 \leq K_2 \leq \cdots \leq K_n$ and let p_i denote the current price of the ith derivative security.

Assume there is no arbitrage involving the n derivatives with payoffs $\Psi_i(S_T)$, for $i = 1, \ldots, n$. The previous reasoning applied to the larger set of $n + 1$ derivatives shows that there is no arbitrage if and only if the following two conditions hold:

- First, $p_{\text{new}} \geq \mathbf{p}^\mathsf{T} \mathbf{x}$ for any portfolio $\mathbf{x} = \begin{bmatrix} x_1 & \cdots & x_n \end{bmatrix}^\mathsf{T}$ such that

$$\Psi_{\text{new}}(S_T) \geq \Psi^\times(S_T) \text{ for all } S_T \geq 0.$$

- Second, $p_{\text{new}} \leq \mathbf{p}^\mathsf{T} \mathbf{x}$ for any portfolio $\mathbf{x} = \begin{bmatrix} x_1 & \cdots & x_n \end{bmatrix}^\mathsf{T}$ such that

$$\Psi_{\text{new}}(S_T) \leq \Psi^\times(S_T) \text{ for all } S_T \geq 0.$$

In words, the first condition states that the price of the new derivative has to be at least as large as the price of any sub-replicating portfolio of the old securities. Likewise, the second condition states that the price of the new derivative has to be at most as large as the price of any super-replicating portfolio of the old securities. The above two conditions automatically yield the following static arbitrage bounds on p.

Lower bound:

$$p_{\text{new}}^\ell := \max \quad \mathbf{p}^\mathsf{T} \mathbf{x}$$
$$\text{s.t.} \quad \Psi_{\text{new}}(S_T) \geq \Psi^\times(S_T) \text{ for all } S_T \geq 0.$$

Upper bound:

$$p_{\text{new}}^u := \min \quad \mathbf{p}^\mathsf{T} \mathbf{x}$$
$$\text{s.t.} \quad \Psi_{\text{new}}(S_T) \leq \Psi^\times(S_T) \text{ for all } S_T \geq 0.$$

The piecewise linearity of $\Psi_{\text{new}}(S_T)$ and $\Psi_i(S_T)$, $i = 1, \ldots, n$, implies that both inequalities $\Psi_{\text{new}}(S_T) \geq \Psi^\times(S_T)$ for all $S_T \geq 0$, and $\Psi_{\text{new}}(S_T) \geq \Psi^\times(S_T)$ for all $S_T \geq 0$ can be formulated as a finite system of linear inequalities. Therefore, both the upper and lower static arbitrage bounds can be formulated as linear programming models. In particular, for the special case where $\Psi_i(S_T) = (S_T - K_i)^+$, for $i = 1, \ldots, n$, and $\Psi_{\text{new}}(S_T) = (S_T - K)^+$ with $K_1 \leq K \leq K_n$ the static arbitrage upper bound p_{new}^u on p can be written as the following linear programming model (Exercise 4.10):

$$p_{\text{new}}^u := \min \quad \mathbf{p}^\mathsf{T} \mathbf{x}$$
$$\text{s.t.} \quad \mathbf{A}\mathbf{x} \geq \mathbf{b}, \tag{4.8}$$

where

$$
\mathbf{A} = \begin{bmatrix} K_2 - K_1 & 0 & 0 & \cdots & 0 \\ K_3 - K_1 & K_3 - K_2 & 0 & \cdots & 0 \\ \vdots & \vdots & \vdots & \ddots & \vdots \\ K_n - K_1 & K_n - K_2 & K_n - K_3 & \cdots & 0 \\ 1 & 1 & 1 & \cdots & 1 \end{bmatrix}, \quad \mathbf{b} = \begin{bmatrix} (K_2 - K)^+ \\ (K_3 - K)^+ \\ \vdots \\ (K_n - K)^+ \\ 0 \end{bmatrix}.
$$

4.5   Tax Clientele Effects in Bond Portfolio Management

This section presents a model proposed by Ronn (1987) to elicit *clientele effects* induced by taxes in the bond market. Related models were also proposed by Hodges and Schaefer (1977) and Schaefer (1982). The crux of the model is to formulate a linear program that exploits the price differential of bonds given their after-tax cash flows. To do so, the model finds a long–short portfolio that simultaneously buys "underpriced" bonds and sells "overpriced" bonds while ensuring non-negative cash flows throughout the lives of the bonds.

Next we describe the details of the model. Assume the bond market includes N bonds with the following characteristics:

- The ask and bid prices of bond j are p_j^a and p_j^b respectively for $j = 1, \ldots, N$.
- Each unit of bond j generates a cash flow a_j^t at date t for $j = 1, \ldots, N$ and $t = 1, \ldots, T$. These cash flows are after-tax coupon and/or principal payments.
- The minimal risk-free reinvestment rate at future dates $t = 1, \ldots, T$ is ρ.

Linear programming model for tax clientele effects in the bond market

Variables:

x_j^a: number of units of bond j bought, for $j = 1, \ldots, N$

x_j^b: number of units of bond j sold, for $j = 1, \ldots, N$

z_t: surplus cash flow at date t, for $t = 1, \ldots, T$.

Objective:

$$
\max \ \sum_{j=1}^{N} p_j^b x_j^b - \sum_{j=1}^{N} p_j^a x_j^a.
$$

Constraints: Cash balance constraints in each date and bounds on $x_j^a, x_j^b,$ and z_t:

$$z_1 = \sum_{j=1}^{N} a_j^1 x_j^a - \sum_{j=1}^{N} a_j^1 x_j^b$$

$$z_t = (1+\rho)z_{t-1} + \sum_{j=1}^{N} a_j^t x_j^a - \sum_{j=1}^{N} a_j^t x_j^b, \quad \text{for } t = 2, \ldots, T$$

$$x_j^a, x_j^b \geq 0, \quad \text{for } j = 1, \ldots, N$$

$$z_t \geq 0, \quad \text{for } t = 1, \ldots, T$$

$$x_j^a, x_j^b \leq 1, \quad \text{for } j = 1, \ldots, N.$$

$$(4.9)$$

Some comments are in order. The above objective function is the net difference between the value of the short positions and long positions of the portfolio. The short positions have to settle at the bid prices whereas the long positions have to settle at the ask prices. Because of this distinction, the constraints $x_j^a, x_j^b \geq 0$ are required. To ensure that the portfolio is risk-free, we require the surplus cash flows z_t to be non-negative for each date t.

The resulting linear program admits two main types of solutions. Either all bonds are priced within the bid–ask spread. In that case the optimal value of the linear program is zero and it is trivially attained by not taking any short or long positions. On the other hand, if there are exploitable price differentials in the bonds, the linear program chooses long and short holdings so as to maximize the difference between the values of the long and short positions. In that case the optimal value is positive. To avoid unbounded values, the model includes the upper bounds $x_j^a, x_j^b \leq 1$ on the long and short holdings.

Note that the model requires bonds with perfectly forecastable cash flows. Thus, non-callable bonds and notes are deemed appropriate, but callable bonds are excluded.

The proposed model explicitly accounts for the taxation of income and capital gains for specific investor classes. This means that the cash flows need to be adjusted for the presence of taxes. For a discount bond (that is, when $p_j^a < 100$), the after-tax cash flow of bond j at date t is

$$a_j^t = c_j^t(1 - \tau),$$

where c_j^t is the coupon payment at date t and τ is the ordinary income tax rate. At maturity, the after-tax cash flow of bond j is

$$a_j^t = (100 - p_j^a)(1 - g) + p_j^a,$$

where g is the capital gains tax rate.

On the other hand, for a premium bond (that is, when $p_j^a > 100$), the premium is amortized against ordinary income over the life of the bond, giving rise to an after-tax coupon payment of

$$a_j^t = \left[c_j^t - \frac{p_j^a - 100}{n_j} \right] (1 - \tau) + \frac{p_j^a - 100}{n_j},$$

where n_j is the number of coupon payments remaining to maturity.

A premium bond also makes a non-taxable repayment of

$$a_j^t = 100$$

at maturity.

Major categories of taxable investors are domestic banks, insurance companies, individuals, non-financial corporations, and foreigners. In each case, one needs to distinguish the tax rates on capital gains versus ordinary income.

As an example, consider tax-exempt investors. For this class of investors, Schaefer (1982) observed that the "purchased" portfolio contains high coupon bonds and the "sold" portfolio is dominated by low coupon bonds. This can be explained as follows: The preferential taxation of capital gains for (most) taxable investors causes them to gravitate towards low coupon bonds. Consequently, for tax-exempt investors, low coupon bonds are "overpriced" and not desirable as investment vehicles.

4.6   Notes

The fundamental theorem of asset pricing is central to the mathematical finance literature. The connection between arbitrage and risk-neutral pricing underlies the classical work of Merton (1973) and Black and Scholes (1973). More explicit and formal statements on the relation between absence of arbitrage and existence of stochastic discount factors in single-period as well as in multi-period settings were developed by Ross (1976), Harrison and Kreps (1979), and Harrison and Pliska (1981). The textbooks by Back (2010), Duffie (2001), and Shreve (2000) give a detailed treatment of this important topic.

4.7   Exercises

Exercise 4.1   The Excel spreadsheet "Exercise 4.1 FX model" gives cross-currency exchange rates among the currencies USD, EUR, GBP, AUD, and JPY. Use a linear programming model to detect if these exchange rates contain an arbitrage opportunity. To do so, use the following decision variables:

x_{ij}: amount of currency i converted to currency j.

y_k: net amount of currency k after all transactions.

Is there an arbitrage opportunity? If the answer is yes, then describe it, for example: "Convert 1000 USD to EUR then to JPY then back to USD to net 1 USD without putting money in."

Exercise 4.2   Let S_0 be the current share price of a "risky" security and assume that there are two possible share prices for this security at a future time T: $S_T(u) = S_0 \cdot u$ and $S_T(d) = S_0 \cdot d$, where $u > d > 0$. Assume there is also a "risk-free" security with current share price 1 and future share price $1 + r$ at time T. Show that there is no arbitrage opportunity involving the risky and risk-free securities if and only if $u > 1 + r > d$.

Exercise 4.3 Assume that the XYZ stock is currently priced at \$40. At the end of the next period, the price of XYZ is expected to be in one of the following two states: $40 \cdot u$ or $40 \cdot d$. We know that $d < 1 < \frac{5}{4} < u$ but we do not know d or u. The interest rate is zero. If a European call option with strike price \$50 is priced at \$10 while a European call option with strike price \$40 is priced at \$13, and we assume that these prices do not contain any arbitrage opportunities, what is the fair price of a European put option with a strike price of \$40?

Exercise 4.4 Assume that the XYZ stock is currently priced at \$40. At the end of the next period, the price of XYZ is expected to be in one of the following two states: $40 \cdot u$ or $40 \cdot d$. We know that $d < 1 < u$ but we do not know d or u. The interest rate is $r = 0$. European call options on XYZ with strike prices of \$30, \$40, \$50, and \$60 are priced at \$10, \$7, \$10/3, and \$0. Which one of these options is mispriced? Why?

Exercise 4.5 Prove the equivalences (ii) $\Leftrightarrow$ (iii) $\Leftrightarrow$ (iv) in Theorem 4.2.

Exercise 4.6 Consider the setting of Proposition 4.1. Assume there is no arbitrage and thus a positive linear pricing rule exists.

(a) Show that the linear pricing rule is unique if and only if the matrix $\mathbf{S}$ has full column rank.
(b) Consider a new asset with payoff $S_1^{m+1}(\omega_j)$ per share in state ω_j at time 1. Show that if the linear pricing rule is unique then there exists a portfolio $\mathbf{y} = \begin{bmatrix} y_1 & \cdots & y_m \end{bmatrix}^{\mathsf{T}}$ of the m old assets that replicates the payoff of the new asset; that is,

$$\sum_{i=1}^{m} S_1^i(\omega_j) y_i = S_1^{m+1}(\omega_j), \ j = 1, \ldots, n.$$

(c) Conclude that to rule out arbitrage, the price S_0^{m+1} at time 0 of the new asset must be equal to

$$\mathbf{S}_0^{\mathsf{T}} \mathbf{y} = \sum_{i=1}^{m} S_0^i y_i.$$

Exercise 4.7 Prove Theorem 4.4.

Exercise 4.8 Both Theorem 4.4 and the linear programming model (4.8) implicitly assume that the ith call can be bought or sold at the same price p_i. In real markets, there is always a gap between the price a buyer pays for a security and the amount the seller collects called the *bid–ask spread*.

Assume that the ask price of the ith call is p_i^a and its bid price is p_i^b with $p_i^a > p_i^b$. Develop analogs of Theorem 4.4 and of (4.8) in the case where we can only purchase the calls at their ask prices or sell them at their bid prices.

Exercise 4.9 Consider all the call options on the S&P 500 index or on a highly traded security that expire on the same day, about three months from today. Their current prices can be downloaded from the website of the Chicago Board of Options Exchange at www.cboe.com or several other market quote websites. Formulate the linear programming problem (4.7) (or, rather, the version you developed for Exercise 4.8 since market quotes will include bid and ask prices) to determine whether these prices contain any arbitrage opportunities.

Sometimes, illiquid securities (those that are not traded very often) can have misleading prices since the reported price corresponds to the last transaction in that security, which may have happened several days ago, and if there were to be a new transaction, this value would change dramatically. As a result, it is quite possible that you will discover false "arbitrage opportunities" because of these misleading prices. Repeat this exercise but this time use only prices of call options that have had a trading volume of at least 100 on the day you downloaded the prices.

Exercise 4.10 Prove that, for the special case where $\Psi_i(S_T) = (S_T - K_i)^+$, with $i = 1, \ldots, n$, and $\Psi_{\text{new}}(S_T) = (S_T - K)^+$ with $K_1 \leq K \leq K_n$, the static arbitrage upper bound p_{new}^u on p can be written as the following linear programming model:

$$p_{\text{new}}^u := \min \quad \mathbf{p}^\top \mathbf{x}$$
$$\text{s.t.} \quad \mathbf{Ax} \geq \mathbf{b},$$

where

$$\mathbf{A} = \begin{bmatrix} K_2 - K_1 & 0 & 0 & \cdots & 0 \\ K_3 - K_1 & K_3 - K_2 & 0 & \cdots & 0 \\ \vdots & \vdots & \vdots & \ddots & \vdots \\ K_n - K_1 & K_n - K_2 & K_n - K_3 & \cdots & 0 \\ 1 & 1 & 1 & \cdots & 1 \end{bmatrix}, \quad \mathbf{b} = \begin{bmatrix} (K_2 - K)^+ \\ (K_3 - K)^+ \\ \vdots \\ (K_n - K)^+ \\ 0 \end{bmatrix}.$$

Exercise 4.11 The purpose of this exercise is to see whether the results observed by Schaefer (1982) (see Section 4.5) occur in the current bond market. Only use non-callable bonds and notes.

Consider first the class of tax-exempt investors. Using current data, form the optimal "purchased" and "sold" bond portfolios using the linear program presented in Section 4.5. Do you observe the same tax clientele effect as documented by Schaefer for British government securities; namely, the "purchased" portfolio contains high coupon bonds and the "sold" portfolio is dominated by low coupon bonds.

Repeat the same analysis with different types of taxable investors.

(a) Is there a clientele effect in the pricing of US government investments, with tax-exempt investors, or those without preferential treatment of capital gains, gravitating towards high coupon bonds?

(b) Do you observe that not all high coupon bonds are desirable to investors without preferential treatment of capital gains? Nor are all low coupon bonds attractive to those with preferential treatment of capital gains. Can you find reasons why this may be the case?

The dual price, say u_t, associated with the cash balance constraint at date t in (4.9) represents the present value of an additional dollar at time t. Explain why. It follows that u_t may be used to compute the term structure of spot interest rates R_t, given by the relation

$$R_t = \left(\frac{1}{u_t}\right)^{1/t} - 1.$$

Compute this week's term structure of spot interest rates for tax-exempt investors.

Part II

Single-Period Models

5 Quadratic Programming: Theory and Algorithms

5.1 Quadratic Programming

A *quadratic program* is an optimization problem whose objective is to minimize or maximize a quadratic function subject to a finite set of linear equality and inequality constraints. By flipping signs if necessary, a quadratic program can be written in the generic form:

$$\min_{\mathbf{x}} \quad \tfrac{1}{2}\mathbf{x}^{\mathsf{T}}\mathbf{Q}\mathbf{x} + \mathbf{c}^{\mathsf{T}}\mathbf{x}$$
$$\text{s.t.} \quad \mathbf{A}\mathbf{x} = \mathbf{b} \tag{5.1}$$
$$\mathbf{D}\mathbf{x} \geq \mathbf{d}$$

for some vectors and matrices $\mathbf{c} \in \mathbb{R}^n$, $\mathbf{b} \in \mathbb{R}^m$, $\mathbf{d} \in \mathbb{R}^p$, $\mathbf{A} \in \mathbb{R}^{m \times n}$, $\mathbf{D} \in \mathbb{R}^{p \times n}$, $\mathbf{Q} \in \mathbb{R}^{n \times n}$. As observed in Chapter 1 we may assume that $\mathbf{Q}$ is a symmetric matrix. The term *quadratic programming model* is also used to refer to a quadratic program. We will use these terms interchangeably throughout the book.

Quadratic programming models arise in a variety of practical contexts. The seminal *mean–variance model* of Markowitz and most of its variants for portfolio selection are quadratic programs as we illustrate in Example 5.1 below and discuss in full detail in Chapter 6. The popular ordinary least-squares and lasso estimation procedures in linear regression are also quadratic programs. Quadratic programs are also often solved as subproblems in the solution of more general nonlinear optimization problems.

Observe that the constraint set in (5.1) is convex since it is a system of linear inequalities. Furthermore, the objective function of (5.1) is convex when $\mathbf{Q}$ is a positive semidefinite matrix. Throughout this chapter we assume that $\mathbf{Q}$ is symmetric and positive semidefinite. Therefore problem (5.1) is a convex program.

A quadratic programming model is in *standard form* if it is written as follows:

$$\min_{\mathbf{x}} \quad \tfrac{1}{2}\mathbf{x}^{\mathsf{T}}\mathbf{Q}\mathbf{x} + \mathbf{c}^{\mathsf{T}}\mathbf{x}$$
$$\text{s.t.} \quad \mathbf{A}\mathbf{x} = \mathbf{b} \tag{5.2}$$
$$\mathbf{x} \geq \mathbf{0}.$$

Example 5.1 (Asset allocation) Assume the one-year returns of the asset classes large stocks, small stocks, and bonds have the following correlations and standard deviations:

	Large	Small	Bonds	Standard deviation
Large	1	0.6	0.2	0.12
Small	0.6	1	0.5	0.20
Bonds	0.2	0.5	1	0.05

Determine the asset allocation of minimum risk, that is, find a portfolio comprised of these three asset classes whose return has the lowest standard deviation. Assume the portfolio can only hold long positions in each of the asset classes.

This problem can be formulated as a quadratic programming model. To that end, first construct the *covariance matrix* $\mathbf{V}$ of asset returns: this is the matrix whose (i,j) entry is the covariance of asset i and asset j; that is, $\rho_{ij} \cdot \sigma_i \cdot \sigma_j$. Using matrix notation and 'o' to denote the componentwise product of matrices, the covariance matrix can be computed as

$$
\mathbf{V} = \begin{bmatrix} 1 & 0.6 & 0.2 \\ 0.6 & 1 & 0.5 \\ 0.2 & 0.5 & 1 \end{bmatrix} \circ \begin{bmatrix} 0.12 \\ 0.20 \\ 0.05 \end{bmatrix} \begin{bmatrix} 0.12 & 0.20 & 0.05 \end{bmatrix}
$$

$$
= \begin{bmatrix} 1 & 0.6 & 0.2 \\ 0.6 & 1 & 0.5 \\ 0.2 & 0.5 & 1 \end{bmatrix} \circ \begin{bmatrix} 0.0144 & 0.024 & 0.006 \\ 0.024 & 0.04 & 0.01 \\ 0.006 & 0.01 & 0.0025 \end{bmatrix}
$$

$$
= \begin{bmatrix} 0.0144 & 0.0144 & 0.0012 \\ 0.0144 & 0.04 & 0.005 \\ 0.0012 & 0.005 & 0.0025 \end{bmatrix}.
$$

We are now ready to describe the quadratic programming formulation for the above asset allocation problem. (A more detailed discussion is given in Chapter 6.)

Quadratic programming model for asset allocation
Variables:
$\qquad x_i$: percentage of the portfolio invested in asset i for $i = 1, 2, 3$.
Objective (minimize the variance of the portfolio return):

$$
\min_{\mathbf{x}} \mathbf{x}^{\mathsf{T}} \mathbf{V} \mathbf{x} = \min_{x_1, x_2, x_3} \left(0.0144 x_1^2 + 0.04 x_2^2 + 0.0025 x_3^2 + 0.0288 x_1 x_2 \right.
$$

$$
\left. + 0.0024 x_1 x_3 + 0.01 x_2 x_3 \right)
$$

Constraints:

$$
\begin{aligned}
x_1 + x_2 + x_3 &= 1 \quad \text{(percentages add up to one)} \\
x_1, x_2, x_3 &\geq 0 \quad \text{(long-only positions).}
\end{aligned}
$$

Observe that even in this small example the quadratic objective is much more concise and easier to write using matrix notation.

We now discuss the special case of a convex quadratic program without constraints. As Example 5.3 below illustrates, this kind of model arises naturally in the ordinary least-squares procedure.

Consider a quadratic program without constraints:

$$\min_{\mathbf{x}} \ \tfrac{1}{2}\mathbf{x}^{\mathsf{T}}\mathbf{Q}\mathbf{x} + \mathbf{c}^{\mathsf{T}}\mathbf{x}. \tag{5.3}$$

The optimality conditions in this case are as follows.

Theorem 5.2 *Let* $\mathbf{c} \in \mathbb{R}^n$, $\mathbf{Q} \in \mathbb{R}^{n \times n}$ *and assume* $\mathbf{Q}$ *is symmetric and positive semidefinite. If* (5.3) *is bounded, then it attains its minimum. Furthermore, a point* $\mathbf{x} \in \mathbb{R}^n$ *is an optimal solution to* (5.3) *if and only if*

$$\mathbf{Q}\mathbf{x} + \mathbf{c} = \mathbf{0}. \tag{5.4}$$

When $\mathbf{Q}$ is positive definite, the problem (5.3) has the unique minimizer $\mathbf{x} = -\mathbf{Q}^{-1}\mathbf{c}$. When $\mathbf{Q}$ is positive semidefinite but not positive definite, the matrix $\mathbf{Q}$ is singular and the problem (5.3) is either unbounded or has multiple solutions.

Example 5.3 (Ordinary least squares) Assume $(\mathbf{x}_i, y_i)$, for $i = 1, \ldots, N$, is a random sample drawn from the joint distribution of X, Y where X, Y are respectively $\mathbb{R}^p$-valued and $\mathbb{R}$-valued random variables. Using the *training data* $(\mathbf{x}_i, y_i)$, with $i = 1, \ldots, N$, estimate a vector of coefficients β for the linear model

$$Y = \beta^{\mathsf{T}} X + \epsilon.$$

The most popular approach to this problem is to find the estimate of β that solves the following least-squares problem:

$$\min_{\beta} \sum_{i=1}^{N} (\beta^{\mathsf{T}}\mathbf{x}_i - y_i)^2.$$

Observe that

$$\sum_{i=1}^{N} (\beta^{\mathsf{T}}\mathbf{x}_i - y_i)^2 = (\mathbf{X}\beta - \mathbf{y})^{\mathsf{T}}(\mathbf{X}\beta - \mathbf{y}) = \beta^{\mathsf{T}}\mathbf{X}^{\mathsf{T}}\mathbf{X}\beta - 2\mathbf{y}^{\mathsf{T}}\mathbf{X}\beta + \mathbf{y}^{\mathsf{T}}\mathbf{y}$$

for

$$\mathbf{X} = \begin{bmatrix} \mathbf{x}_1^{\mathsf{T}} \\ \vdots \\ \mathbf{x}_N^{\mathsf{T}} \end{bmatrix}, \quad \mathbf{y} = \begin{bmatrix} y_1 \\ \vdots \\ y_N \end{bmatrix}.$$

Hence the least-squares problem can be formulated as follows.

Quadratic programming formulation for least-squares estimation.
Variables:
   β: vector of coefficients in the linear model $Y = \beta^{\mathsf{T}} X + \epsilon$.
Objective:

$$\min_{\beta} \ \tfrac{1}{2}\beta^{\mathsf{T}}\mathbf{Q}\beta - \mathbf{b}^{\mathsf{T}}\beta,$$

   where $\mathbf{Q} := \mathbf{X}^{\mathsf{T}}\mathbf{X}$, $\mathbf{b} = \mathbf{X}^{\mathsf{T}}\mathbf{y}$.
Constraints: None.

By applying Theorem 5.2 we obtain the widely known solution to the least-squares problem:

$$\hat{\beta} := \mathbf{Q}^{-1}\mathbf{b} = (\mathbf{X}^{\mathsf{T}}\mathbf{X})^{-1}\mathbf{X}^{\mathsf{T}}\mathbf{y},$$

provided the $N \times p$ matrix $\mathbf{X}$ has full column rank. This latter condition usually holds in the typical practical situation when there are more observations than predictor variables; that is, when $N > p$. However, the case $N < p$ occurs as well. In this kind of situation the matrix $\mathbf{X}$ is never full column rank so the ordinary least-squares approach is not appropriate. Section 5.6.2 describes two popular variants for this kind of situation, namely *ridge regression* and *lasso regression,* both of which can be seen as modifications of the ordinary least-squares procedure.

5.2   Numerical Quadratic Programming Solvers

As with linear programming, there are a variety of highly efficient, fast, and reliable commercial and open-source software packages for convex quadratic programming. Most of these packages implement versions of the algorithms sketched in Section 5.5 below. We illustrate two of these solvers by applying them to Example 5.1.

Excel Solver

Figure 5.1 displays a printout of an Excel spreadsheet implementation of the quadratic programming model for Example 5.1 as well as the dialog box obtained when we run the Excel add-in `Solver`. The spreadsheet model contains the three components of the quadratic program. The decision variables are in the range **B20:D20**. The objective function is in cell **E22**. The Excel formula in this cell, using matrix operations, is as follows:

MMULT(B20 : D20, MMULT(B16 : D18, TRANSPOSE(B20 : D20))).

The left-hand and right-hand sides of the equality constraint are in cells **E20** and **G20** respectively.

MATLAB CVX

Figure 5.2 displays a CVX script for the same problem. The script can be run provided the freely available CVX toolbox is installed.

Using either of these solvers we obtain the optimal solution to the problem in Example 5.1:

$$\mathbf{x}^* = \begin{bmatrix} 0.0897 \\ 0 \\ 0.9103 \end{bmatrix}.$$

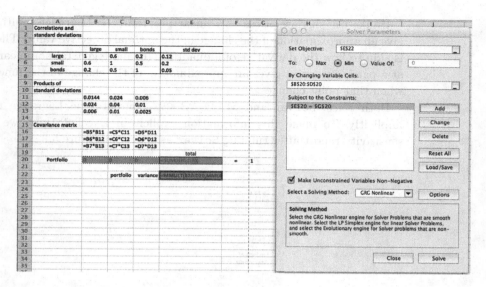

Figure 5.1 Spreadsheet implementation and the Solver dialog box for the asset allocation model

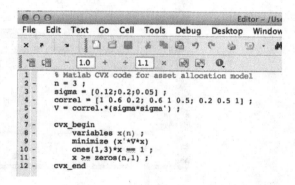

Figure 5.2 MATLAB CVX code for the asset allocation model

5.3 Sensitivity Analysis

As is the case for linear programming, the process of solving a quadratic program also generates some interesting *sensitivity information* via the so-called *Lagrange multipliers* associated with the constraints. Assume the constraints of a quadratic program, and hence the Lagrange multipliers, are indexed by $i = 1, \ldots, m$. The *Lagrange multiplier* y_i^* of the ith constraint has the following sensitivity interpretation:

> If the right-hand side of the ith constraint changes by Δ, then the optimal value of the quadratic program changes by approximately $\Delta \cdot y_i^*$ for small Δ.

Unlike the shadow prices of a linear program, the Lagrange multipliers only give an approximation of the change in the optimal objective value. The situation is akin to how the derivative of a quadratic (or more general nonlinear) function at a particular point gives an approximation of the change in the function value when that point changes.

Both Excel `Solver` and MATLAB `CVX` compute the Lagrange multipliers implicitly. To make this information explicit in Excel `Solver`, we request a sensitivity report after running `Solver` as shown in Figure 5.3.

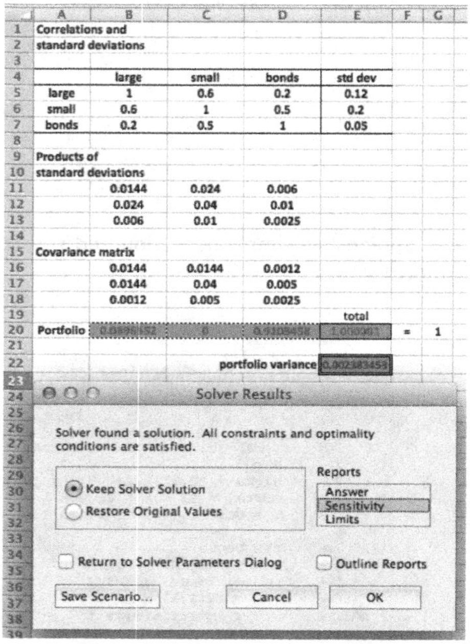

Figure 5.3 Requesting sensitivity report in Solver

Figure 5.4 displays the sensitivity report for Example 5.1. The values y_i^* can be found in the column labeled "Lagrange Multiplier". In CVX this information can also be obtained by including a line of code to save the dual information y as shown in Figure 5.5. Both solvers yield the dual value $\mathbf{y}^* = 0.0047669$.

5.4   *Duality and Optimality Conditions

As in linear programming, there is a *dual* quadratic program associated with every *primal* quadratic programming problem, and this dual can be obtained via the Lagrangian function. Throughout this section consider the *primal* quadratic

Variable Cells

Cell	Name	Final Value	Reduced Gradient
B20	Portfolio large	0.089655197	0
C20	Portfolio small	0	0.006918652
D20	Portfolio bonds	0.910345803	0

Constraints

Cell	Name	Final Value	Lagrange Multiplier
E20	Portfolio total	1.000001	0.004766915

Figure 5.4 Sensitivity report

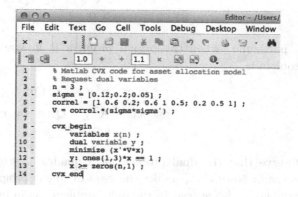

Figure 5.5 MATLAB CVX code with dual variables

program

$$\min_{\mathbf{x}} \quad \tfrac{1}{2}\mathbf{x}^{\mathsf{T}}\mathbf{Q}\mathbf{x} + \mathbf{c}^{\mathsf{T}}\mathbf{x}$$
$$\text{s.t.} \quad \mathbf{A}\mathbf{x} = \mathbf{b} \tag{5.5}$$
$$\mathbf{D}\mathbf{x} \geq \mathbf{d},$$

where $\mathbf{c} \in \mathbb{R}^n$, $\mathbf{Q} \in \mathbb{R}^{n \times n}$, $\mathbf{A} \in \mathbb{R}^{m \times n}$, $\mathbf{b} \in \mathbb{R}^m$, $\mathbf{D} \in \mathbb{R}^{p \times n}$, $\mathbf{d} \in \mathbb{R}^p$, and $\mathbf{Q}$ is symmetric and positive semidefinite.

The *Lagrangian function* associated with (5.5) is

$$L(\mathbf{x}, \mathbf{y}, \mathbf{s}) := \tfrac{1}{2}\mathbf{x}^{\mathsf{T}}\mathbf{Q}\mathbf{x} + \mathbf{c}^{\mathsf{T}}\mathbf{x} + \mathbf{y}^{\mathsf{T}}(\mathbf{b} - \mathbf{A}\mathbf{x}) + \mathbf{s}^{\mathsf{T}}(\mathbf{d} - \mathbf{D}\mathbf{x}).$$

The constraints of (5.5) can be encoded via the Lagrangian function through the following observation: For a given vector $\mathbf{x}$

$$\max_{\substack{\mathbf{y}, \mathbf{s} \\ \mathbf{s} \geq 0}} L(\mathbf{x}, \mathbf{y}, \mathbf{s}) = \begin{cases} \tfrac{1}{2}\mathbf{x}^{\mathsf{T}}\mathbf{Q}\mathbf{x} + \mathbf{c}^{\mathsf{T}}\mathbf{x} & \text{if } \mathbf{A}\mathbf{x} = \mathbf{b} \text{ and } \mathbf{D}\mathbf{x} \geq \mathbf{d} \\ +\infty & \text{otherwise.} \end{cases}$$

Therefore the primal problem (5.5) can be written as

$$\min_{\mathbf{x}} \max_{\substack{\mathbf{y},\mathbf{s} \\ \mathbf{s} \geq 0}} L(\mathbf{x},\mathbf{y},\mathbf{s}).$$

The dual problem is obtained by flipping the order of the min and max operations:

$$\max_{\substack{\mathbf{y},\mathbf{s} \\ \mathbf{s} \geq 0}} \min_{\mathbf{x}} L(\mathbf{x},\mathbf{y},\mathbf{s}).$$

It is easy to see that the dual problem can be written as follows:

$$\begin{array}{ll} \max_{\mathbf{x},\mathbf{y},\mathbf{s}} & \mathbf{b}^{\mathsf{T}}\mathbf{y} + \mathbf{d}^{\mathsf{T}}\mathbf{s} - \frac{1}{2}\mathbf{x}^{\mathsf{T}}\mathbf{Q}\mathbf{x} \\ \text{s.t.} & \mathbf{A}^{\mathsf{T}}\mathbf{y} + \mathbf{D}^{\mathsf{T}}\mathbf{s} - \mathbf{Q}\mathbf{x} = \mathbf{c} \\ & \mathbf{s} \geq \mathbf{0}. \end{array} \qquad (5.6)$$

In particular, when the primal problem is in standard form (5.2), the dual problem is

$$\begin{array}{ll} \max_{\mathbf{x},\mathbf{y},\mathbf{s}} & \mathbf{b}^{\mathsf{T}}\mathbf{y} - \frac{1}{2}\mathbf{x}^{\mathsf{T}}\mathbf{Q}\mathbf{x} \\ \text{s.t.} & \mathbf{A}^{\mathsf{T}}\mathbf{y} - \mathbf{Q}\mathbf{x} + \mathbf{s} = \mathbf{c} \\ & \mathbf{s} \geq \mathbf{0}. \end{array}$$

Observe that the dual problem of a quadratic program is again a quadratic program. Note that, unlike the case of linear programming, some primal-like variables $\mathbf{x}$ also appear in the dual problem. As in linear programming, there is a deep connection between the primal problem (5.5) and its dual (5.6). The next result follows by construction.

Theorem 5.4 (Weak duality) *Assume* $\mathbf{x}$ *is a feasible point for* (5.5) *and* $(\tilde{\mathbf{x}},\mathbf{y},\mathbf{s})$ *is a feasible point for* (5.6). *Then*

$$\mathbf{b}^{\mathsf{T}}\mathbf{y} + \mathbf{d}^{\mathsf{T}}\mathbf{s} - \frac{1}{2}\tilde{\mathbf{x}}^{\mathsf{T}}\mathbf{Q}\tilde{\mathbf{x}} \leq \frac{1}{2}\mathbf{x}^{\mathsf{T}}\mathbf{Q}\mathbf{x} + \mathbf{c}^{\mathsf{T}}\mathbf{x}.$$

Proof If $\mathbf{x}$ and $(\tilde{\mathbf{x}},\mathbf{y},\mathbf{s})$ satisfy the above assumptions then

$$\begin{aligned} \mathbf{b}^{\mathsf{T}}\mathbf{y} + \mathbf{d}^{\mathsf{T}}\mathbf{s} - \frac{1}{2}\tilde{\mathbf{x}}^{\mathsf{T}}\mathbf{Q}\tilde{\mathbf{x}} &\leq (\mathbf{A}\mathbf{x})^{\mathsf{T}}\mathbf{y} + (\mathbf{D}\mathbf{x})^{\mathsf{T}}\mathbf{s} - \frac{1}{2}\tilde{\mathbf{x}}^{\mathsf{T}}\mathbf{Q}\tilde{\mathbf{x}} \\ &= (\mathbf{A}^{\mathsf{T}}\mathbf{y} + \mathbf{D}^{\mathsf{T}}\mathbf{s})^{\mathsf{T}}\mathbf{x} - \frac{1}{2}\tilde{\mathbf{x}}^{\mathsf{T}}\mathbf{Q}\tilde{\mathbf{x}} \\ &= (\mathbf{c} + \mathbf{Q}\tilde{\mathbf{x}})^{\mathsf{T}}\mathbf{x} - \frac{1}{2}\tilde{\mathbf{x}}^{\mathsf{T}}\mathbf{Q}\tilde{\mathbf{x}} \\ &= \frac{1}{2}\mathbf{x}^{\mathsf{T}}\mathbf{Q}\mathbf{x} + \mathbf{c}^{\mathsf{T}}\mathbf{x} - \frac{1}{2}(\mathbf{x} - \tilde{\mathbf{x}})^{\mathsf{T}}\mathbf{Q}(\mathbf{x} - \tilde{\mathbf{x}}) \\ &\leq \frac{1}{2}\mathbf{x}^{\mathsf{T}}\mathbf{Q}\mathbf{x} + \mathbf{c}^{\mathsf{T}}\mathbf{x}. \end{aligned}$$

$\square$

The following much deeper result also holds.

Theorem 5.5 (Strong duality)   *Assume one of the problems* (5.5) *or* (5.6) *is feasible. Then this problem is bounded if and only if the other one is feasible. In that case both problems have optimal solutions and their optimal values are the same.*

We refer the reader to Güler (2010) or Nocedal and Wright (2006) for a proof of Theorem 5.5. This result is closely tied to certain kinds of *separation theorems* for convex sets. For details see Güler (2010, chapters 6 and 11). A powerful consequence of Theorem 5.5 is the following characterization of the solutions to both (5.5) and (5.6).

Theorem 5.6 (Optimality conditions)   *The vectors* $\mathbf{x} \in \mathbb{R}^n$ *and* $(\tilde{\mathbf{x}}, \mathbf{y}, \mathbf{s}) \in \mathbb{R}^n \times \mathbb{R}^m \times \mathbb{R}^p$ *are optimal solutions to* (5.5) *and* (5.6) *respectively if and only if* $\mathbf{Q}\mathbf{x} = \mathbf{Q}\tilde{\mathbf{x}}$ *and*

$$
\begin{aligned}
\mathbf{Q}\mathbf{x} + \mathbf{c} - \mathbf{A}^\mathsf{T}\mathbf{y} - \mathbf{D}^\mathsf{T}\mathbf{s} &= \mathbf{0} \\
\mathbf{A}\mathbf{x} - \mathbf{b} &= \mathbf{0} \\
\mathbf{D}\mathbf{x} - \mathbf{d} &\geq \mathbf{0} \\
\mathbf{s} &\geq \mathbf{0} \\
(\mathbf{D}\mathbf{x} - \mathbf{d})_i s_i &= 0, \quad i = 1, \ldots, p.
\end{aligned}
\tag{5.7}
$$

For a quadratic program in standard form (5.2), the optimality conditions (5.7) can be written as follows:

$$
\begin{aligned}
-\mathbf{Q}\mathbf{x} + \mathbf{A}^\mathsf{T}\mathbf{y} + \mathbf{s} &= \mathbf{c} \\
\mathbf{A}\mathbf{x} &= \mathbf{b} \\
\mathbf{x} &\geq \mathbf{0} \\
\mathbf{s} &\geq \mathbf{0} \\
x_i s_i &= 0, \quad i = 1, \ldots, n.
\end{aligned}
\tag{5.8}
$$

Observe that (5.8) nicely extends the optimality conditions (2.9) for linear programming in standard form.

The optimality conditions (5.7) can be seen as "saddle-point" conditions for the Lagrangian function

$$
L(\mathbf{x}, \mathbf{y}, \mathbf{s}) = \tfrac{1}{2}\mathbf{x}^\mathsf{T}\mathbf{Q}\mathbf{x} + \mathbf{c}^\mathsf{T}\mathbf{x} + \mathbf{y}^\mathsf{T}(\mathbf{b} - \mathbf{A}\mathbf{x}) + \mathbf{s}^\mathsf{T}(\mathbf{d} - \mathbf{D}\mathbf{x}).
$$

We next discuss the special case of a quadratic program with equality constraints only. Consider the problem

$$
\begin{aligned}
\min_{\mathbf{x}} \quad & \tfrac{1}{2}\mathbf{x}^\mathsf{T}\mathbf{Q}\mathbf{x} + \mathbf{c}^\mathsf{T}\mathbf{x} \\
\text{s.t.} \quad & \mathbf{A}\mathbf{x} = \mathbf{b},
\end{aligned}
\tag{5.9}
$$

where $\mathbf{c} \in \mathbb{R}^n$, $\mathbf{Q} \in \mathbb{R}^{n \times n}$, $\mathbf{A} \in \mathbb{R}^{m \times n}$, $\mathbf{b} \in \mathbb{R}^m$, and $\mathbf{Q}$ is symmetric and positive semidefinite. In this case the optimality conditions (5.7) simplify to

$$
\begin{aligned}
\mathbf{Q}\mathbf{x} + \mathbf{c} - \mathbf{A}^\mathsf{T}\mathbf{y} &= \mathbf{0} \\
\mathbf{A}\mathbf{x} - \mathbf{b} &= \mathbf{0}.
\end{aligned}
\tag{5.10}
$$

The optimality conditions (5.10) in turn can be stated in terms of the Lagrangian function of (5.9):

$$L(\mathbf{x}, \mathbf{y}) = \tfrac{1}{2}\mathbf{x}^\mathsf{T}\mathbf{Q}\mathbf{x} + \mathbf{c}^\mathsf{T}\mathbf{x} + \mathbf{y}^\mathsf{T}(\mathbf{b} - \mathbf{A}\mathbf{x}).$$

Indeed observe that (5.10) can be succinctly written as

$$\nabla L(\mathbf{x}, \mathbf{y}) = \mathbf{0}.$$

When $\mathbf{Q}$ is positive definite and $\mathbf{A}$ has full row rank, problem (5.9) has a unique minimizer $\mathbf{x}$ and a unique Lagrange multiplier $\mathbf{y}$ given by

$$\begin{bmatrix} \mathbf{x} \\ \mathbf{y} \end{bmatrix} = \begin{bmatrix} \mathbf{Q} & -\mathbf{A}^\mathsf{T} \\ \mathbf{A} & 0 \end{bmatrix}^{-1} \begin{bmatrix} -\mathbf{c} \\ \mathbf{b} \end{bmatrix}.$$

In particular, if $\mathbf{Q}$ is positive definite and $\mathbf{A}$ has full row rank, then the minimizer and vector of Lagrange multipliers for the problem

$$\begin{aligned} \min_{\mathbf{x}} \quad & \tfrac{1}{2}\mathbf{x}^\mathsf{T}\mathbf{Q}\mathbf{x} \\ \text{s.t.} \quad & \mathbf{A}\mathbf{x} = \mathbf{b} \end{aligned} \tag{5.11}$$

are respectively

$$\mathbf{x}^* = \mathbf{Q}^{-1}\mathbf{A}^\mathsf{T}(\mathbf{A}\mathbf{Q}^{-1}\mathbf{A}^\mathsf{T})^{-1}\mathbf{b}$$
$$\mathbf{y}^* = (\mathbf{A}\mathbf{Q}^{-1}\mathbf{A}^\mathsf{T})^{-1}\mathbf{b}.$$

Example 5.7 (Asset allocation)   Consider the same problem as in Example 5.1 but assume this time that the portfolio is allowed to hold short positions.

The formulation for this modification of Example 5.1 is straightforward: just drop the non-negativity constraint on the variables. Thus we obtain the quadratic programming model

$$\begin{aligned} \min_{\mathbf{x}} \quad & \tfrac{1}{2}\mathbf{x}^\mathsf{T}\mathbf{V}\mathbf{x} \\ \text{s.t.} \quad & \mathbf{1}^\mathsf{T}\mathbf{x} = 1. \end{aligned}$$

From the above discussion it readily follows that the optimal solution and Lagrange multiplier are

$$\mathbf{x}^* = \frac{1}{\mathbf{1}^\mathsf{T}\mathbf{V}^{-1}\mathbf{1}}\mathbf{V}^{-1}\mathbf{1}$$
$$y^* = \frac{1}{\mathbf{1}^\mathsf{T}\mathbf{V}^{-1}\mathbf{1}}.$$

For the particular value of $\mathbf{V}$ in Example 5.1 we get the following optimal solution and Lagrange multiplier

$$\mathbf{x}^* = \begin{bmatrix} 0.1934 \\ -0.1406 \\ 0.9472 \end{bmatrix}, \quad y^* = 0.001897074.$$

5.5 *Algorithms

We next sketch extensions of the two main algorithmic schemes for linear programming discussed in Chapter 2. The first scheme, namely *active-set methods*, can be seen as an analog of the simplex method. The second scheme, namely interior-point methods, is a straightforward extension from the linear programming to the quadratic programming context.

5.5.1 Active-Set Methods

Active-set methods are based on the following key observation. Assume $\bar{x}$ is an optimal solution to (5.5) and

$$I := \{i = 1, \ldots, p : (D\bar{x} - d)_i = 0\}.$$

Then the optimality conditions (5.7) can be rewritten as

$$
\begin{aligned}
Qx + c - A^\mathsf{T}y - D_I^\mathsf{T}s_I &= 0 \\
Ax - b &= 0 \\
D_I x - d_I &= 0 \\
s_I &\geq 0.
\end{aligned}
\tag{5.12}
$$

If we ignore the last constraint $s_I \geq 0$, the remaining conditions in (5.12) are precisely the optimality conditions of the problem

$$
\begin{aligned}
\min_{x} \quad & \tfrac{1}{2}x^\mathsf{T}Qx + c^\mathsf{T}x \\
\text{s.t.} \quad & Ax = b \\
& D_I x = d_I.
\end{aligned}
\tag{5.13}
$$

This suggests an algorithmic strategy to solve (5.5): guess the *active set I* and solve the subproblem (5.13). If the solution $\bar{x}$ to this subproblem satisfies the other conditions in (5.7) then stop. Otherwise, make a new guess for I. Algorithm 5.1 gives a possible version of this strategy.

Each main iteration of Algorithm 5.1 requires solving the following subproblem for some current trial solution $\bar{x}$ and trial active set I:

$$
\begin{aligned}
\min_{\Delta x} \quad & \tfrac{1}{2}(\Delta x)^\mathsf{T}Q\Delta x + (Q\bar{x} + c)^\mathsf{T}\Delta x \\
\text{s.t.} \quad & A\Delta x = 0 \\
& D_I \Delta x = 0.
\end{aligned}
\tag{5.14}
$$

To update the trial solution we also need to compute the step length

$$
\alpha := \min\left\{ 1, \; \min_{\substack{i \notin I \\ D_i \Delta x < 0}} \frac{d_i - D_i \Delta x}{D_i \Delta x} \right\}.
\tag{5.15}
$$

Algorithm 5.1 Active-set method

1: choose $\mathbf{x}_0$ feasible for (5.5) and $I_0 \subseteq \{i : \mathbf{D}_i \mathbf{x}_0 = \mathbf{d}_i, \ i = 1, \ldots, p\}$
2: **for** $k = 0, 1, \ldots$ **do**
3:       solve (5.14) for $I = I_k$ and $\bar{\mathbf{x}} = \mathbf{x}_k$
4:       **if** $\Delta \mathbf{x} = 0$ **then**
5:             compute the Lagrange multipliers $\bar{\mathbf{s}}_I$ of (5.14) for $I = I_k$ and $\bar{\mathbf{x}} = \mathbf{x}_k$
6:             **if** $\bar{\mathbf{s}}_I \geq \mathbf{0}$ **then** HALT $\bar{\mathbf{x}}$ is an optimal solution to (5.5)
7:             **else**
8:                   let $j := \arg\min_{i \in I} \bar{\mathbf{s}}_i$, $I_{k+1} := I_k \backslash \{j\}$, and $\mathbf{x}_{k+1} := \mathbf{x}_k$
9:             **end if**
10:      **else**
11:            compute α via (5.15) for $I = I_k$ and let $\mathbf{x}_{k+1} := \mathbf{x}_k + \alpha \Delta \mathbf{x}$
12:            **if** α_k has a blocking constraint j **then** $I_{k+1} := I_k \cup \{j\}$
13:            **else** $I_{k+1} := I_k$
14:            **end if**
15:      **end if**
16: **end for**

We say that the step length α computed in (5.15) has a *blocking constraint*, $j \notin I$, if

$$\alpha = \min_{\substack{i \notin I \\ \mathbf{D}_i \Delta \mathbf{x} < 0}} \frac{\mathbf{d}_i - \mathbf{D}_i \Delta \mathbf{x}}{\mathbf{D}_i \Delta \mathbf{x}} = \frac{\mathbf{d}_j - \mathbf{D}_j \Delta \mathbf{x}}{\mathbf{D}_j \Delta \mathbf{x}} < 1.$$

And we say that α has no blocking constraints when

$$\alpha = 1 < \min_{\substack{i \notin I \\ \mathbf{D}_i \Delta \mathbf{x} < 0}} \frac{\mathbf{d}_i - \mathbf{D}_i \Delta \mathbf{x}}{\mathbf{D}_i \Delta \mathbf{x}}.$$

5.5.2   Interior-Point Methods

For notational convenience and without loss of generality we assume that the problem of interest is in standard form (5.2).

As in the linear programming case (Section 2.7.3), interior-point methods generate a sequence of iterates that satisfy $\mathbf{x}, \mathbf{s} > \mathbf{0}$. Each iteration of the algorithm aims to make progress towards satisfying $-\mathbf{Q}\mathbf{x} + \mathbf{A}^\mathsf{T}\mathbf{y} + \mathbf{s} = \mathbf{c}$, $\mathbf{A}\mathbf{x} = \mathbf{b}$, and $x_i s_i = 0$, with $i = 1, \ldots, n$.

As before we use the following notational convention: Given a vector $\mathbf{x} \in \mathbb{R}^n$, let $\mathbf{X} \in \mathbb{R}^{n \times n}$ denote the diagonal matrix defined by $X_{ii} = x_i$, with $i = 1, \ldots, n$, and let $\mathbf{1} \in \mathbb{R}^n$ denote the vector whose components are all 1s. The optimality conditions (5.8) can be restated as

$$\begin{bmatrix} -\mathbf{Q}\mathbf{x} + \mathbf{A}^\mathsf{T}\mathbf{y} + \mathbf{s} - \mathbf{c} \\ \mathbf{A}\mathbf{x} - \mathbf{b} \\ \mathbf{X}\mathbf{S}\mathbf{1} \end{bmatrix} = \begin{bmatrix} \mathbf{0} \\ \mathbf{0} \\ \mathbf{0} \end{bmatrix}, \quad \mathbf{x}, \mathbf{s} \geq \mathbf{0}.$$

Given $\mu > 0$, let $(\mathbf{x}(\mu), \mathbf{y}(\mu), \mathbf{s}(\mu))$ be the solution to the following perturbed version of the above optimality conditions:

$$\begin{bmatrix} -\mathbf{Q}\mathbf{x} + \mathbf{A}^\mathsf{T}\mathbf{y} + \mathbf{s} - \mathbf{c} \\ \mathbf{A}\mathbf{x} - \mathbf{b} \\ \mathbf{X}\mathbf{S}\mathbf{1} \end{bmatrix} = \begin{bmatrix} \mathbf{0} \\ \mathbf{0} \\ \mu\mathbf{1} \end{bmatrix}, \quad \mathbf{x}, \mathbf{s} > \mathbf{0}.$$

The first condition above can be written as $\mathbf{r}_\mu(\mathbf{x}, \mathbf{y}, \mathbf{s}) = \mathbf{0}$ for the *residual vector*

$$\mathbf{r}_\mu(\mathbf{x}, \mathbf{y}, \mathbf{s}) := \begin{bmatrix} -\mathbf{Q}\mathbf{x} + \mathbf{A}^\mathsf{T}\mathbf{y} + \mathbf{s} - \mathbf{c} \\ \mathbf{A}\mathbf{x} - \mathbf{b} \\ \mathbf{X}\mathbf{S}\mathbf{1} - \mu\mathbf{1} \end{bmatrix}.$$

The *central path* is the set $\{(\mathbf{x}(\mu), \mathbf{y}(\mu), \mathbf{s}(\mu)) : \mu > 0\}$. It is intuitively clear that $(\mathbf{x}(\mu), \mathbf{y}(\mu), \mathbf{s}(\mu))$ converges to an optimal solution to both (5.2) and its dual. This suggests the following algorithmic strategy. Suppose $(\mathbf{x}, \mathbf{y}, \mathbf{s})$ is "near" $(\mathbf{x}(\mu), \mathbf{y}(\mu), \mathbf{s}(\mu))$ for some $\mu > 0$. Use $(\mathbf{x}, \mathbf{y}, \mathbf{s})$ to move to a better point $(\mathbf{x}^+, \mathbf{y}^+, \mathbf{s}^+)$ "near" $(\mathbf{x}(\mu^+), \mathbf{y}(\mu^+), \mathbf{s}(\mu^+))$ for some $\mu^+ < \mu$.

It can be shown that if a point $(\mathbf{x}, \mathbf{y}, \mathbf{s})$ is on the central path, then the corresponding value of μ satisfies $\mathbf{x}^\mathsf{T}\mathbf{s} = n\mu$. Likewise, given $\mathbf{x}, \mathbf{s} > \mathbf{0}$, define

$$\mu(\mathbf{x}, \mathbf{s}) := \frac{\mathbf{x}^\mathsf{T}\mathbf{s}}{n}.$$

To move from a current point $(\mathbf{x}, \mathbf{y}, \mathbf{s})$ to a new point, we use the so-called *Newton step*; that is, the solution to the system of equations

$$\begin{bmatrix} -\mathbf{Q} & \mathbf{A}^\mathsf{T} & \mathbf{I} \\ \mathbf{A} & \mathbf{0} & \mathbf{0} \\ \mathbf{S} & \mathbf{0} & \mathbf{X} \end{bmatrix} \begin{bmatrix} \Delta\mathbf{x} \\ \Delta\mathbf{y} \\ \Delta\mathbf{s} \end{bmatrix} = \begin{bmatrix} \mathbf{c} + \mathbf{Q}\mathbf{x} - \mathbf{A}^\mathsf{T}\mathbf{y} - \mathbf{s} \\ \mathbf{b} - \mathbf{A}\mathbf{x} \\ \mu\mathbf{1} - \mathbf{X}\mathbf{S}\mathbf{1} \end{bmatrix}. \tag{5.16}$$

Algorithm 5.2 presents a template for an interior-point method.

Algorithm 5.2 Interior-point method for quadratic programming

---

1: choose $\mathbf{x}^0, \mathbf{s}^0 > \mathbf{0}$
2: **for** $k = 0, 1, \ldots$ **do**
3: solve the Newton system (5.16) for $(\mathbf{x}, \mathbf{y}, \mathbf{s}) = (\mathbf{x}^k, \mathbf{y}^k, \mathbf{s}^k)$ and $\mu := 0.1\mu(\mathbf{x}^k, \mathbf{s}^k)$
4: choose a step length $\alpha \in (0, 1]$ and set $(\mathbf{x}^{k+1}, \mathbf{y}^{k+1}, \mathbf{s}^{k+1}) = (\mathbf{x}^k, \mathbf{y}^k, \mathbf{s}^k) + \alpha(\Delta\mathbf{x}, \Delta\mathbf{y}, \Delta\mathbf{s})$
5: **end for**

The step length α in step 4 should be chosen so that $\mathbf{x}^{k+1}, \mathbf{s}^{k+1} > \mathbf{0}$ and the size of $\mathbf{r}_\mu(\mathbf{x}^{k+1}, \mathbf{y}^{k+1}, \mathbf{s}^{k+1})$ is sufficiently smaller than $\mathbf{r}_\mu(\mathbf{x}^k, \mathbf{y}^k, \mathbf{s}^k)$. A line-search procedure such as the one described in Algorithm 2.4 in Chapter 2 can be used for choosing the step length α.

5.6   Applications to Machine Learning

We next discuss some iconic applications of quadratic programming to machine learning. We must note that the literature on optimization models in machine learning is vast and continues to grow at a rapid pace. For a more detailed discussion on this timely subject, we refer the reader to the excellent textbooks by Friedman et al. (2001), Sra et al. (2012), and Vapnik (2013).

5.6.1   Binary Classification and Support Vector Machines

Classification problems constitute an important class of problems in financial mathematics that can be solved using optimization models and techniques. In a classification problem we have a vector of features describing an entity and the goal is to analyze these features to determine which class each entity belongs to, among two (or more) classes. For example, the classes might be "growth stocks" and "value stocks", and the entities (stocks) may be described by a feature vector that contains elements such as stock price, price–earnings ratio, growth rate for the previous periods, growth estimates, etc.

Mathematical approaches to classification often start with a training exercise. One is supplied with a list of entities, their feature vectors, and the classes they belong to. From this information, one tries to extract a mathematical structure for the entity classes so that additional entities can be classified using this mathematical structure and their feature vectors. For two-class classification, a hyperplane is probably the simplest mathematical structure that can be used to separate the feature vectors of these two different classes. Of course, there may not be any hyperplane that separates two sets of vectors. When such a hyperplane exists, we say that the two sets can be linearly separated.

Consider feature vectors $\mathbf{a}_i \in \mathbb{R}^n$ for $i = 1, \ldots, k_1$ corresponding to class 1, and vectors $\mathbf{b}_i \in \mathbb{R}^n$ for $i = 1, \ldots, k_2$ corresponding to class 2. If these two vector sets can be linearly separated, a hyperplane $\mathbf{w}^\mathsf{T}\mathbf{x} = \gamma$ exists with $\mathbf{w} \in \mathbb{R}^n, \gamma \in \mathbb{R}$ such that

$$\mathbf{w}^\mathsf{T}\mathbf{a}_i \geq \gamma, \text{ for } i = 1, \ldots, k_1$$
$$\mathbf{w}^\mathsf{T}\mathbf{b}_i \leq \gamma, \text{ for } i = 1, \ldots, k_2.$$

To have a "strict" separation, we often prefer to obtain w and γ such that

$$\mathbf{w}^\mathsf{T}\mathbf{a}_i \geq \gamma + 1, \text{ for } i = 1, \ldots, k_1$$
$$\mathbf{w}^\mathsf{T}\mathbf{b}_i \leq \gamma - 1, \text{ for } i = 1, \ldots, k_2.$$

In this manner, we find two parallel lines ($\mathbf{w}^\mathsf{T}\mathbf{x} = \gamma + 1$ and $\mathbf{w}^\mathsf{T}\mathbf{x} = \gamma - 1$) that form the boundaries of the class 1 and class 2 portions of the vector space; see Figure 5.6.

There may be several such parallel lines that separate the two classes. Which one should one choose? A good criterion is to choose the lines that have the

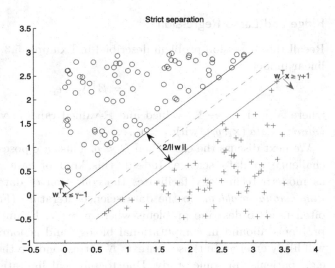

Figure 5.6 Linear separation of two classes of data points

largest margin (distance between the lines). In machine learning, this type of classification model is known as a *support vector machine* (Friedman et al., 2001; Vapnik, 2013).

(i) Consider the following quadratic problem:

$$\min_{\mathbf{w},\gamma} \quad \|\mathbf{w}\|_2^2$$
$$\begin{aligned} \mathbf{a}_i^{\mathsf{T}}\mathbf{w} &\geq \gamma + 1, \text{ for } i = 1, \ldots, k_1 \\ \mathbf{b}_i^{\mathsf{T}}\mathbf{w} &\leq \gamma - 1, \text{ for } i = 1, \ldots, k_2. \end{aligned} \tag{5.17}$$

The objective function of this problem is equivalent to maximizing the margin between the lines $\mathbf{w}^{\mathsf{T}}\mathbf{x} = \gamma + 1$ and $\mathbf{w}^{\mathsf{T}}\mathbf{x} = \gamma - 1$ (see Exercise 5.6).

(ii) The linear separation idea we presented above can be used even when the two vector sets $\{\mathbf{a}_i\}$ and $\{\mathbf{b}_i\}$ are not linearly separable. (Note that linearly inseparable sets will result in an infeasible problem in formulation (5.17).) This is achieved by introducing a non-negative violation variable for each constraint of (5.17). Then, one has two objectives: to minimize the total of the constraint violations and to maximize the margin. One can formulate a quadratic programming model that combines these two objectives using an adjustable parameter that can be chosen in a way to put more weight on violations or margin, depending on one's preference (see Exercise 5.7).

5.6.2   Ridge and Lasso Regression

Recall the regression problem described in Example 5.3, namely to estimate the linear model

$$Y = \beta^\mathsf{T} X + \epsilon,$$

where X and Y are $\mathbb{R}^p$-valued and $\mathbb{R}$-valued random variables, by using some *training data* $(\mathbf{x}_i, y_i)$, with $i = 1, \ldots, N$.

We next discuss the case when $N < p$. This case poses a classical and modern challenge in data science. Indeed, this kind of case is increasingly common as modern technology facilitates the collection of data. The expression *high-dimensional problems* in the data science literature (Friedman et al., 2001) is often used to describe problems where $p \gg N$. Examples of high-dimensional problems abound in computational biology and genomics, and other instances will likely emerge. In those contexts N corresponds to the number of individuals, e.g., patients, in some study. Due to physical limitations, N may only be of the order of a few hundred. In contrast, the number of features p that can be gathered, e.g., gene measurements, could be of the order of tens of thousands.

When $p < N$ the $p \times p$ matrix $\mathbf{X}^\mathsf{T}\mathbf{X}$ has rank at most $N < p$ and thus the least-squares approach

$$\min_{\beta} \|\mathbf{X}^\mathsf{T}\beta - \mathbf{y}\|_2^2$$

is inadequate because the optimality conditions lead to an underdetermined system of equations

$$(\mathbf{X}^\mathsf{T}\mathbf{X})\beta = \mathbf{X}^\mathsf{T}\mathbf{y}.$$

We next describe two popular modifications to the ordinary least-squares approach that aim to rectify this difficulty, namely *ridge regression* and *lasso regression*.

Ridge regression adds a quadratic penalty term to the objective function in the least-squares model

$$\min_{\beta} \ \|\mathbf{X}^\mathsf{T}\beta - \mathbf{y}\|_2^2 + \lambda\|\beta\|_2^2, \tag{5.18}$$

where $\lambda > 0$ is a tuning parameter. The effect of the penalty term is to shrink the regression coefficients towards zero. The magnitude of λ determines the shrinking effect. In the limit when $\lambda \to \infty$ the solution to the ridge regression model is $\beta = \mathbf{0}$. On the other hand, when $\lambda = 0$ ridge regression and ordinary least squares coincide.

The optimality conditions for (5.18) yield the following system of equations:

$$(\mathbf{X}^\mathsf{T}\mathbf{X} + \lambda\mathbf{I})\beta - \mathbf{X}^\mathsf{T}\mathbf{y} = \mathbf{0}.$$

Thus the solution to (5.18) is

$$\beta = (\mathbf{X}^\mathsf{T}\mathbf{X} + \lambda\mathbf{I})^{-1}\mathbf{X}^\mathsf{T}\mathbf{y}.$$

On the other hand, the lasso regression model, proposed in a seminal paper by Tibshirani (1996), adds a 1-norm penalty term to the objective function in the least-squares model

$$\min_{\beta} \ \|\mathbf{X}^{\mathsf{T}}\beta - \mathbf{y}\|_2^2 + \lambda\|\beta\|_1, \tag{5.19}$$

where $\lambda > 0$ is a tuning parameter. The effect of the penalty term is again to shrink the regression coefficients towards zero. However, the properties of the 1-norm have a far more interesting effect. The penalty term $\lambda\|\beta\|_1$ makes some of the regression coefficients be *equal to* zero. In particular, the solutions to the lasso regression model (5.19) are typically sparse and the level of sparsity is controlled by the tuning parameter λ. Lasso regression can be formulated as a quadratic program (see Exercise 5.9). Unlike ridge regression, there is no closed-form formula for the solution to lasso regression.

5.7   Exercises

Exercise 5.1   Assume $\mathbf{c} \in \mathbb{R}^n$ and $\mathbf{Q} \in \mathbb{R}^{n \times n}$ is symmetric. Show that the function

$$f(\mathbf{x}) = \tfrac{1}{2}\mathbf{x}^{\mathsf{T}}\mathbf{Q}\mathbf{x} + \mathbf{c}^{\mathsf{T}}\mathbf{x}$$

is convex if and only if $\mathbf{Q}$ is positive semidefinite.

Assume $\mathbf{c} \in \mathbb{R}^n$ and $\mathbf{Q} \in \mathbb{R}^{n \times n}$ is symmetric and positive semidefinite but not positive definite. Show that the problem

$$\min_{\mathbf{x}} \ \tfrac{1}{2}\mathbf{x}^{\mathsf{T}}\mathbf{Q}\mathbf{x} + \mathbf{c}^{\mathsf{T}}\mathbf{x}$$

is either bounded or has infinitely many optimal solutions.

Exercise 5.2   Let $\mathbf{c} \in \mathbb{R}^n$. Show that the solution to

$$\begin{aligned} \min_{\mathbf{x}} \quad & \tfrac{1}{2}\|\mathbf{x}\|_2^2 - \mathbf{c}^{\mathsf{T}}\mathbf{x} \\ \text{s.t.} \quad & \mathbf{1}^{\mathsf{T}}\mathbf{x} = 1 \\ & \mathbf{x} \geq \mathbf{0} \end{aligned}$$

is

$$\mathbf{x} = (\lambda\mathbf{1} + \mathbf{c})^+,$$

where $\lambda \in \mathbb{R}$ is a suitable threshold value such that $\mathbf{1}^{\mathsf{T}}(\lambda\mathbf{1} + \mathbf{c})^+ = 1$.

Exercise 5.3   Let $\mathbf{c}, \mathbf{d} \in \mathbb{R}^n$. Assume $\mathbf{d} > \mathbf{0}$ and $\mathbf{D} = (\mathrm{Diag}(\mathbf{d}))$. Show that the solution to

$$\begin{aligned} \min_{\mathbf{x}} \quad & \tfrac{1}{2}\mathbf{x}^{\mathsf{T}}\mathbf{D}^{-1}\mathbf{x} - \mathbf{c}^{\mathsf{T}}\mathbf{x} \\ \text{s.t.} \quad & \mathbf{1}^{\mathsf{T}}\mathbf{x} = 1 \\ & \mathbf{x} \geq \mathbf{0} \end{aligned}$$

is

$$\mathbf{x} = (\lambda \mathbf{d} + \mathbf{Dc})^+,$$

where $\lambda \in \mathbb{R}$ is a suitable threshold value such that $\mathbf{1}^\mathsf{T}(\lambda \mathbf{d} + \mathbf{Dc})^+ = 1$.

Exercise 5.4   Write a CVX MATLAB script that takes as inputs $\mathbf{c} \in \mathbb{R}^n$, $\mathbf{Q} \in \mathbb{R}^{n \times n}$, $\mathbf{A} \in \mathbb{R}^{m \times n}$, $\mathbf{b} \in \mathbb{R}^m$ and solves the optimization problem

$$
\begin{array}{ll}
\min & \frac{1}{2}\mathbf{x}^\mathsf{T}\mathbf{Qx} + \mathbf{c}^\mathsf{T}\mathbf{x} \\
\text{s.t.} & \mathbf{Ax} = \mathbf{b} \\
& \mathbf{x} \geq \mathbf{0}.
\end{array}
$$

Test your script on instances generated as follows:

```
>> m=1, n=5, c=randn(n,1), Q=eye(n), A=ones(m,n), b=1;
```

and

```
>> m=1, n=5, c=randn(n,1), Q=diag(rand(n,1)), A=ones(m,n), b=1;
```

Are the results consistent with Exercises 5.2 and 5.3?

Exercise 5.5   Consider a quadratic program with non-negativity inequality constraints only:

$$
\begin{array}{ll}
\min_{\mathbf{x}} & \frac{1}{2}\mathbf{x}^\mathsf{T}\mathbf{Qx} + \mathbf{c}^\mathsf{T}\mathbf{x} \\
\text{s.t.} & \mathbf{x} \geq \mathbf{0}.
\end{array} \tag{5.20}
$$

There is some intuition behind the optimality conditions: at an optimal solution the non-negativity constraints split into binding and non-binding constraints. The former behave like equality constraints whereas the latter can be treated as if they did not exist. Suppose $I \subseteq \{1, \ldots, n\}$ is the set of binding constraints and $J = \{1, \ldots, n\} \setminus I$. This reasoning suggests that we think of the problem

$$
\begin{array}{ll}
\min_{\mathbf{x}} & \frac{1}{2}\mathbf{x}^\mathsf{T}\mathbf{Qx} + \mathbf{c}^\mathsf{T}\mathbf{x} \\
\text{s.t.} & \mathbf{x}_I = \mathbf{0},
\end{array}
$$

whose optimality conditions are

$$
\begin{array}{rcl}
(\mathbf{Qx} + \mathbf{c})_I - \mathbf{s}_I & = & \mathbf{0} \\
(\mathbf{Qx} + \mathbf{c})_J & = & \mathbf{0} \\
\mathbf{x}_I & = & \mathbf{0}.
\end{array} \tag{5.21}
$$

Prove that this intuition is indeed correct.

Exercise 5.6   Consider the quadratic problem (5.17) presented in Section 5.6.1.

(a) Show that the objective function of this problem is equivalent to maximizing the margin between the lines $\mathbf{w}^\mathsf{T}\mathbf{x} = \gamma + 1$ and $\mathbf{w}^\mathsf{T}\mathbf{x} = \gamma - 1$.
(b) Write the optimality conditions for problem (5.17).
(c) Write the dual.

Exercise 5.7   The linear separation idea presented in Section 5.6.1 can be used even when the two vector sets $\{\mathbf{a}_i\}$ and $\{\mathbf{b}_i\}$ are not linearly separable. This is achieved by introducing a non-negative violation variable for each constraint of (5.17). Then, one has two objectives: to minimize the total of the constraint violations and to maximize the margin. Develop a quadratic programming model that combines these two objectives using an adjustable parameter that can be chosen in a way to put more weight on violations or margin, depending on one's preference.

Exercise 5.8   The classification problems discussed in the two previous exercises can also be formulated as linear programming problems, if one agrees to use the 1-norm rather than the 2-norm of $\mathbf{w}$ in the objective function. Recall that $\|\mathbf{w}\|_1 = \sum_i |w_i|$. Show that if we replace $\|\mathbf{w}\|_2^2$ by $\|\mathbf{w}\|_1$ in the objective function of (5.17), we can write the resulting problem as a linear program. Show also that this new objective function is equivalent to maximizing the distance between $\mathbf{w}^\mathsf{T}\mathbf{x} = \gamma + 1$ and $\mathbf{w}^\mathsf{T}\mathbf{x} = \gamma - 1$ if one measures the distance using the ∞-norm $\|\mathbf{g}\|_\infty = \max_i |g_i|$.

Exercise 5.9   Show that the lasso regression model (5.19) can be equivalently formulated as

$$\min_{\mathbf{x}} \quad \tfrac{1}{2}\mathbf{x}^\mathsf{T}\mathbf{Q}\mathbf{x} + \mathbf{c}^\mathsf{T}\mathbf{x}$$
$$\text{s.t.} \quad \mathbf{D}\mathbf{x} \geq \mathbf{d}$$

for some suitable $\mathbf{Q}, \mathbf{c}, \mathbf{D}, \mathbf{d}$.

6 Quadratic Programming Models: Mean–Variance Optimization

6.1 Portfolio Return

Consider an investment environment where there is a universe of n risky assets. In the next few chapters we will be concerned with a *one-period* model of the problem of investing in these n risky assets. Assume a portfolio must be selected at some initial time t_0 and held until time t. Let $\mathbf{v}_0 = \begin{bmatrix} v_{1,0} & \cdots & v_{n,0} \end{bmatrix}^{\mathsf{T}}$ and $\mathbf{v} = \begin{bmatrix} v_1 & \cdots & v_n \end{bmatrix}^{\mathsf{T}}$ denote the vectors of asset prices at times t_0 and t respectively. The vector $\mathbf{v}_0$ is known whereas $\mathbf{v}$ is a vector of random variables. A vector $\mathbf{h} \in \mathbb{R}^n$ of share holdings in each of the assets defines a portfolio whose values at time t_0 and t are $W_0 := \mathbf{v}_0^{\mathsf{T}} \mathbf{h}$ and $W := \mathbf{v}^{\mathsf{T}} \mathbf{h}$ respectively. The value W_0 is known at time t_0 whereas W is a random variable. The gist of portfolio construction is to choose $\mathbf{h}$ to optimize some measure of satisfaction on the random variable W.

It is customary to use the initial portfolio value W_0 as a reference and to write the above problem in terms of the portfolio return

$$r_P = \frac{W - W_0}{W_0}.$$

The return of asset i, which is the same as that of a portfolio entirely invested in asset i, is similarly defined as

$$r_i = \frac{v_i - v_{i,0}}{v_{i,0}}.$$

Instead of the vector of holdings $\mathbf{h} \in \mathbb{R}^n$, the portfolio construction problem is often stated in terms of percentage holdings $\mathbf{x} \in \mathbb{R}^n$ where

$$x_i = \frac{h_i v_{i,0}}{W_0} = \frac{h_i v_{i,0}}{\sum_{j=1}^n h_j v_{j,0}}.$$

Observe that $W = \mathbf{v}^{\mathsf{T}} \mathbf{h}$ can be equivalently written as

$$r_P = \sum_{i=1}^n r_i x_i = \mathbf{r}^{\mathsf{T}} \mathbf{x}.$$

In spite of its wide popularity, this convention runs into difficulties in some cases. For example, the above quantity r_P does not make sense for a long–short portfolio associated with a pairs trading strategy. More broadly, the quantity r_P does not make sense for a situation where the initial value of a portfolio W_0 is zero as when one enters a futures contract or constructs a long–short portfolio with equal long and short cash positions.

As Meucci (2005, 2010) nicely puts it, this difficulty can be amended by assuming that returns are measured relative to some predefined *basis value b* as opposed to the initial portfolio value W_0. In some cases, it is natural to choose $b = W_0$ but it is more proper to think of b as a general reference point. To make this idea more precise, we associate with each asset and portfolio a basis b that satisfies the following four properties:

- The basis b for a long position of an asset is positive.
- The basis b is measured in the same unit as the asset values.
- The basis is homogeneous: the basis of k shares of an asset is k times the basis of one share.
- The basis is known at time t_0.

Equipped with this concept, we get a formal and unambiguous definition of asset and portfolio returns:

$$r_i = \frac{v_i - v_{i,0}}{b_i}, \quad r_P = \frac{W - W_0}{b_P}.$$

Likewise, we obtain a formal and unambiguous definition of percentage holdings:

$$x_i = \frac{h_i b_i}{b_P}.$$

Once again, the identity $W = \mathbf{v}^\mathsf{T}\mathbf{h}$ can be equivalently written as

$$r_P = \mathbf{r}^\mathsf{T}\mathbf{x}.$$

Throughout this chapter $\mathbf{x} = \begin{bmatrix} x_1 & \cdots & x_n \end{bmatrix}^\mathsf{T}$ will denote the vector of percentage holdings of a portfolio in a universe of n risky assets. When it is applicable and evident from the context, we shall assume the usual basis values $b_i = v_{i,0}$ and $b_P = W_0$ respectively.

6.2 Markowitz Mean–Variance (Basic Model)

Markowitz's key insight into the above one-period investment problem was to consider the expected value and standard deviation of the return as measures of performance and risk respectively. The portfolio selection problem can then be formally stated as a quadratic programming model. To simplify our discussion of this model, we will proceed in three incremental steps. First, we will look at the case when there are only two assets; second, we will look at the case when there are three risky assets; and finally, we will see the general case with any number of risky assets.

Two Assets

Suppose we are combining two assets whose random returns are r_1 and r_2. Let

$$\mu_1 := \mathbb{E}(r_1), \quad \mu_2 := \mathbb{E}(r_2),$$

and

$$\sigma_1^2 := \text{var}(r_1), \quad \sigma_2^2 = \text{var}(r_2), \quad \sigma_{12} = \text{cov}(r_1, r_2) = \rho \cdot \sigma_1 \cdot \sigma_2.$$

In this case a *portfolio* of these two assets is determined by the proportion invested in one of the two assets. Let x denote the proportion in asset 1. Thus the portfolio return is

$$r_P = x \cdot r_1 + (1 - x) \cdot r_2,$$

the portfolio expected return is

$$
\begin{aligned}
\mu_P := \mathbb{E}(r_P) \quad &= \quad x \cdot \mathbb{E}(r_1) + (1 - x) \cdot \mathbb{E}(r_2) \\
&= \quad x \cdot \mu_1 + (1 - x) \cdot \mu_2,
\end{aligned}
$$

and the portfolio variance is

$$\sigma_P^2 = x^2 \sigma_1^2 + (1 - x)^2 \sigma_2^2 + 2 \cdot x(1 - x) \cdot \rho \cdot \sigma_1 \cdot \sigma_2.$$

In the special case when one of the assets, say asset 2, is the asset with risk-free return r_f we get

$$\mu_P = x \cdot \mu_1 + (1 - x) \cdot r_f = r_f + (\mu_1 - r_f)x, \quad \sigma_P^2 = x^2 \sigma_1^2.$$

In this case the portfolio selection is particularly simple: a target level of expected return μ_P corresponds to one particular portfolio obtained by choosing $x = (\mu_P - r_f)/(\mu_1 - r_f)$. The situation with three assets leads to a more interesting situation.

Three Risky Assets

Suppose now that there are three assets with random returns r_1, r_2, and r_3. As before, let

$$\mu_j = \mathbb{E}(r_j), \quad \sigma_j^2 := \text{var}(r_j) \quad \text{for } j = 1, 2, 3,$$

and

$$\sigma_{ij} := \text{cov}(r_i, r_j) = \rho_{ij} \cdot \sigma_i \cdot \sigma_j \quad \text{for } i, j = 1, 2, 3.$$

Now a portfolio determines the holdings in the three assets. Let x_j denote the proportion (weight) invested in asset j, for $j = 1, 2, 3$. Notice that these proportions should add up to one if the portfolio is fully invested in the three assets:

$$x_1 + x_2 + x_3 = 1.$$

Similar to what we did before, the portfolio return is

$$r_P = r_1 x_1 + r_2 x_2 + r_3 x_3.$$

So the portfolio expected return is

$$\mu_P = \mu_1 x_1 + \mu_2 x_2 + \mu_3 x_3,$$

and the portfolio variance is

$$\sigma_P^2 = \sigma_1^2 x_1^2 + \sigma_2^2 x_2^2 + \sigma_3^2 x_3^2 + 2(\sigma_{12} x_1 x_2 + \sigma_{23} x_2 x_3 + \sigma_{13} x_1 x_3).$$

Observe that now there are multiple portfolios that can achieve a target expected level of return. A portfolio is *efficient* if it has minimum risk for a given target return, or equivalently, if it has the maximum expected return for a given target risk. This naturally leads to the following quadratic programming formulation.

To find a portfolio of minimum risk (variance) with expected return *at least* $\bar{\mu}$ solve the following *mean–variance optimization model*:

$$\min_{\mathbf{x}} \quad \sum_{i=1}^{3} \sigma_{ii} x_i^2 + 2 \sum_{i=1}^{3} \sum_{j=i+1}^{3} \sigma_{ij} x_i x_j$$
$$\text{s.t.} \quad \mu_1 x_1 + \mu_2 x_2 + \mu_3 x_3 \geq \bar{\mu}$$
$$x_1 + x_2 + x_3 = 1.$$

The *efficient frontier* is the set of efficient portfolios. The efficient frontier is often "visualized" by plotting the expected return against the standard deviation of the efficient portfolios. To generate portfolios on the efficient frontier, we can minimize variance, for varying target return $\bar{\mu}$:

$$\min_{\mathbf{x}} \quad \sum_{i=1}^{3} \sigma_{ii} x_i^2 + 2 \sum_{i=1}^{3} \sum_{j=i+1}^{3} \sigma_{ij} x_i x_j$$
$$\text{s.t.} \quad \mu_1 x_1 + \mu_2 x_2 + \mu_3 x_3 \geq \bar{\mu}$$
$$x_1 + x_2 + x_3 = 1.$$

We can also maximize return, for varying target variance $\bar{\sigma}^2 > 0$:

$$\max_{\mathbf{x}} \quad \mu_1 x_1 + \mu_2 x_2 + \mu_3 x_3$$
$$\text{s.t.} \quad \sum_{i=1}^{3} \sigma_{ii} x_i^2 + 2 \sum_{i=1}^{3} \sum_{j=i+1}^{3} \sigma_{ij} x_i x_j \leq \bar{\sigma}^2$$
$$x_1 + x_2 + x_3 = 1.$$

Or we can maximize quadratic *utility*, for varying risk aversion $\gamma > 0$:

$$\max_{\mathbf{x}} \quad \mu_1 x_1 + \mu_2 x_2 + \mu_3 x_3 - \frac{\gamma}{2} \left(\sum_{i=1}^{3} \sigma_{ii} x_i^2 + 2 \sum_{i=1}^{3} \sum_{j=i+1}^{3} \sigma_{ij} x_i x_j \right)$$
$$\text{s.t.} \quad x_1 + x_2 + x_3 = 1.$$

Any Number of Risky Assets

Let us now take a leap to the most general case. Assume we have n risky assets. Let $\mathbf{r} \in \mathbb{R}^n$ be the n-dimensional random vector of returns, i.e., r_i denotes the return of asset i between times t_0 and t. Let $\boldsymbol{\mu} \in \mathbb{R}^n$ denote the vector of expected returns, and $\mathbf{V} \in \mathbb{R}^{n \times n}$ denote the return covariance matrix. More precisely,

$$
\boldsymbol{\mu} = \begin{bmatrix} \mu_1 \\ \vdots \\ \mu_n \end{bmatrix}, \quad \mathbf{V} = \begin{bmatrix} \sigma_{11} & \cdots & \sigma_{1n} \\ \vdots & \ddots & \vdots \\ \sigma_{n1} & \cdots & \sigma_{nn} \end{bmatrix},
$$

where $\mu_i := \mathbb{E}(r_i)$, $\sigma_{ij} := \mathrm{cov}(r_i, r_j)$, $i, j = 1, \ldots, n$.

From the linearity properties of expectation, it follows that the expected return and variance of a given portfolio $\mathbf{x} = \begin{bmatrix} x_1 & \cdots & x_n \end{bmatrix}^\mathsf{T}$ of the risky assets are respectively

$$
\boldsymbol{\mu}^\mathsf{T} \mathbf{x} = \sum_{j=1}^{n} \mu_j x_j
$$

and

$$
\mathbf{x}^\mathsf{T} \mathbf{V} \mathbf{x} = \sum_{i=1}^{n} \sum_{j=1}^{n} \sigma_{ij} x_i x_j = \sum_{i=1}^{n} \sigma_{ii} x_i^2 + 2 \sum_{i=1}^{n} \sum_{j=i+1}^{n} \sigma_{ij} x_i x_j.
$$

The problem of selecting a portfolio can be formally stated as a tradeoff between these two components. A fully invested portfolio is *efficient* if it has minimum risk for a given level of return, or equivalently if it has maximum expected return for a given level of risk.

A fully invested efficient portfolio can then be characterized as the solution to the following quadratic program:

$$
\begin{aligned}
\max_{\mathbf{x}} \quad & \boldsymbol{\mu}^\mathsf{T} \mathbf{x} - \tfrac{1}{2} \gamma \cdot \mathbf{x}^\mathsf{T} \mathbf{V} \mathbf{x} \\
& \mathbf{1}^\mathsf{T} \mathbf{x} = 1
\end{aligned}
\tag{6.1}
$$

for some risk-aversion coefficient $\gamma > 0$.

The set of efficient portfolios can also be obtained as the set of solutions to the quadratic program:

$$
\begin{aligned}
\min_{\mathbf{x}} \quad & \mathbf{x}^\mathsf{T} \mathbf{V} \mathbf{x} \\
\text{s.t.} \quad & \boldsymbol{\mu}^\mathsf{T} \mathbf{x} \geq \bar{\mu} \\
& \mathbf{1}^\mathsf{T} \mathbf{x} = 1,
\end{aligned}
\tag{6.2}
$$

and also as the set of solutions to

$$
\begin{aligned}
\max_{\mathbf{x}} \quad & \boldsymbol{\mu}^\mathsf{T} \mathbf{x} \\
\text{s.t.} \quad & \mathbf{x}^\mathsf{T} \mathbf{V} \mathbf{x} \leq \bar{\sigma}^2 \\
& \mathbf{1}^\mathsf{T} \mathbf{x} = 1
\end{aligned}
\tag{6.3}
$$

by varying $\bar{\mu}$ and $\bar{\sigma}$ respectively. The exercises at the end of the chapter sketch how to give a formal proof of the equivalence of the above three models.

We shall refer to the equivalent mean–variance models (6.1), (6.2), and (6.3) as the *basic mean–variance models* as they include only the following three essential components: mean and variance of return, and the full investment constraint. Observe that these three optimization models are convex because the quadratic function $\mathbf{x} \mapsto \mathbf{x}^\mathsf{T} \mathbf{V} \mathbf{x}$ is convex as the covariance matrix $\mathbf{V}$ is positive semidefinite.

Section 6.3 below details several interesting insights that can be gained from the solution to these basic mean–variance models.

As we discuss later in this chapter, the types of mean–variance models used in portfolio construction typically include a number of additional constraints.

Asset Allocation and Security Selection

There are two distinct levels of portfolio analysis that are amenable to mean–variance models. The conventional *top-down* investment approach to portfolio construction consists of two main steps, namely *asset allocation* and *security selection*.

On the one hand, the *asset allocation* decision is concerned with portfolio choices among broad asset classes. At the coarsest level, these asset classes could be stocks, bonds, and cash. At a more refined level, some of these broad asset classes could be subdivided. For instance, stocks can be divided according to geography or market capitalization. The asset allocation decision involves only a small number of assets, typically ranging from a handful to a dozen or so. It generally involves simple constraints such as budget constraints and upper and lower bounds on individual positions.

On the other hand, the *security selection* decision is concerned with the specific securities within each particular asset class. For instance, if the relevant asset class is equities in the S&P 500 market index, then the security selection problem is concerned with the specific portfolio holdings at the individual stock level. The security selection problem typically involves a large number of securities, ranging from a few hundred to potentially thousands. It also involves a myriad of constraints and is often formulated relative to a predefined *benchmark,* as we discuss in more detail in Section 6.5.

6.3 Analytical Solutions to Basic Mean–Variance Models

The solution to the basic mean–variance models described in Section 6.2 can be characterized by relying on the tools introduced in Chapter 5. Throughout this section we assume that the covariance matrix of asset returns $\mathbf{V}$ is positive definite. In particular, $\mathbf{V}^{-1}$ exists.

Minimum Risk and Characteristic Portfolios

Consider the simplified version of (6.1) that is obtained in the limit when $\gamma \to \infty$:

$$\min_{\mathbf{x}} \quad \mathbf{x}^{\mathsf{T}}\mathbf{V}\mathbf{x}$$
$$\mathbf{1}^{\mathsf{T}}\mathbf{x} = 1. \tag{6.4}$$

The model (6.4) corresponds to the problem of finding the minimum-risk fully invested portfolio. We discussed this problem in Example 5.7 where the optimal solution was shown to be

$$\mathbf{x}^* = \frac{1}{\mathbf{1}^\mathsf{T}\mathbf{V}^{-1}\mathbf{1}}\mathbf{V}^{-1}\mathbf{1}.$$

A related problem that is often of interest is to find the minimum-risk portfolio with unit exposure to a vector of *attributes* $\mathbf{a}$ associated with the assets. As we will see later, some interesting attributes could be the betas of the assets relative to a benchmark, the asset volatilities, or the asset expected returns. The *characteristic portfolio* of a vector of attributes $\mathbf{a}$ is the solution to the problem

$$\begin{aligned}\min_{\mathbf{x}} \quad & \mathbf{x}^\mathsf{T}\mathbf{V}\mathbf{x} \\ & \mathbf{a}^\mathsf{T}\mathbf{x} = 1.\end{aligned} \tag{6.5}$$

Using the solution of (5.9) obtained in Chapter 5, it follows that the solution to (6.5) is

$$\mathbf{x}^* = \frac{1}{\mathbf{a}^\mathsf{T}\mathbf{V}^{-1}\mathbf{a}}\mathbf{V}^{-1}\mathbf{a}.$$

Observe that a characteristic portfolio $\mathbf{x}^* = (1/\mathbf{a}^\mathsf{T}\mathbf{V}^{-1}\mathbf{a})\mathbf{V}^{-1}\mathbf{a}$ is not necessarily fully invested as its components may not necessarily add up to one. Observe that the variance of the characteristic porfolio $\mathbf{x}^* = (1/\mathbf{a}^\mathsf{T}\mathbf{V}^{-1}\mathbf{a})\mathbf{V}^{-1}\mathbf{a}$ is

$$(\mathbf{x}^*)^\mathsf{T}\mathbf{V}\mathbf{x}^* = \frac{1}{\mathbf{a}^\mathsf{T}\mathbf{V}^{-1}\mathbf{a}}.$$

Two-Fund Separation Theorem

Consider the basic mean–variance model

$$\begin{aligned}\max_{\mathbf{x}} \quad & \boldsymbol{\mu}^\mathsf{T}\mathbf{x} - \tfrac{1}{2}\gamma \cdot \mathbf{x}^\mathsf{T}\mathbf{V}\mathbf{x} \\ & \mathbf{1}^\mathsf{T}\mathbf{x} = 1\end{aligned} \tag{6.6}$$

for some risk-aversion coefficient $\gamma > 0$. We next derive an interesting result often called the *two-fund separation theorem*. The theorem states that every fully invested efficient portfolio is a combination of two particular efficient portfolios.

Applying the optimality conditions (5.10) from Theorem 5.6 to problem (6.6) we obtain the solution

$$\mathbf{x}^* = \lambda \cdot \frac{1}{\mathbf{1}^\mathsf{T}\mathbf{V}^{-1}\boldsymbol{\mu}}\mathbf{V}^{-1}\boldsymbol{\mu} + (1-\lambda) \cdot \frac{1}{\mathbf{1}^\mathsf{T}\mathbf{V}^{-1}\mathbf{1}}\mathbf{V}^{-1}\mathbf{1}$$

where $\lambda = \mathbf{1}^\mathsf{T}\mathbf{V}^{-1}\boldsymbol{\mu}/\gamma$. The following *two-fund theorem* readily follows.

Theorem 6.1 (Two-fund theorem)   *Consider model (6.6) for some $\gamma > 0$. There exist two efficient portfolios (funds), namely*

$$\frac{1}{\mathbf{1}^\mathsf{T}\mathbf{V}^{-1}\boldsymbol{\mu}}\mathbf{V}^{-1}\boldsymbol{\mu} \quad and \quad \frac{1}{\mathbf{1}^\mathsf{T}\mathbf{V}^{-1}\mathbf{1}}\mathbf{V}^{-1}\mathbf{1},$$

such that every efficient portfolio, that is, every solution to (6.6), *is a combination of these two portfolios.*

Observe that one of the two portfolios in the two-fund theorem is the minimum-risk portfolio $(1/\mathbf{1}^\mathsf{T}\mathbf{V}^{-1}\mathbf{1})\mathbf{V}^{-1}\mathbf{1}$ and the other one is a multiple of the characteristic portfolio $(1/\boldsymbol{\mu}^\mathsf{T}\mathbf{V}^{-1}\boldsymbol{\mu})\mathbf{V}^{-1}\boldsymbol{\mu}$ of the vector of attributes $\boldsymbol{\mu}$.

One-Fund Separation Theorem

We next derive the *one-fund* or *mutual fund* separation theorem. This result is similar in spirit to the two-fund separation theorem. It states that if there is a risk-free asset, then every efficient portfolio is a combination of the risk-free asset and a particular fund.

Consider the case when, in addition to the universe of n risky assets, there is an additional asset $n+1$ with risk-free return r_f. In this case, problem (6.1) extends as follows

$$\max_{\mathbf{x},x_{n+1}} \quad \boldsymbol{\mu}^\mathsf{T}\mathbf{x} + r_f \cdot x_{n+1} - \tfrac{1}{2}\gamma \cdot \mathbf{x}^\mathsf{T}\mathbf{V}\mathbf{x}$$
$$\mathbf{1}^\mathsf{T}\mathbf{x} + x_{n+1} = 1. \tag{6.7}$$

By substituting $x_{n+1} = 1 - \mathbf{1}^\mathsf{T}\mathbf{x}$ in the objective and dropping the constraint, problem (6.7) can be rewritten as the following unconstrained optimization problem:

$$\max_{\mathbf{x}} \ (\boldsymbol{\mu} - r_f\mathbf{1})^\mathsf{T}\mathbf{x} - \tfrac{1}{2}\gamma \cdot \mathbf{x}^\mathsf{T}\mathbf{V}\mathbf{x}.$$

Applying the optimality conditions (5.4) from Theorem 5.2, we obtain the following solution to (6.7):

$$\mathbf{x}^* = \frac{1}{\gamma} \cdot \mathbf{V}^{-1}(\boldsymbol{\mu} - r_f\mathbf{1}) = \lambda \cdot \frac{1}{\mathbf{1}^\mathsf{T}\mathbf{V}^{-1}(\boldsymbol{\mu} - r_f\mathbf{1})}\mathbf{V}^{-1}(\boldsymbol{\mu} - r_f\mathbf{1}), \ x_{n+1}^* = 1 - \mathbf{1}^\mathsf{T}\mathbf{x}^*,$$

where $\lambda = \mathbf{1}^\mathsf{T}\mathbf{V}^{-1}(\boldsymbol{\mu} - r_f\mathbf{1})/\gamma$. The following *one-fund theorem* readily follows.

Theorem 6.2 (One-fund theorem) *Suppose the investment universe includes n risky assets and a risk-free asset. Then there exists a fully invested efficient portfolio (fund) namely*

$$\frac{1}{\mathbf{1}^\mathsf{T}\mathbf{V}^{-1}(\boldsymbol{\mu} - r_f\mathbf{1})}\mathbf{V}^{-1}(\boldsymbol{\mu} - r_f\mathbf{1})$$

such that every efficient portfolio – that is, every solution to (6.7) *for some $\gamma > 0$ – is a combination of this portfolio and the risk-free asset.*

The portfolio $[1/\mathbf{1}^\mathsf{T}\mathbf{V}^{-1}(\boldsymbol{\mu} - r_f\mathbf{1})]\mathbf{V}^{-1}(\boldsymbol{\mu} - r_f\mathbf{1})$ is called the *tangency portfolio*. This name is motivated by the geometric interpretation illustrated in Figure 6.1. Consider the plot of expected return versus standard deviation for the efficient frontier portfolios. The portfolio $[1/\mathbf{1}^\mathsf{T}\mathbf{V}^{-1}(\boldsymbol{\mu} - r_f\mathbf{1})]\mathbf{V}^{-1}(\boldsymbol{\mu} - r_f\mathbf{1})$ lies exactly at the tangency point on this frontier defined by the straight line emerging from the point $(0, r_f)$. The point $(0, r_f)$ corresponds to the expected return versus standard deviation of the risk-free asset. The tangency line is also known as the *capital allocation line* (CAL) as it corresponds to portfolios with different allocations of capital between the tangency portfolio and the risk-free asset.

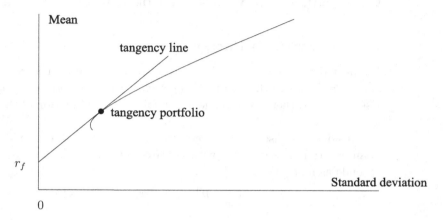

Figure 6.1 Tangency portfolio

Capital Asset Pricing Model (CAPM)

Under suitable equilibrium assumptions the tangency portfolio discussed above yields the main mathematical foundation for the capital asset pricing model (CAPM), a fundamental asset pricing model in financial economics. The key step in this derivation is that, in equilibrium, the tangency portfolio is precisely the market portfolio $\mathbf{x}_M$. That is,

$$\mathbf{x}_M = \frac{1}{\mathbf{1}^\mathsf{T}\mathbf{V}^{-1}(\boldsymbol{\mu} - r_f\mathbf{1})}\mathbf{V}^{-1}(\boldsymbol{\mu} - r_f\mathbf{1}). \tag{6.8}$$

From (6.8) we readily obtain

$$\mathbf{V}\mathbf{x}_M = \frac{1}{\mathbf{1}^\mathsf{T}\mathbf{V}^{-1}(\boldsymbol{\mu} - r_f\mathbf{1})}(\boldsymbol{\mu} - r_f\mathbf{1}), \tag{6.9}$$

and

$$\mathbf{x}_M^\mathsf{T}\mathbf{V}\mathbf{x}_M = \frac{(\boldsymbol{\mu} - r_f\mathbf{1})^\mathsf{T}\mathbf{x}_M}{\mathbf{1}^\mathsf{T}\mathbf{V}^{-1}(\boldsymbol{\mu} - r_f\mathbf{1})} = \frac{\mu_M - r_f}{\mathbf{1}^\mathsf{T}\mathbf{V}^{-1}(\boldsymbol{\mu} - r_f\mathbf{1})}, \tag{6.10}$$

where $\mu_M = \boldsymbol{\mu}^\mathsf{T}\mathbf{x}_M$ is the expected value of the market portfolio return.

Combining (6.9) and (6.10) we get

$$\boldsymbol{\mu} - r_f\mathbf{1} = \mathbf{1}^\mathsf{T}\mathbf{V}^{-1}(\boldsymbol{\mu} - r_f\mathbf{1})\mathbf{V}\mathbf{x}_M = \left(\frac{1}{\mathbf{x}_M^\mathsf{T}\mathbf{V}\mathbf{x}_M}\mathbf{V}\mathbf{x}_M\right)(\mu_M - r_f) = \boldsymbol{\beta}\cdot(\mu_M - r_f), \tag{6.11}$$

where $\boldsymbol{\beta} = (1/\mathbf{x}_M^\mathsf{T} \mathbf{V} \mathbf{x}_M) \mathbf{V} \mathbf{x}_M$. The above can be equivalently stated as

$$\mu_j - r_f = \beta_j(\mu_M - r_f), \quad \text{where} \quad \beta_j = \frac{\sigma_{j,M}}{\sigma_M^2} \quad \text{for } j = 1, \ldots, n. \tag{6.12}$$

Equation (6.11) or its equivalent (6.12) is the formal statement of the capital asset pricing model (CAPM). The CAPM postulates that the excess return of asset j is determined entirely by its beta coefficient times the excess return of the market.

In the expression (6.12), $\sigma_{j,M}$ denotes the covariance between the return of asset j and the return of the market portfolio, and σ_M^2 denotes the variance of the market portfolio return. The last two quantities in turn have the following expressions in terms of the covariance matrix $\mathbf{V}$:

$$\sigma_{j,M} = \mathrm{cov}(r_j, r_M) = (\mathbf{V} \mathbf{x}_M)_j, \quad \sigma_M^2 = \mathrm{var}(r_M) = \mathbf{x}_M^\mathsf{T} \mathbf{V} \mathbf{x}_M.$$

6.4 More General Mean–Variance Models

The basic mean–variance model discussed in the previous section provides the foundation of modern portfolio theory. However, when mean–variance models are used as a normative tool in portfolio construction, it is common to use modifications of the basic model by including additional constraints and possibly additional terms in the objective.

Common Constraints

Aside from a target expected return or a target variance, the only portfolio constraint in the basic mean–variance model is the full investment constraint

$$\mathbf{1}^\mathsf{T} \mathbf{x} = 1.$$

Furthermore, this constraint disappears if the portfolio is allowed to include holdings in a risk-free asset. In both cases the individual portfolio holdings could in principle take arbitrary positive and negative values as there is no explicit restriction on them. This motivates the following types of constraints that are often included in a mean–variance model:

- Budget constraints, such as fully invested portfolios.
- Upper and/or lower bounds on the size of individual positions.
- Upper and/or lower bounds on exposure to industries or sectors.
- Leverage constraints such as long-only, or 130/30 constraints.
- Turnover constraints.

The above types of constraints replace the single portfolio constraint

$$\mathbf{1}^\mathsf{T} \mathbf{x} = 1$$

by a more elaborate set of constraints of the form

$$\mathbf{Ax} = \mathbf{b}$$
$$\mathbf{Dx} \geq \mathbf{d}.$$

Consequently, we get the following general version of the basic mean–variance model (6.1):

$$\max_{\mathbf{x}} \quad \boldsymbol{\mu}^\mathsf{T}\mathbf{x} - \tfrac{1}{2}\gamma \cdot \mathbf{x}^\mathsf{T}\mathbf{Vx}$$
$$\mathbf{Ax} = \mathbf{b} \tag{6.13}$$
$$\mathbf{Dx} \geq \mathbf{d}.$$

The set of portfolios obtained via the model (6.13) can also be obtained via the following two equivalent models. The first one enforces a target expected return:

$$\min_{\mathbf{x}} \quad \mathbf{x}^\mathsf{T}\mathbf{Vx}$$
$$\text{s.t.} \quad \boldsymbol{\mu}^\mathsf{T}\mathbf{x} \geq \bar{\mu}$$
$$\mathbf{Ax} = \mathbf{b} \tag{6.14}$$
$$\mathbf{Dx} \geq \mathbf{d}.$$

The second one enforces a target variance of return:

$$\max_{\mathbf{x}} \quad \boldsymbol{\mu}^\mathsf{T}\mathbf{x}$$
$$\text{s.t.} \quad \mathbf{x}^\mathsf{T}\mathbf{Vx} \leq \bar{\sigma}^2$$
$$\mathbf{Ax} = \mathbf{b} \tag{6.15}$$
$$\mathbf{Dx} \geq \mathbf{d}.$$

The models (6.13), (6.14), and (6.15) are still convex quadratic optimization models. Unlike the basic mean–variance model, they generally do not have an analytical closed-form solution due to the additional inequality constraints. However, they can be solved numerically very efficiently via optimization solvers.

We next discuss how some of the above five types of constraints can be incorporated into a mean–variance model. The first three types of constraints have straightforward formulations. We concentrate on the last two, namely, leverage constraints and turnover constraints. A long-only constraint can readily be enforced via $\mathbf{x} \geq 0$. A relaxed version of this constraint, popular in certain contexts, is not to rule out leverage altogether but to limit it. For instance, a "130/30" leverage constraint means that the total value of the holdings in short positions must be at most 30% of the portfolio value. In general, suppose that we want the value of the total short positions to be at most L. This means that we want to enforce the following restriction:

$$\sum_{j=1}^{n} \min(x_j, 0) \geq -L \quad \Leftrightarrow \quad \sum_{j=1}^{n} \max(-x_j, 0) \leq L.$$

Although this is a correct mathematical formulation of the constraint, it is not ideal for computational purposes because of the non-smooth terms $\max(-x_j, 0)$.

In particular, if a constraint were written in this form the resulting mean–variance model would not be a quadratic program. To formulate this constraint efficiently in the quadratic optimization model, we trade terms of the form $\max(-x_j, 0)$ for new terms involving possibly new variables and linear inequalities. To that end, add the new vector of variables $\mathbf{y} = \begin{bmatrix} y_1 & \cdots & y_n \end{bmatrix}^\mathsf{T}$ and constraints

$$\mathbf{x} \geq -\mathbf{y}$$

$$\sum_{j=1}^{n} y_j \leq L$$

$$\mathbf{y} \geq \mathbf{0}.$$

A *turnover* constraint is a constraint on the total change in the portfolio positions. This constraint is generally included as a way to limit certain kinds of costs such as taxes and transaction costs. Suppose that we have an initial portfolio $\mathbf{x}^0 = \begin{bmatrix} x_1^0 & \cdots & x_n^0 \end{bmatrix}^\mathsf{T}$ and we want to ensure that the new portfolio incurs a total turnover no larger than h. This means that we want to enforce the restriction

$$\sum_{j=1}^{n} |x_j^0 - x_j| \leq h.$$

To formulate this constraint efficiently in the quadratic optimization model, add the new vector of variables $\mathbf{y} = \begin{bmatrix} y_1 & \cdots & y_n \end{bmatrix}^\mathsf{T}$ and constraints

$$x_j - x_j^0 \leq y_j$$

$$x_j^0 - x_j \leq y_j$$

$$\sum_{j=1}^{n} y_j \leq h$$

(see Exercise 6.3). The total turnover $\sum_{j=1}^{n} |x_j^0 - x_j|$ is also sometimes called the *two-sided* turnover.

Maximizing the Sharpe Ratio

The three equivalent mean–variance models (6.13), (6.14), and (6.15) define a frontier of efficient portfolios. These portfolios are determined by some optimal tradeoff of expected return and variance, or equivalently, standard deviation of return. The ratio of expected return to standard deviation, called *Sharpe ratio* or *reward-to-risk ratio,* singles out the efficient portfolio that offers the highest reward per measure of risk.

Definition 6.3 (Sharpe ratio)   *The* Sharpe ratio *of a given portfolio* $\mathbf{x} = \begin{bmatrix} x_1 & \cdots & x_n \end{bmatrix}^\mathsf{T}$ *is the ratio of its expected return to its volatility (standard deviation) of return:*

$$Sharpe\ ratio := \frac{\boldsymbol{\mu}^\mathsf{T}\mathbf{x}}{\sqrt{\mathbf{x}^\mathsf{T}\mathbf{V}\mathbf{x}}}.$$

As we further elaborate in the next sections, sometimes $\boldsymbol{\mu}$ may not necessarily stand for the vector of expected *absolute* returns but instead it may make sense for $\boldsymbol{\mu}$ to stand for the vector of expected *relative* returns. In particular, if there is a risk-free asset, in the above definition of the Sharpe ratio it is usual to assume that $\boldsymbol{\mu}$ stands for the vector of expected *excess* returns. The *excess return* of an asset is simply the difference of its return and the risk-free return.

As an alternative or a complement to the equivalent mean–variance models (6.13), (6.14), and (6.15), consider the problem of finding the efficient portfolio with maximum Sharpe ratio. The natural formulation for this problem is the following:

$$\begin{array}{cl} \max_{\mathbf{x}} & \dfrac{\boldsymbol{\mu}^\mathsf{T}\mathbf{x}}{\sqrt{\mathbf{x}^\mathsf{T}\mathbf{V}\mathbf{x}}} \\ \text{s.t.} & \mathbf{A}\mathbf{x} = \mathbf{b} \\ & \mathbf{D}\mathbf{x} \geq \mathbf{d}. \end{array} \tag{6.16}$$

This natural formulation is evidently not a quadratic optimization model. Furthermore, the formulation is not convex as the objective function is not convex. We next show that this problem can be recast as a quadratic convex optimization problem via a suitable *homogenization*. To this end, make the following mild assumptions:

- There is a feasible portfolio $\mathbf{x}$ such that $\boldsymbol{\mu}^\mathsf{T}\mathbf{x} > 0$.

- The matrices $\mathbf{A}, \mathbf{D}$ and vector $\boldsymbol{\mu}$ satisfy the following technical condition:

$$\mathbf{A}\mathbf{z} = \mathbf{0}, \ \mathbf{D}\mathbf{z} \geq \mathbf{0} \quad \Rightarrow \quad \boldsymbol{\mu}^\mathsf{T}\mathbf{z} \leq 0.$$

The latter condition readily holds when the following stronger but easier to verify condition holds:

$$\mathbf{A}\mathbf{z} = \mathbf{0}, \ \mathbf{D}\mathbf{z} \geq \mathbf{0} \quad \Rightarrow \quad \mathbf{z} = \mathbf{0}.$$

The above assumptions ensure the soundness of the approach described next. To see what goes wrong when these assumptions do not hold, see the exercises at the end of the chapter.

The gist of the reformulation of (6.16) as a quadratic optimization problem is the following *homogenization*. Consider the change of variables obtained by putting $\mathbf{z} := \kappa\mathbf{x}$, where $\kappa > 0$ is a new scalar variable. The problem (6.16) can be rewritten as

$$\max_{\mathbf{z},\kappa} \quad \frac{\boldsymbol{\mu}^\mathsf{T}\mathbf{z}}{\sqrt{\mathbf{z}^\mathsf{T}\mathbf{V}\mathbf{z}}}$$
$$\text{s.t.} \quad \mathbf{A}\frac{\mathbf{z}}{\kappa} = \mathbf{b}$$
$$\mathbf{D}\frac{\mathbf{z}}{\kappa} \geq \mathbf{d}$$
$$\kappa > 0.$$

(6.17)

The assumption $\boldsymbol{\mu}^\mathsf{T}\mathbf{x} > 0$ for some feasible $\mathbf{x}$ implies that we can choose $\kappa > 0$ such that $\boldsymbol{\mu}^\mathsf{T}\mathbf{z} = 1$. Using this together with the second assumption, it follows that the problem (6.17) is equivalent to

$$\min_{\mathbf{z},\kappa} \quad \mathbf{z}^\mathsf{T}\mathbf{V}\mathbf{z}$$
$$\text{s.t.} \quad \boldsymbol{\mu}^\mathsf{T}\mathbf{z} \quad = \quad 1$$
$$\mathbf{A}\mathbf{z} - \mathbf{b}\kappa \quad = \quad \mathbf{0}$$
$$\mathbf{D}\mathbf{z} - \mathbf{d}\kappa \quad \geq \quad \mathbf{0}$$
$$\kappa \quad \geq \quad 0.$$

(6.18)

As the exercises at the end of the chapter detail, this approach also yields the following characterization of the portfolio with maximum Sharpe ratio in the case when we only include the full investment constraint $\mathbf{1}^\mathsf{T}\mathbf{x} = 1$.

Proposition 6.4 *Suppose the minimum-risk portfolio $(1/\mathbf{1}^\mathsf{T}\mathbf{V}^{-1}\mathbf{1})\mathbf{V}^{-1}\mathbf{1}$ has positive expected return; that is, $\boldsymbol{\mu}^\mathsf{T}\mathbf{V}^{-1}\mathbf{1} > 0$. Then the solution to the following maximum Sharpe ratio problem*

$$\max_{\mathbf{x}} \quad \frac{\boldsymbol{\mu}^\mathsf{T}\mathbf{x}}{\sqrt{\mathbf{x}^\mathsf{T}\mathbf{V}\mathbf{x}}}$$
$$\text{s.t.} \quad \mathbf{1}^\mathsf{T}\mathbf{x} = 1$$

(6.19)

is the tangency portfolio

$$\mathbf{x}^* = \frac{1}{\mathbf{1}^\mathsf{T}\mathbf{V}^{-1}\boldsymbol{\mu}}\mathbf{V}^{-1}\boldsymbol{\mu}.$$

6.5 Portfolio Management Relative to a Benchmark

In an investment portfolio, the *security selection* problem is concerned with determining the holdings of specific securities within a given asset class. It is customary to manage and evaluate the portfolio of securities relative to some predefined *benchmark portfolio* that represents a particular asset class. The benchmark portfolio provides a reference point. It serves the role of the *market portfolio* if the investment universe is restricted to the particular asset class that the benchmark represents. The management of a portfolio of securities relative to a benchmark could be *passive* or *active*. The goal of the former is to replicate the benchmark whereas the goal of the latter is to beat the benchmark.

Systematic (Beta) and Individual (Alpha) Returns

Both passive and active management rely on a fundamental decomposition of individual securities return into *systematic* and *individual* (or *residual*) components. The former is the component of return that can be explained by the security exposure to the benchmark. The latter is the component of return that is idiosyncratic to the individual security.

To make the above decomposition more precise, assume the investment universe determined by a particular asset class includes n individual securities. Let r_i denote the excess return of security i for $i = 1, \ldots, n$. Let r_B denote the excess return of the benchmark.

The return of security i can be decomposed via the following linear regression model:

$$r_i = \beta_i r_B + \theta_i,$$

where θ_i is the component of return uncorrelated to r_B; that is, $\text{cov}(r_B, \theta_i) = 0$. The coefficient β_i is the *beta* of security i relative to the benchmark B and is given by

$$\beta_i := \frac{\text{cov}(r_i, r_B)}{\text{var}(r_B)}.$$

The term $\beta_i r_B$ is the *systematic* component of return of security i. The term θ_i is the *residual* component of return of security i. The *alpha* of security i is the expected value of the residual return θ_i:

$$\alpha_i = \mathbb{E}(\theta_i).$$

Consider a portfolio of securities with percentage holdings $\mathbf{x} = \begin{bmatrix} x_1 & \cdots & x_n \end{bmatrix}^\mathsf{T}$. The above type of decomposition also applies to the portfolio return

$$r_P := \mathbf{r}^\mathsf{T} \mathbf{x} = r_1 x_1 + \cdots + r_n x_n.$$

That is, we can decompose the portfolio return r_P as

$$r_P = \beta_P r_B + \theta_P,$$

where the systematic and residual components of the portfolio return are respectively

$$\beta_P r_B = (\boldsymbol{\beta}^\mathsf{T} \mathbf{x}) r_B = (\beta_1 x_1 + \cdots + \beta_n x_n) r_B$$

and

$$\theta_P = \boldsymbol{\theta}^\mathsf{T} \mathbf{x} = \theta_1 x_1 + \cdots + \theta_n x_n.$$

Furthermore, it is easy to see that the beta and alpha of the portfolio are respectively

$$\beta_P = \boldsymbol{\beta}^\mathsf{T} \mathbf{x} = \beta_1 x_1 + \cdots + \beta_n x_n$$

and

$$\alpha_P = \mathbb{E}(\theta_P) = \boldsymbol{\alpha}^\mathsf{T} \mathbf{x} = \alpha_1 x_1 + \cdots + \alpha_n x_n.$$

Active Return, Tracking Error, Information Ratio

Consider a portfolio with percentage holdings $\mathbf{x} = \begin{bmatrix} x_1 & \cdots & x_n \end{bmatrix}^\mathsf{T}$. The *active return* of the portfolio is the difference between the portfolio return and the benchmark return:

$$\mathbf{r}^\mathsf{T}\mathbf{x} - r_B.$$

If the portfolio of benchmark holdings is $\mathbf{x}^B = \begin{bmatrix} x_1^B & \cdots & x_n^B \end{bmatrix}$, then $r_B = \mathbf{r}^\mathsf{T}\mathbf{x}^B$ and thus the active return can also be written as

$$\mathbf{r}^\mathsf{T}\mathbf{x} - r_B = \mathbf{r}^\mathsf{T}(\mathbf{x} - \mathbf{x}^B).$$

The vector $\mathbf{x} - \mathbf{x}^B$ is the vector of *active holdings* of the portfolio.

The *active risk* or *tracking error* ψ^2 of a portfolio is the standard deviation of the portfolio active return. In other words,

$$\psi^2 := \mathrm{var}(\mathbf{r}^\mathsf{T}(\mathbf{x} - \mathbf{x}^B)).$$

Some straightforward matrix calculations show that if $\mathbf{V}$ is the covariance matrix of securities returns, then

$$\psi^2 = \mathrm{var}(\mathbf{r}^\mathsf{T}(\mathbf{x} - \mathbf{x}^B)) = (\mathbf{x} - \mathbf{x}^B)^\mathsf{T}\mathbf{V}(\mathbf{x} - \mathbf{x}^B).$$

A straightforward calculation also shows that the active risk can be decomposed as

$$\psi^2 = (\beta_P - 1)^2 \sigma_B^2 + \omega_P^2,$$

where $\sigma_B^2 = \mathrm{var}(r_B)$ and $\omega_P^2 = \mathrm{var}(\theta_P)$. The first term $(\beta_P - 1)^2 \sigma_B^2$ is the component of active risk due to the *active beta* $\beta_P - 1$ of the portfolio. The second term ω_P^2 is the portfolio *residual risk*. Observe that the active risk and residual risk are the same when $\beta_P = 1$.

The *information ratio* is a cousin of the Sharpe ratio defined in Section 6.2.

Definition 6.5 (Information ratio)   The information ratio (IR) of a portfolio P is the ratio of expected residual return to volatility (standard deviation) of residual return:

$$IR_P := \frac{\alpha_P}{\omega_P}.$$

Portfolio Optimization with Benchmark Considerations

The consideration of a benchmark in portfolio construction typically leads to mean–variance models that include some adjustments and constraints induced by the benchmark.

The following are some of the most common adjustments and constraints when a mean–variance model is used for portfolio construction relative to a benchmark:

- Use expected residual returns $\alpha^\mathsf{T}\mathbf{x}$ instead of expected total return $\mu^\mathsf{T}\mathbf{x}$.
- Use active risk $\psi^2 = (\mathbf{x} - \mathbf{x}^B)^\mathsf{T}\mathbf{V}(\mathbf{x} - \mathbf{x}^B)$ instead of total risk $\mathbf{x}^\mathsf{T}\mathbf{V}\mathbf{x}$.

- Bounds on the size of *active positions*. These adjustments and constraints are typically of the form

$$L_i \leq x_i - x_i^B \leq U_i, \ i = 1, \ldots, n,$$

  that restrict the deviations between the portfolio holdings and the benchmark holdings.
- Bounds on the beta of the portfolio. Again this type of constraint is typically of the form

$$L \leq \boldsymbol{\beta}^{\mathsf{T}} \mathbf{x} - 1 \leq U.$$

As an example, the optimization problem might be

$$
\begin{aligned}
\max_{\mathbf{x}} \quad & \boldsymbol{\alpha}^{\mathsf{T}} \mathbf{x} \\
\text{s.t.} \quad & (\mathbf{x} - \mathbf{x}^B)^{\mathsf{T}} \mathbf{V} (\mathbf{x} - \mathbf{x}^B) \leq \bar{\psi}^2 \\
& \mathbf{1}^{\mathsf{T}} \mathbf{x} = 1 \\
& L \leq \boldsymbol{\beta}^{\mathsf{T}} \mathbf{x} - 1 \leq U.
\end{aligned}
\tag{6.20}
$$

6.6 Estimation of Inputs to Mean–Variance Models

The estimation of input parameters, namely the covariance matrix of returns $\mathbf{V}$ and the vector of total expected returns $\boldsymbol{\mu}$ or residual expected returns $\boldsymbol{\alpha}$, is one of the most critical and challenging steps in the use of mean–variance models. We next describe some of the central ideas that underlie most popular approaches to this fundamental problem. A comprehensive treatment of this subject is well beyond the scope of this book. Thus we only describe the key building blocks of *factor models*. We refer the reader to the textbooks of Grinold and Kahn (1999) and Litterman (2003) and to the articles by Rosenberg (1974) and Ledoit and Wolf (2003, 2004) as well as the references therein for further details on the vast variety of techniques and approaches that can be used for estimating the mean–variance input parameters $\mathbf{V}$ and $\boldsymbol{\mu}$.

Throughout this section assume the investment universe has n assets and let r_i denote the *excess return* of asset i for $i = 1, \ldots, n$. Let $\mathbf{r} \in \mathbb{R}^n$ denote the vector of excess returns. A rudimentary approach to estimate $\boldsymbol{\mu}$ and $\mathbf{V}$ via sample means and sample covariances is based on historical data. More precisely, given a time series of realized excess returns $\mathbf{r}(1), \mathbf{r}(2), \ldots, \mathbf{r}(T)$, the vectors of sample means and sample covariance are respectively

$$\hat{\boldsymbol{\mu}} := \frac{1}{T} \sum_{t=1}^{T} \mathbf{r}(t), \quad \hat{\mathbf{V}} := \frac{1}{T-1} \sum_{t=1}^{T} (\mathbf{r}(t) - \hat{\boldsymbol{\mu}})(\mathbf{r}(t) - \hat{\boldsymbol{\mu}})^{\mathsf{T}}.$$

The vector $\hat{\boldsymbol{\mu}}$ and matrix $\hat{\mathbf{V}}$ provide estimates of $\boldsymbol{\mu}$ and $\mathbf{V}$. However, these estimators have three major shortcomings:

- The sample mean and sample covariance do not incorporate other data that could contain useful forecasting information.

- For an investment universe with n assets, there are a total of $n + \frac{1}{2}n(n+1)$ $= \frac{1}{2}n(n+3)$ different parameters to estimate. Although this could be manageable for a small asset allocation model, it is not viable for an equity portfolio management model, as the number of securities n in a stock universe could easily range in the hundreds or thousands.
- The sample mean and sample covariance inevitably contain a fair amount of estimation errors, which, as we further explain in the next chapter, are magnified by the mean–variance optimizer.

The first two shortcomings above can be largely mitigated by assuming some kind of structure in the portfolio returns $\mathbf{r}$, as the following subsections detail. The next chapter is devoted entirely to the third shortcoming.

Single-Factor Model

The task of estimating a risk model can be drastically simplified by assuming that each asset has two components of risk: market risk and residual risk. This is a *single-factor risk model*. Historically this model was introduced by Sharpe as an intellectual precursor of the capital asset pricing model (CAPM). The model assumes that excess returns are decomposed as in the following regression model:

$$r_i = \beta_i r_M + \theta_i.$$

Here β_i is the beta of asset i, and θ_i is its residual return, uncorrelated with r_M. The model also assumes that the residual returns θ_i are uncorrelated with each other. The rationale for the model is that a single common factor r_M, typically the return of the market portfolio, accounts for all of the common shocks between pairs of assets. The parameter β_i is also called the *factor loading* or *factor exposure* of asset i. The component θ_i is also called the *residual* or *specific* return of asset i, as it is the portion of r_i not accounted for by the common factor r_M.

A bit of algebra shows that in this model the expected return of asset i is

$$\mathbb{E}(r_i) = \beta_i \mathbb{E}(r_M) + \mathbb{E}(\theta_i),$$

the covariance between two different assets i and j is

$$\mathrm{cov}(r_i, r_j) = \beta_i \beta_j \sigma_M^2,$$

and the variance of asset i is

$$\mathrm{var}(r_i) = \beta_i^2 \sigma_M^2 + \omega_i^2,$$

where $\sigma_M^2 = \mathrm{var}(r_M)$, $\omega_i^2 = \mathrm{var}(\theta_i)$.

Using matrix–vector notation, the single-factor risk model assumption can be succinctly written as

$$\mathbf{r} = \boldsymbol{\beta} r_M + \boldsymbol{\theta}$$

and the vector of expected returns and covariance matrix can be written as

$$\mathbb{E}(\mathbf{r}) = \boldsymbol{\beta}\mathbb{E}(r_M) + \mathbb{E}(\boldsymbol{\theta}), \quad \mathbf{V} = \sigma_M^2\boldsymbol{\beta}\boldsymbol{\beta}^\mathsf{T} + \mathbf{D},$$

where $\mathbf{D}$ is the diagonal matrix $\mathbf{D} = \mathrm{diag}(\omega_1^2, \ldots, \omega_n^2) = \mathrm{cov}(\boldsymbol{\theta})$.

We observe that under the single-factor model, the estimation of the covariance matrix only requires the estimation of $\boldsymbol{\beta}, \sigma_M^2$, and $\mathbf{D}$. That is a total of $n+1+n = 2n + 1$ parameters in contrast to the $\frac{1}{2}n(n + 1)$ parameters for a non-structured covariance matrix. The particular structure of the covariance matrix for a single-factor risk model also enables the derivation of some interesting properties of minimum-risk portfolios. (See the exercises at the end of the chapter.)

A basic estimation of the parameters of a single-factor model can be performed as follows. Assume we have some historical data of realized returns $\mathbf{r}(1), \ldots, \mathbf{r}(T)$ as well as the corresponding returns for the factor $r_M(1), \ldots, r_M(T)$. Use these data to run n simple linear regressions

$$r_i = \alpha_i + \beta_i r_M + \epsilon_i, \quad i = 1, \ldots, n.$$

Each of these linear regressions yields estimates $\hat{\beta}_i$ of β_i, $\hat{\alpha}_i$ of $\mathbb{E}(\theta_i)$, and $\hat{\omega}_i$ of $\mathrm{var}(\epsilon_i) = \mathrm{var}(\theta_i)$. Using the historical data $r_M(1), \ldots, r_M(T)$ for the factor, we can also obtain an estimate $\hat{\sigma}_M^2$ of $\mathrm{var}(r_M)$.

The above basic regression method can be enhanced to produce more accurate estimates. In particular, it is known that the quality of the estimates of $\boldsymbol{\beta}$ can be improved via a *shrinkage procedure* as explained by Blume (1975). The basic idea, which can be traced back to the classical work of Stein (1956), is that improved estimates on $\boldsymbol{\beta}$ can be obtained by taking a convex combination of the raw estimates $\hat{\boldsymbol{\beta}}$ and $\mathbf{1}$:

$$(1 - \tau)\hat{\boldsymbol{\beta}} + \tau\mathbf{1},$$

for some shrinkage factor τ. The articles of Ledoit and Wolf (2003, 2004) elaborate further on using shrinkage for improved estimates of the covariance matrix. Efron and Morris (1977) present a related and entertaining discussion of shrinkage estimation applied to baseball statistics.

The estimates of σ_M and of ω_i can also be improved by using techniques such as exponential smoothing and generalized autoregressive conditional heteroskedasticity (GARCH) (Campbell et al., 1997; Engle, 1982).

The CAPM is related to, although not the same as, a single-factor risk model. In the context of a single-factor model where the factor is the market portfolio r_M, the CAPM postulates

$$\mathbb{E}(r_i) = \beta_i\mathbb{E}(r_M).$$

In other words, the expected value of the asset-specific return is zero. The CAPM thus gives a straightforward estimation procedure for the vector of expected returns $\boldsymbol{\mu} = \mathbb{E}(\mathbf{r})$, namely $\hat{\boldsymbol{\mu}} := \hat{\boldsymbol{\beta}}\hat{\mu}_M$, where $\hat{\boldsymbol{\beta}}$ and $\hat{\mu}_M$ are estimates of $\boldsymbol{\beta}$ and $\mathbb{E}(r_M)$ respectively. As we discuss in Section 6.6 below, other alternatives for estimating expected returns are often used in equity portfolio management.

Constant Correlation Models

A second way of imposing structure on the asset returns is to assume that the correlation between any two different assets in the investment universe is the same. Under this assumption, the estimation of the covariance matrix only requires an estimate of each individual asset volatility σ_i and the average correlation ρ between different pairs of assets. This yields a "quick and dirty" estimate of the covariance matrix given by

$$\text{cov}(r_i, r_j) = \rho \sigma_i \sigma_j, \quad i \neq j.$$

In this model the estimation of the covariance matrix only requires estimates of σ and ρ. That is a total of $n+1$ parameters.

Under the reasonable assumption that $\rho > 0$, the constant correlation model can be seen as the following kind of single-factor model with predetermined factor loadings. Assume the following single-factor model for volatility *scaled* excess returns:

$$\frac{r_i}{\sigma_i} = f + \theta_i,$$

where f is a common factor to all scaled returns and θ_i is a specific scaled return on asset i. It is easy to see that this particular single-factor model yields a constant correlation model with ρ being the variance of the single factor f.

Using matrix notation, the constant correlation covariance matrix can be written as

$$\mathbf{V} = \rho \boldsymbol{\sigma} \boldsymbol{\sigma}^\mathsf{T} + (1 - \rho)\text{diag}(\boldsymbol{\sigma})^2.$$

A basic estimation procedure for this model is straightforward: first, using historical data, compute estimates $\hat{\sigma}_i$ of σ_i and estimates $\hat{\rho}_{ij}$ of each correlation ρ_{ij} for all $i \neq j$. Finally, take the average

$$\hat{\rho} := \frac{1}{n(n-1)} \sum_{i \neq j} \hat{\rho}_{ij}$$

as an estimate of ρ.

Multiple-Factor Models

Multiple-factor models are a generalization of the single-factor model discussed above. These models are based on the assumption that the return of each asset can be explained by a small collection of common factors in addition to some other specific return. Aside from simplifying the estimation task, multiple-factor models provide a useful breakdown of risk, incorporate some economic logic, and are fairly flexible. The majority of quantitative money managers rely on multi-factor models provided by third-party vendors such as MSCI, Axioma, Northfield, etc. for the management of equity portfolios.

A multi-factor model assumes that excess returns are as follows:

$$r_i = \sum_{k=1}^{K} B_{ik} f_k + u_i,$$

where

- r_i: excess return of asset i
- B_{ik}: exposure of asset i to factor k
- f_k: rate of return of factor k
- u_i: specific (or residual) return of asset i.

It is convenient to rewrite the relation above in matrix form as

$$\mathbf{r} = \mathbf{Bf} + \mathbf{u}.$$

A bit of matrix algebra shows that the expected value and covariance of $\mathbf{r}$ are respectively

$$\mathbb{E}[\mathbf{r}] = \mathbf{B}\mathbb{E}[\mathbf{f}] + \mathbb{E}[\mathbf{u}], \quad \mathbf{V} = \mathbf{BFB}^{\mathsf{T}} + \Delta,$$

where $\mathbf{F} = \mathrm{cov}(\mathbf{f})$ and $\Delta = \mathrm{cov}(\mathbf{u})$. Observe that Δ is diagonal since the u_i are assumed to be uncorrelated with each other.

The construction and estimation of a multi-factor model hinges on the choice of factors. For an equity universe, the following three main classes of factors are commonly used:

- Macroeconomic factors: inflation, economic growth, etc.
- Fundamental factors: earning/price, dividend yield, market cap, etc.
- Statistical factors: principal component analysis, hidden factors.

Empirical evidence suggests that the second type of fundamental factors works better than the other two (Connor, 1995). This is also the prevalent class of factors used by most risk model providers. In this approach we have

$$\mathbf{r} = \mathbf{Bf} + \mathbf{u},$$

where the matrix of factor loadings $\mathbf{B}$ is predetermined. The estimation of the corresponding covariance matrix is as follows. Using historical data for the asset returns, infer the corresponding historical data for factor returns by solving each of the weighted least-squares problems

$$\min(\mathbf{r}(t) - \mathbf{Bf}(t))^{\mathsf{T}} \mathbf{D}^{-1}(\mathbf{r}(t) - \mathbf{Bf}(t)).$$

The matrix $\mathbf{D}$ is a diagonal matrix whose entries are estimates of the asset variances. A common proxy is to use instead the reciprocal of the market capitalizations of the assets. The solution to this weighted least-squares problem is

$$\mathbf{f}(t) = (\mathbf{BD}^{-1}\mathbf{B}^{\mathsf{T}})^{-1}\mathbf{B}^{\mathsf{T}}\mathbf{D}^{-1}\mathbf{r}(t).$$

Each row of the matrix $(\mathbf{B}\mathbf{D}^{-1}\mathbf{B}^{\mathsf{T}})^{-1}\mathbf{B}^{\mathsf{T}}\mathbf{D}^{-1}$ can be interpreted as a *factor mimicking* portfolio.

Equipped with this historical data of factor returns, we can estimate the factor covariance matrix. The residuals $\mathbf{u}(t) := \mathbf{r}(t) - \mathbf{B}\mathbf{f}(t)$ can then be used to estimate the covariance matrix Δ of asset-specific returns.

The connection between the CAPM and single-factor models has an analogous counterpart in the context of multi-factor models, namely the *arbitrage pricing theory* (APT). A combination of an arbitrage argument and the assumption that the set of factors $\mathbf{f}$ account for all of the common shocks to the returns of all assets in the investment universe implies that

$$\mathbb{E}(\mathbf{r}) = \mathbf{B}\mathbb{E}(\mathbf{f}).$$

Like the CAPM, the APT model also yields a straightforward estimation procedure for $\boldsymbol{\mu} = \mathbb{E}(\mathbf{r})$.

Estimation of Alpha

In a benchmark-relative context, an estimate of expected residual returns $\boldsymbol{\alpha}$ is typically the relevant estimate instead of an estimate of expected total return $\boldsymbol{\mu}$.

According to the CAPM or the more general APT model, the expected residual returns are zero. However, numerous articles have documented certain *anomalies* that are systematically associated with the over- and underperformance of the return of securities after controlling for their systematic component of return. Some of these anomalies include the SMB (small minus big market capitalization) and HML (high minus low book-to-price) factors introduced in the classical article by Fama and French (1992).

A generic approach for generating alpha is to rely on *signals* unveiled via a judicious type of analysis. A signal could be an empirical observation such as *momentum* that suggests that the recent performance (good or bad) of individual securities will persist in the near term. A signal could also be a financial principle such as "firms with low book-to-price ratio will outperform" or "firms with higher earnings per share will outperform".

The following is a reasonable and popular rule of thumb for transforming a signal into a forecast of alpha (for a detailed discussion see Grinold and Kahn (1999)):

$$\text{alpha} = (\text{residual volatility}) \cdot \text{IC} \cdot \text{score}.$$

Here the *residual volatility* is the standard deviation of residual return. The *score* is a numerical score associated with the signal. The score is assumed to be scaled so that its cross-sectional mean and standard deviation are respectively 0 and 1. Finally, the *information coefficient* IC is a measure of the forecasting quality of the signal; that is, the correlation between the raw signal score and the residual return.

In addition to proper scaling, the signal score should be *neutralized* so that the alphas do not include biases or undesirable bets on the benchmark or on risk factors. As we illustrate in the exercises at the end of the chapter, neutralization can be achieved in various ways, as there are multiple portfolios that hedge out a bet on the benchmark or on other risk factors.

6.7   Performance Analysis

How can the performance of a portfolio manager be evaluated? Are the *ex post* results due to skill or luck? The goal of performance analysis is to answer these questions. The efficient market hypothesis suggests that skillful active management is impossible. However, there is considerable evidence against the efficient market hypothesis (Shleifer, 2000).

Empirical results also suggest that an *average* active fund manager underperforms their benchmark on a risk-adjusted basis. Furthermore, empirical evidence also shows that good performance does not persist: The winners this year are almost as likely to be winners or losers next year. These are bleak conclusions about asset management. So how could we tell which asset managers are the good ones?

The fundamental goal of performance analysis is to separate skill from luck. The simplest type of performance analysis is a cross-sectional comparison of returns over some time period. This would distinguish winners from losers. However, these kinds of comparisons have several drawbacks. First, they typically do not represent the complete universe of investment managers but only those in existence during a specific time period. They generally contain survivorship bias. Perhaps worst of all, cross-sectional comparisons do not adjust for risk. By contrast, time-series analysis of returns can do a better job at separating skill from luck by measuring both return and risk. An even more complete picture can be obtained via time-series analysis of returns and portfolio holdings.

Return-Based Performance Analysis (Basic)

The development of the CAPM and the notion of market efficiency in the 1960s encouraged academics to tackle the problem of performance analysis. According to the CAPM, consistent exceptional returns are unlikely. Academics devised tests to check if the theory was correct. As a byproduct the first performance analysis techniques emerged. One approach, proposed by Jensen, consists of regressing the time series of *realized* portfolio excess returns against benchmark excess return:

$$r_P(t) = \alpha_P + \beta_P r_B(t) + \epsilon_P(t).$$

Jensen's alpha is simply the intercept α_P of this regression. According to the CAPM, this intercept is zero. The regression yields not only alpha and beta, but t-statistics that give information about their statistical significance. The *t-statistic* for α_P is

$$t\text{-stat} = \frac{\alpha_P}{\text{SE}(\alpha_P)}.$$

As a rule of thumb, a t-statistic of 2 or more indicates that the performance of the portfolio is due to skill rather than luck. Assuming normality, the probability of observing such a large t-statistic purely by chance is smaller than 5%.

The t-statistic and the information ratio are closely related. The main difference between them is that the information ratio is annualized. By contrast, the t-statistic scales with the number of years of data. If we observe returns over a period of T years, the information ratio is approximately the t-statistic divided by the square root of the number of years of observation:

$$IR \approx \frac{t\text{-stat}}{\sqrt{T}}.$$

The standard error of the information ratio is approximately

$$\text{SE}(IR) \approx \frac{1}{\sqrt{T}}.$$

A simple alternative to Jensen's approach is to compare Sharpe ratios for the portfolio and the benchmark. A portfolio with

$$\frac{\bar{r}_P}{\sigma_P} > \frac{\bar{r}_B}{\sigma_B},$$

where $\bar{r}$ denotes mean excess return over the period, has demonstrated positive performance. Once again, the statistical significance of this relationship is relevant for distinguishing luck from skill. If we assume that the standard errors of the portfolio and benchmark volatilities are fairly small compared to $\bar{r}$ standard errors, then the standard error of the Sharpe ratio is approximately $1/\sqrt{N}$, where N is the number of observations. Hence a statistically significant demonstration of skill occurs when

$$\frac{\bar{r}_P}{\sigma_P} - \frac{\bar{r}_B}{\sigma_B} > 2\sqrt{\frac{2}{N}}.$$

Return-Based Style Analysis

Style analysis was developed by Nobel laureate William Sharpe (1992). The popularity of this concept was aided by a study (Brinson et al., 1991) concluding that 91.5% of the variation in returns of 82 mutual funds could be explained by the allocation to bills, stocks, and bonds. Later studies considering asset allocation across a broader range of asset classes have shown that as much as 97% of fund returns can be explained by asset allocation alone.

Style analysis attempts to determine the effective asset mix of a fund using only the time series of returns for the fund and for a number of carefully chosen asset classes. Like a factor model approach, style analysis assumes that portfolio returns have the form

$$r_P(t) = \sum_{j=1}^{m} w_j f_j(t) + u_P(t),$$

where the $f_j(t)$ are the returns of m benchmark asset classes. The holdings w_j, $j = 1, \ldots, m$, represent the *style* of the portfolio. That is, the effective allocation to the m asset classes that could be replicated via a passive portfolio. The term $u_P(t)$ represents the *selection return*; that is, the portion of the portfolio return that style cannot explain. The effective holdings can be estimated via the quadratic program

$$
\begin{aligned}
\min_{\mathbf{w}} \quad & \mathrm{var}(u_P(t)) \\
\text{s.t.} \quad & \sum_{j=1}^{m} w_j = 1 \\
& w_j \geq 0, \ j = 1, \ldots, m.
\end{aligned}
\tag{6.21}
$$

Notice that there are two key differences between this model and conventional multiple regression. First, the weights are constrained to be non-negative and to add up to 1. Second, instead of minimizing the sum of squared errors $\sum_{t=1}^{T} u_P(t)^2$ we minimize the variance of these quantities. The reason for the first restriction is that the w_j are to be interpreted as an effective asset allocation representing the style of the fund. In essence, they create a fund-specific benchmark. The reason for the second restriction is that we want to allow for a non-zero selection effect by the fund manager. The model finds the style that minimizes the variance of this effect. Once the optimal weights are determined, the average value of $u_P(t)$ gives the value added by the manager's selection skills, which can be negative or positive.

Assume the data available for style analysis are the return time series $r_P(t)$, $f_1(t), \ldots, f_m(t)$ for $t = 1, \ldots, T$. For ease of notation, put

$$
\mathbf{r} := \begin{bmatrix} r_P(1) \\ \vdots \\ r_P(T) \end{bmatrix}, \quad
\mathbf{F} := \begin{bmatrix} f_1(1) & \cdots & f_m(1) \\ \vdots & \ddots & \vdots \\ f_1(T) & \cdots & f_m(T) \end{bmatrix}, \quad
\mathbf{1} := \begin{bmatrix} 1 \\ \vdots \\ 1 \end{bmatrix}.
$$

Then the objective function in (6.21) can be written as

$$
\begin{aligned}
\mathrm{var}\,(\mathbf{r} - \mathbf{F}\mathbf{w}) &= \frac{1}{T}\|\mathbf{r} - \mathbf{F}\mathbf{w}\|^2 - \frac{1}{T^2}(\mathbf{1}^{\mathsf{T}}(\mathbf{r} - \mathbf{F}\mathbf{w}))^2 \\
&= \left(\frac{\|\mathbf{r}\|^2}{T} - \frac{(\mathbf{1}^{\mathsf{T}}\mathbf{r})^2}{T^2} \right) - 2\left(\frac{\mathbf{r}^{\mathsf{T}}\mathbf{F}}{T} - \frac{\mathbf{1}^{\mathsf{T}}\mathbf{r}}{T^2}\mathbf{1}^{\mathsf{T}}\mathbf{F} \right) \mathbf{w} \\
&\quad + \mathbf{w}^{\mathsf{T}} \left(\frac{1}{T}\mathbf{F}^{\mathsf{T}}\left(I - \frac{1}{T}\mathbf{1}\mathbf{1}^{\mathsf{T}} \right) \mathbf{F} \right) \mathbf{w}.
\end{aligned}
$$

Style analysis provides an improvement tool for measuring performance. The constructed style usually tracks the performance of the fund more accurately than a predefined benchmark. Style analysis has also some limitations. For instance, the weights may not necessarily match the style disclosed by the fund manager. However, as Sharpe puts it: "If it acts like a duck, it is ok to assume it is a duck." Style analysis also makes the simplifying assumptions that the weights are constant. This is clearly not the case in actively managed funds, even without active trading. There exist some variations of style analysis that allow for weights to change. The model gets a bit more technical because it needs to incorporate some "regularization" term that prevents the weights from changing too much too often.

6.8 Notes

The mean–variance model was introduced in the seminal article of Markowitz (1952). The CAPM was developed by Treynor[1], Sharpe (1964), Lintner (1965), and Mossin (1966), by building on the mean–variance approach of Markowitz. In recognition of their work on portfolio choice and the CAPM, Sharpe and Markowitz were jointly awarded the 1990 Nobel Prize in Economics. Both Lintner and Mossin passed away before 1990 and Treynor's manuscript was never published.

The textbook by Grinold and Kahn (1999) is a classical reference in active portfolio management. In their textbook, Grinold and Kahn developed and relied extensively on characteristic portfolios.

6.9 Exercises

Exercise 6.1 The purpose of this exercise is to prove the two-fund theorem (Theorem 6.1).

(a) Find the Lagrangian function $L(\mathbf{x}, \theta)$ for (6.1).
(b) Solve the optimality conditions $\nabla L(\mathbf{x}, \theta) = \mathbf{0}$ to conclude that the optimal solution to (6.1) is

$$\mathbf{x}^* = \lambda \cdot \frac{1}{\mathbf{1}^\mathsf{T} \mathbf{V}^{-1} \boldsymbol{\mu}} \mathbf{V}^{-1} \boldsymbol{\mu} + (1 - \lambda) \cdot \frac{1}{\mathbf{1}^\mathsf{T} \mathbf{V}^{-1} \mathbf{1}} \mathbf{V}^{-1} \mathbf{1}$$

where $\lambda = \mathbf{1}^\mathsf{T} \mathbf{V}^{-1} \boldsymbol{\mu} / \gamma$.

Exercise 6.2 Assume $\boldsymbol{\mu}$ and $\mathbf{V}$ are respectively the vector of expected returns and covariance matrix of n risky assets. Assume V is non-singular and $\bar{\mu} > \boldsymbol{\mu}^\mathsf{T} \mathbf{V}^{-1} \mathbf{1} / \mathbf{1}^\mathsf{T} \mathbf{V}^{-1} \mathbf{1}$. Consider the mean–variance optimization problem

[1] "Toward a theory of market value of risky assets". Unpublished manuscript, 1961.

$$\text{min} \quad \mathbf{x}^{\mathsf{T}}\mathbf{V}\mathbf{x}$$
$$\text{s.t.} \quad \boldsymbol{\mu}^{\mathsf{T}}\mathbf{x} \geq \bar{\mu} \qquad\qquad (6.22)$$
$$\mathbf{1}^{\mathsf{T}}\mathbf{x} = 1.$$

Now consider the following variations:

$$\text{max} \quad \boldsymbol{\mu}^{\mathsf{T}}\mathbf{x}$$
$$\text{s.t.} \quad \mathbf{x}^{\mathsf{T}}\mathbf{V}\mathbf{x} \leq \bar{\sigma}^2 \qquad\qquad (6.23)$$
$$\mathbf{1}^{\mathsf{T}}\mathbf{x} = 1,$$

and

$$\text{max} \quad \boldsymbol{\mu}^{\mathsf{T}}\mathbf{x} - \tfrac{1}{2}\gamma \cdot \mathbf{x}^{\mathsf{T}}\mathbf{V}\mathbf{x}$$
$$\text{s.t.} \quad \mathbf{1}^{\mathsf{T}}\mathbf{x} = 1. \qquad\qquad (6.24)$$

Let $\mathbf{x}^*$ be the optimal solution to (6.22). Find appropriate values of $\bar{\sigma}$ and γ so that the optimal solutions to (6.23) and (6.24) are also $\mathbf{x}^*$.

Exercise 6.3   Prove that

$$\sum_{j=1}^{n} |x_j^0 - x_j| \leq h$$

if and only if there exists a vector $\mathbf{y} = \begin{bmatrix} y_1 & \cdots & y_n \end{bmatrix}^{\mathsf{T}}$ such that

$$x_j - x_j^0 \leq y_j$$
$$x_j^0 - x_j \leq y_j$$
$$\sum_{j=1}^{n} y_j \leq h.$$

Exercise 6.4   Prove that under the two assumptions made in Section 6.4, the maximum Sharpe ratio problem (6.16) is indeed equivalent to (6.18).

Exercise 6.5   The purpose of this exercise is to prove Proposition 6.4. Assume the covariance matrix of asset returns $\mathbf{V}$ is positive definite and the minimum- risk portfolio $(1/\mathbf{1}^{\mathsf{T}}\mathbf{V}^{-1}\mathbf{1})\mathbf{V}^{-1}\mathbf{1}$ has positive expected return; that is, $\boldsymbol{\mu}^{\mathsf{T}}\mathbf{V}^{-1}\mathbf{1} > 0$.

(a) Show that (6.19) can be rewritten as follows:

$$\begin{aligned} \min_{\mathbf{z},\kappa} \quad & \mathbf{z}^{\mathsf{T}}\mathbf{V}\mathbf{z} \\ \text{s.t.} \quad & \boldsymbol{\mu}^{\mathsf{T}}\mathbf{z} && = 1 \\ & \mathbf{1}^{\mathsf{T}}\mathbf{z} - \kappa && = 0 \\ & \kappa && > 0. \end{aligned} \qquad (6.25)$$

(b) Show that the solution to (6.25) is

$$\mathbf{z}^* = \frac{1}{\boldsymbol{\mu}^{\mathsf{T}}\mathbf{V}^{-1}\boldsymbol{\mu}}\mathbf{V}^{-1}\boldsymbol{\mu}$$
$$\kappa^* = \mathbf{1}^{\mathsf{T}}\mathbf{z}^*.$$

(c) Use part (b) to conclude that the solution to (6.19) is indeed

$$\mathbf{x}^* = \frac{1}{\mathbf{1}^\mathsf{T}\mathbf{V}^{-1}\boldsymbol{\mu}}\mathbf{V}^{-1}\boldsymbol{\mu}.$$

(d) *Show that if $\boldsymbol{\mu}^\mathsf{T}\mathbf{V}^{-1}\mathbf{1} < 0$ then (6.19) is bounded but does not attain its maximum value. Use this fact to illustrate why the two assumptions made in Section 6.4 cannot simply be dropped without making some other assumptions.

Exercise 6.6   The Excel spreadsheet "Exercise 6.6 Six Stocks" provides hypothetical estimates of the expected return and variance–covariance matrix for a set of six stocks.

(a) Set up a quadratic programming model to determine the long-only minimum-variance portfolio that can be constructed with the six stocks. What is the expected return of your minimum-variance portfolio?

(b) Set up the classical Markowitz model with long-only constraints. Solve your model for at least six different levels of expected return ranging from the level found in part (a) up to the largest expected return level for which there are feasible portfolios. What is the value of such largest return level?

     Use your results to generate the expected return versus standard deviation plot for the efficient frontier.

(c) Assume the "benchmark" is a portfolio equally divided among the six stocks. Compute the beta of each stock (with respect to this benchmark) and the consensus (i.e., CAPM) returns assuming the risk-free rate is zero.

(d) Assume that your current portfolio is the benchmark, i.e., it is equally divided among the six stocks. Include an additional total turnover constraint of 70% in the model from part (b). Determine the new optimal portfolio for a desired expected return somewhere in the middle of the range used in part (b). Is it possible to find portfolios with any return level in the range in part (b)? If it is not, can you explain why?

(e) Find the portfolio with maximum Sharpe ratio subject to all constraints in part (d). Again, assume the risk-free rate is zero.

Exercise 6.7   The Excel spreadsheet "Exercise 6.7 Twenty Stocks" contains estimated expected values, standard deviations, and correlations of monthly returns for a set of 20 large-capitalization stocks from the S&P 500.

(a) Find the fully invested long-only portfolio with minimum variance. Find the numerical values of the first two and last two positions in your portfolio (i.e., those of BOL, NE, and XTO, ABC). These numbers are between 0 and 1.

     Find the numerical value of the variance of the portfolio (in bps^2).

(b) Assume the benchmark is an equally weighted portfolio of the 20 assets Determine the beta of each asset relative to this benchmark.

Find the numerical values of the beta of the first two stocks and last stock

(c) Find the fully invested long-only portfolio with highest expected return that satisfies the following constraints:

- The size of every position is at most 10%.
- The portfolio has beta equal to 1.

Find the numerical values of the first and last positions in your portfolio

Find the numerical value of the expected return of the portfolio (in bps)

(d) Assume the risk-free rate is zero. Find the fully invested long-only portfolio with highest Sharpe ratio that satisfies the following constraints:

- The size of every position is at most 10%.
- The portfolio has beta equal to 1.

Find the numerical values of the positions 15 and 16 in your portfolio (i.e. those of LH and R).

Find the numerical value of the Sharpe ratio of the portfolio.

Exercise 6.8 Suppose M is the market portfolio in a universe of securities According to the CAPM, the excess return of each security is given by

$$r_i = \beta_i r_M + \epsilon_i,$$

where ϵ_i is the zero-mean, security-specific risk, and r_M is the market excess return. For simplicity assume the risk-free rate is zero.

Suppose that via a thorough security analysis a manager identifies an active portfolio A whose return is

$$r_A = \alpha_A + \beta_A r_M + \epsilon_A;$$

let $w_A^2 = \text{var}(\epsilon_A)$ denote the residual variance of the active portfolio A.

Consider a portfolio P obtained by investing a proportion w in the active portfolio A and the remaining proportion $1 - w$ in the market portfolio:

$$r_P(w) = w r_A + (1 - w) r_M.$$

(a) Find the expressions for the expected return and variance of the portfolio P.

(b) Assume $\beta_A = 1$, $\alpha_A > 0$, and $\mu_M > 0$. Show that the portfolio P with highest Sharpe ratio is attained for the following proportion value:

$$w_0 = \frac{\alpha_A / w_A^2}{\mu_M / \sigma_M^2}.$$

Furthermore, show that, for this proportion value, the Sharpe ratio of the portfolio is

$$S_P^2 = S_M^2 + IR_A^2 = \left(\frac{\mu_M}{\sigma_M}\right)^2 + \left(\frac{\alpha_A}{w_A}\right)^2.$$

Exercise 6.9 Suppose the covariance matrix of a universe of N stocks has the following *single-factor risk model* form:

$$\mathbf{V} = \sigma_M^2 \boldsymbol{\beta}\boldsymbol{\beta}^\mathsf{T} + \mathbf{D}.$$

Here σ_M^2 is the single-factor risk, $\boldsymbol{\beta} = \begin{bmatrix} \beta_1 & \cdots & \beta_N \end{bmatrix}^\mathsf{T} \in \mathbb{R}^N$ is the vector of stock loadings on that factor, and $\mathbf{D} = \mathrm{diag}(\omega_1^2, \ldots, \omega_N^2)$ where each ω_i^2 is the idiosyncratic risk of stock i for $i = 1, \ldots, N$.

(a) Recall that the Sherman–Morrison–Woodbury matrix inverse formula is

$$(\mathbf{A} + \mathbf{u}\mathbf{v}^\mathsf{T})^{-1} = \mathbf{A}^{-1} - \frac{\mathbf{A}^{-1}\mathbf{u}\mathbf{v}^\mathsf{T}\mathbf{A}^{-1}}{1 + \mathbf{v}^\mathsf{T}\mathbf{A}^{-1}\mathbf{u}}$$

provided $\mathbf{A}^{-1}$ exists and $1 + \mathbf{v}^\mathsf{T}\mathbf{A}^{-1}\mathbf{u} \neq 0$. Use this formula to show that

$$\mathbf{V}^{-1} = \mathbf{D}^{-1} - \frac{\sigma_M^2}{1 + \sigma_M^2 \boldsymbol{\beta}^\mathsf{T}\mathbf{D}^{-1}\boldsymbol{\beta}} \cdot \mathbf{D}^{-1}\boldsymbol{\beta}\boldsymbol{\beta}^\mathsf{T}\mathbf{D}^{-1}.$$

(b) *Using (a) conclude that the holdings of the minimum-variance fully invested portfolio $(1/\mathbf{1}^\mathsf{T}\mathbf{V}^{-1}\mathbf{1}) \cdot \mathbf{V}^{-1}\mathbf{1}$ are given by

$$x_i = \frac{\sigma_{MV}^2}{\omega_i^2}\left(1 - \frac{\beta_i}{\beta_{LS}}\right), \quad i = 1, \ldots, N,$$

where

$$\sigma_{MV}^2 = \frac{1}{\mathbf{1}^\mathsf{T}\mathbf{V}^{-1}\mathbf{1}}$$

is the variance of the minimum-variance portfolio and β_{LS} is the following *long–short threshold beta*:

$$\beta_{LS} = \frac{1 + \sigma_M^2\boldsymbol{\beta}^\mathsf{T}\mathbf{D}^{-1}\boldsymbol{\beta}}{\sigma_M^2\boldsymbol{\beta}^\mathsf{T}\mathbf{D}^{-1}\mathbf{1}}.$$

(c) *Show that the holdings of the *long-only* minimum-variance portfolio of the N stocks are given by an expression similar to that in part (b) above:

$$x_i = \frac{\sigma_{LMV}^2}{\omega_i^2}\left(1 - \frac{\beta_i}{\beta_L}\right)^+, \quad i = 1, \ldots, N,$$

where σ_{LMV}^2 is the variance of the long-only minimum-variance portfolio and β_L is a suitable long-only threshold beta.

Exercise 6.10 Suppose the covariance matrix of a set of assets has the following *constant-correlation* form: for some $\rho \in (0, 1)$

$$\mathbf{V}_{ii} = \sigma_i^2, \ \mathbf{V}_{ij} = \rho\sigma_i\sigma_j, \ \text{for} \ i = 1, \ldots, n, \ \text{and} \ j = 1, \ldots, n, \ \text{with} \ i \neq j.$$

In matrix form, we can write the above constant-correlation matrix as follows:

$$\mathbf{V} = \rho\boldsymbol{\sigma}\boldsymbol{\sigma}^\mathsf{T} + (1 - \rho)\mathrm{Diag}(\boldsymbol{\sigma})^2,$$

where $\boldsymbol{\sigma}$ is the vector with components σ_i, $i = 1, \ldots, n$.

(a) Use the Sherman–Morrison–Woodbury formula to show that

$$\mathbf{V}^{-1} = \frac{1}{1-\rho}\mathrm{Diag}(\boldsymbol{\theta})^2 - \frac{\rho}{(1-\rho)(1+(n-1)\rho)}\boldsymbol{\theta}\boldsymbol{\theta}^\mathsf{T},$$

where $\boldsymbol{\theta}$ is the vector with components $\theta_i = 1/\sigma_i$, $i = 1,\ldots,n$.

(b) Conclude that the holdings x_i, $i = 1,\ldots,n$, of the fully invested, minimum risk portfolio are as follows:

$$x_i = \frac{y_i}{\sum_{j=1}^n y_j},$$

where

$$y_i = \frac{1}{(1-\rho)\sigma_i^2}\left[1 - \frac{\rho}{(1-\rho)(1+(n-1)\rho)}\sum_{j=1}^n \frac{\sigma_i}{\sigma_j}\right].$$

(c) Assume all of the assets have the same volatility: that is, $\sigma_1 = \sigma_2 = \cdots = \sigma_n = \sigma$. Prove that the variance of the fully invested portfolio of minimum variance is

$$\sigma_{\min}^2 = \frac{\sigma^2(1+(n-1)\rho)}{n}.$$

Exercise 6.11　The purpose of this exercise is to detail the derivation of *factor portfolios* used in the construction of risk models. Consider the following factor model for a vector of returns

$$\mathbf{r} = \mathbf{B}\mathbf{f} + \mathbf{u},$$

where $\mathbf{B}$ is a given matrix of factor loadings and the factors $\mathbf{f}$ are to be constructed. A common approach to construct the factors $\mathbf{f}$ is to solve the following kind of weighted least-squares problem:

$$\min_{\mathbf{f}} \ (\mathbf{r} - \mathbf{B}\mathbf{f})^\mathsf{T}\mathbf{D}^{-1}(\mathbf{r} - \mathbf{B}\mathbf{f}), \tag{6.26}$$

where $\mathbf{D}$ is a symmetric (often diagonal) positive definite matrix.

(a) Show that the gradient of the multivariate function $\mathbf{f} \mapsto (\mathbf{r} - \mathbf{B}\mathbf{f})^\mathsf{T}\mathbf{D}^{-1}(\mathbf{r} - \mathbf{B}\mathbf{f})$ is

$$2(\mathbf{B}^\mathsf{T}\mathbf{D}^{-1}\mathbf{B})\mathbf{f} - 2\mathbf{B}^\mathsf{T}\mathbf{D}^{-1}\mathbf{r}.$$

(b) Conclude that the solution to (6.26) is

$$\mathbf{f} = (\mathbf{B}^\mathsf{T}\mathbf{D}^{-1}\mathbf{B})^{-1}\mathbf{B}^\mathsf{T}\mathbf{D}^{-1}\mathbf{r}.$$

(c) Consider the special case when $\mathbf{B} = \mathbf{b}$ has only one column, i.e., there is only one factor f. Conclude that in this case the above optimal f is the return of the following "characteristic portfolio":

$$\frac{1}{\mathbf{b}^\mathsf{T}\mathbf{D}^{-1}\mathbf{b}}\mathbf{D}^{-1}\mathbf{b}.$$

(d) Consider again part (c) and the very special case when the entries of $\mathbf{b}$ are 1 (a "buy" list) and -1 (a "sell" list) and $\mathbf{D}$ is a diagonal matrix. Show that in this case the characteristic portfolio in part (c) is a long–short portfolio with long holdings in the "buy list" and short holdings in the "sell list". Describe the values of the portfolio holdings when $\mathbf{D} = \mathbf{I}$.

Exercise 6.12 Consider two portfolio managers. One has 25 years of performance history, with a realized Sharpe ratio of 0.5. The other one has only four years of performance history but with a realized Sharpe ratio of 0.75. Which one would you prefer to invest in and why? The objective is to minimize the likelihood that you will lose money. Returns can be assumed to be stationary and normally distributed.

6.10   Case Studies

Asset Allocation

The goal of this case study is to apply and test mean–variance optimization models as a tool for asset allocation.

(1) Choose between four and ten asset classes and collect their monthly, quarterly, or annual historical returns over a meaningful horizon (several years or decades). Collect also any other relevant data that may help you forecast expected returns. Briefly discuss why you would like to choose these assets and why the selected horizon is appropriate.

(2) Use the first 67% portion of your data to compute the expected returns and the variance–covariance matrix for these assets. (Use the remaining 33% for out-of-sample testing.)

(3) Set up the classical Markowitz model without short sales and solve it in Excel **Solver** or MATLAB for various levels of expected return.

(4) Evaluate, compare, and report your results in-sample and out-of-sample.

(5) Discuss the results of your model.

Covariance Estimation

The goal of this case study is to compare various approaches to covariance estimation and risk diversification.

(1) Select a universe of at least 25 stocks. Some possible choices are the Dow Jones Industrial Average, the S&P 100, and the Nasdaq 100. If you feel ambitious, you may choose a larger universe. Your purpose is to construct the most diversified fully invested portfolio in this universe. Collect weekly or monthly historical returns for securities in your universe over a horizon of a few years.

(2) Use the first 67% portion of historical data for "model calibration" (estimates of covariance matrix) and the remaining 33% for out-of-sample testing.

(3) Use the in-sample data to generate the following two estimates of the covariance matrix: the sample covariance, and a single-factor model covariance. For the latter, you need to choose a suitable benchmark portfolio. Some

reasonable choices are a value-weighted portfolio and an equally weighted portfolio.

(4) Using the two estimates of the covariance matrices computed in (3), find minimum-risk fully invested portfolios (both long–short and long-only).

(5) Compare the results of your models on out-of-sample data. Generate plots of the value of the different portfolios on out-of-sample data.

(6) Repeat (5) using a rolling-time window assuming that all portfolios are rebalanced monthly. Compare these models with value-weighted and equally weighted portfolios.

Report statistics such as out-of-sample mean and standard deviation of results, and average portfolio turnover. Comment on your results.

Active Portfolio Management

The goal of this case study is to apply mean–variance optimization as a tool for active portfolio management. If you are well versed with the Bloomberg terminal, you may use Bloomberg's portfolio analytics capabilities PORT.

(1) Choose at least 20 securities within an asset class (e.g., stocks in the Dow Jones, the S&P 500, the NASDAQ, or the Russell 3000) and find their weekly or monthly historical returns over a meaningful horizon. Collect also relevant additional data for alpha estimation. For instance, Fama–French factors (book-to-market ratio, size, momentum), or any other factors that you can use to rank your stocks. Briefly discuss your selection of securities and data.

(2) Choose a suitable "benchmark portfolio". For instance, if you chose stocks from the S&P 500, a reasonable benchmark would be a value-weighted portfolio of the sets of selected assets.

(3) Use the first 67% portion of historical data for "model calibration" (estimates of covariance matrix, betas, alphas, etc.) and the remaining 33% for out-of-sample testing.

(4) Use the in-sample data to estimate the covariance matrix, betas and alphas of your stocks. The most straightforward way to estimate the betas of your stocks is via linear regression. This would also give you a rudimentary estimate of the alphas. However, these are "realized" estimates, i.e., they are backward looking. Instead you may try to forecast alphas (i.e., be forward looking) via one of the following approaches:

- Momentum factor: rank stocks according to how they have performed in the recent three to twelve months.
- Other factors: rank stocks according to other factors such as the Fama–French factors, price-to-earnings ratio, debt-to-equity ratio, or some combination of these.

(5) Set up an optimization model with the goal of constructing portfolios that outperform the benchmark. Discuss your selection of objective and constraints in your model.

(6) Test the results of your model on out-of-sample data. You may want to do this for various combinations of constraint levels (e.g., small and large levels of tracking errors, small and large levels of active positions, turnover constraint). The most interesting way of doing this is via a "rolling-time window". To that end, proceed as follows:

(a) Partition the out-of-sample data into m equally sized time intervals, e.g., month-long intervals.

(b) Using the estimates from (4), find the optimal portfolio. Assume you hold this portfolio over the first of the m out-of-sample time intervals.

(c) Next, shift the in-sample time window used in step (b). Keep the length of the in-sample time window unchanged. Use this new in-sample time window to update your estimates of covariance matrix, betas, and alphas. Find the new optimal portfolio. Assume you will hold this portfolio over the next out-of-sample time interval.

(d) Repeat step (c) until you reach the last (mth) out-of-sample time interval.

Report and comment on your results.

7 Sensitivity of Mean–Variance Models to Input Estimation

One of the most salient drawbacks of mean–variance optimization is its high sensitivity to the estimation of input parameters. The sensitivity is due to the very nature of the optimization process: if there are assets whose returns appear to be superior, the portfolios generated by an optimization procedure will try to take advantage of these apparently superior assets by overweighting the holdings on those positions. Unfortunately in a practical setting there is inevitable noise in the estimation of inputs to a mean–variance model. Small perturbations in the values of the inputs may lead to large swings in the composition of the portfolio. This unfortunate phenomenon is basically due to the fact that the optimizer is overly responsive given the quality of the inputs typical in portfolio construction. A related phenomenon is the fact that the composition of portfolios is often non-intuitive. Theoretical and empirical evidence indicates that the estimate of expected returns is more critical than the estimate of the covariance matrix.

The sensitivity of mean–variance models to input estimation manifests itself in the differences among the *true*, *estimated*, and *actual* efficient frontiers, terms coined by Broadie (1993). The true efficient frontier is the one computed with the true (unobservable) expected returns and covariance matrix. The estimated frontier is the one computed with estimates of these parameters. The actual frontier is defined as follows: take the portfolios in the estimated frontier and calculate their true expected returns and variances. The actual frontier always lies below the true frontier. In principle the estimated frontier may lie anywhere with respect to the true frontier. However, due to the optimization process, if the estimation errors have zero mean, the estimated frontier is likely to lie above the true frontier. In that case the actual frontier would be well below the estimated frontier. Equivalently, the *ex post* performance of estimated efficient portfolios would typically be substantially worse than their *ex ante* performance suggested by the mean–variance model. Figure 7.1 illustrates the typing relative placement of the three frontiers.

The following specific example illustrates the sensitivity of mean–variance models to the quality of the inputs.

Consider a simple portfolio optimization problem with three assets whose expected returns and covariance matrix are

$$\mu = \begin{bmatrix} 0.11 \\ 0.10 \\ 0.05 \end{bmatrix}, \quad V = \begin{bmatrix} 0.250 & 0.225 & 0.045 \\ 0.225 & 0.250 & 0.045 \\ 0.045 & 0.045 & 0.090 \end{bmatrix}. \tag{7.1}$$

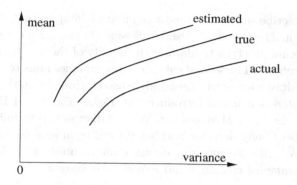

Figure 7.1 Efficient frontiers

Figure 7.2 displays the composition of long-only efficient portfolios for this problem.

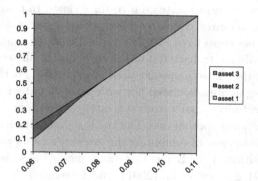

Figure 7.2 Area chart of long-only efficient portfolios for μ and $\mathbf{V}$ as in (7.1)

The picture makes sense from the pure optimization standpoint: assets 1 and 2 are similar but the expected return of asset 1 is slightly larger. Hence for higher target expected returns, the efficient portfolios have a much larger holding in asset 1 than in asset 2. However, from the portfolio construction standpoint this is unintuitive: assets 1 and 2 are very similar and, for all practical purposes, exchangeable because the slight difference could easily be due to estimation error. Therefore, it would be more intuitive for the positions of these two assets to be roughly the same. We can also look at the problem in a different way: Suppose the expected returns of assets 1 and 2 were slightly perturbed so that they are swapped. Then the composition of the efficient portfolios would change drastically. This again is a fairly counterintuitive and unnatural behavior.

The input sensitivity of mean–variance models is a central issue in portfolio management and has been a subject of intense study. There is a tremendous upside potential in finding appropriate ways of harnessing the power of portfolio optimization without getting caught on this major shortcoming. We will next describe some of the most popular techniques that aim at mitigating this problem. The techniques can be classified in two main categories. The first category of techniques tries to improve the quality of the inputs to the portfolio optimization problem. The second category of techniques aims to tweak the optimization procedure. One of the most widely used techniques in the first category is the *Black–Litterman model* introduced by Fisher Black and Bob Litterman at Goldman Sachs Asset Management. We will discuss this technique in some detail. We will also briefly describe another related technique based on Bayesian adjustments. We will subsequently discuss some techniques in the second category, namely *resampled efficiency* and *robust optimization*.

7.1   Black–Litterman Model

The basic idea of the Black–Litterman model is to tilt the market equilibrium returns to incorporate an investor's views. In principle a classical mean–variance model requires estimates of expected returns for all assets in the investment universe considered. This is typically an enormous task. Investment managers are unlikely to have detailed knowledge of all securities at their disposal. Typically, they have a specific area of expertise. Furthermore, some modern trading strategies are associated not with *absolute* but with *relative* rankings of securities. For instance, a pairs trading strategy corresponds to a forecast that one stock will outperform another one. The key insight of Black and Litterman was that there is a suitable way of combining the investor's views with the market equilibrium. The exposition of the Black–Litterman model below is based on Black and Litterman (1992), Fabozzi et al. (2007), and Litterman (2003).

Basic Assumption and Starting Point

The Black–Litterman model is an equilibrium-based model, meaning that the expected returns of the assets should be consistent with the market equilibrium unless the investor has some specific views. In other words, an investor without any views on the market should hold the market. We shall let π denote the equilibrium return vector, and $\mathbf{V}$ the covariance matrix of the asset returns.

The true expected return vector μ is unknown. As a starting point, we assume that the equilibrium return vector serves as a reasonable *prior* estimate of the true return vector in the sense that

$$\mu \sim N(\pi, \mathbf{Q}).$$

That is, μ is a multi-normal random vector with expected value π and covariance matrix $\mathbf{Q}$. The matrix $\mathbf{Q}$ represents the confidence on the equilibrium returns as an estimate of expected returns.

Expressing Investors' Views

A key ingredient of the Black–Litterman model is to incorporate investors' views on the expected returns. The framework is fairly flexible. An investor may have a few different views, each of them involving either a single asset (an absolute view) or several assets (a relative view). Formally, a collection of views is expressed as

$$\mathbf{P}\boldsymbol{\mu} = \mathbf{q} + \boldsymbol{\epsilon}, \quad \boldsymbol{\epsilon} \sim N(\mathbf{0}, \boldsymbol{\Omega}).$$

Each row in the equation $\mathbf{P}\boldsymbol{\mu} = \mathbf{q} + \boldsymbol{\epsilon}$ is a view that represents a forecast. The term $\boldsymbol{\epsilon}$ represents the degree of confidence in the views. The covariance matrix $\boldsymbol{\Omega}$ is typically a diagonal matrix. A weak view is a view with large variance; a strong view is a view with small variance. In the extreme, a certain view is a view with zero variance. Each of the views can be *absolute* or *relative* as described above. For a concrete example, consider an asset allocation problem with seven asset classes: Australia, Canada, France, Germany, Japan, United Kingdom, United States. Suppose we have two views:

- Return on Germany will be 12%.
- UK will outperform US by 2%.

These views can be expressed as

$$\mu_4 = 12\% + \epsilon_1$$
$$\mu_6 - \mu_7 = 2\% + \epsilon_2.$$

In matrix notation this corresponds to $\mathbf{P}\boldsymbol{\mu} = \mathbf{q} + \boldsymbol{\epsilon}$ for

$$\mathbf{P} = \begin{bmatrix} 0 & 0 & 0 & 1 & 0 & 0 & 0 \\ 0 & 0 & 0 & 0 & 0 & 1 & -1 \end{bmatrix}, \quad \mathbf{q} = \begin{bmatrix} 12\% \\ 2\% \end{bmatrix}, \quad \boldsymbol{\epsilon} = \begin{bmatrix} \epsilon_1 \\ \epsilon_2 \end{bmatrix}.$$

Merging Investors' Views and Market Equilibrium

The key insight of the Black–Litterman model is a proper way to combine the investor's views with the prior market equilibrium. First, consider the simpler case when the views are assumed to be certain; that is, $\boldsymbol{\Omega} = \mathbf{0}$ or equivalently the vector of expected returns must be tilted to satisfy the views $\mathbf{P}\boldsymbol{\mu} = \mathbf{q}$. In this case the *posterior* estimate of $\boldsymbol{\mu}$ given the prior $\boldsymbol{\mu} \sim N(\boldsymbol{\pi}, \mathbf{Q})$ and the views $\mathbf{P}\boldsymbol{\mu} = \mathbf{q}$ is

$$\hat{\boldsymbol{\mu}} = \boldsymbol{\pi} + \mathbf{Q}\mathbf{P}^{\mathsf{T}}(\mathbf{P}\mathbf{Q}\mathbf{P}^{\mathsf{T}})^{-1}(\mathbf{q} - \mathbf{P}\boldsymbol{\pi}). \tag{7.2}$$

Some matrix algebra shows that indeed $\mathbf{P}\hat{\boldsymbol{\mu}} = \mathbf{q}$. That is, the posterior estimate satisfies the views $\mathbf{P}\boldsymbol{\mu} = \mathbf{q}$.

In the more general case when the views are not certain, the *posterior* estimate of $\boldsymbol{\mu}$ given the prior $\boldsymbol{\mu} \sim N(\boldsymbol{\pi}, \mathbf{Q})$ and the views $\mathbf{P}\boldsymbol{\mu} = \mathbf{q} + \boldsymbol{\epsilon}$, with $\boldsymbol{\epsilon} \sim N(\mathbf{0}, \boldsymbol{\Omega})$, is

$$\hat{\boldsymbol{\mu}} = \boldsymbol{\pi} + \mathbf{Q}\mathbf{P}^{\mathsf{T}}(\mathbf{P}\mathbf{Q}\mathbf{P}^{\mathsf{T}} + \boldsymbol{\Omega})^{-1}(\mathbf{q} - \mathbf{P}\boldsymbol{\pi}). \tag{7.3}$$

When $\boldsymbol{\Omega}$ is non-singular, $\hat{\boldsymbol{\mu}}$ also has the following equivalent expression:

$$\hat{\boldsymbol{\mu}} = (\mathbf{Q}^{-1} + \mathbf{P}^{\mathsf{T}}\boldsymbol{\Omega}^{-1}\mathbf{P})^{-1}(\mathbf{Q}^{-1}\boldsymbol{\pi} + \mathbf{P}^{\mathsf{T}}\boldsymbol{\Omega}^{-1}\mathbf{q}). \tag{7.4}$$

Observe that the expression (7.2) for certain views can be recovered from (7.3) by taking $\boldsymbol{\Omega} = \mathbf{0}$.

We next give a derivation of the formula (7.4) for the posterior estimate of $\boldsymbol{\mu}$ when $\boldsymbol{\Omega}$ is non-singular. The exercises at the end of the chapter show how the derivation can be tweaked for any $\boldsymbol{\Omega}$. Stack the two equations for the market equilibrium $\boldsymbol{\pi} = \boldsymbol{\mu} + \boldsymbol{\epsilon}_{\boldsymbol{\pi}}$, $\boldsymbol{\epsilon}_{\boldsymbol{\pi}} \sim N(\mathbf{0}, \mathbf{Q})$, and for the investor's views $\mathbf{q} = \mathbf{P}\boldsymbol{\mu} + \boldsymbol{\epsilon}_{\mathbf{q}}$, $\boldsymbol{\epsilon}_{\mathbf{q}} \sim N(\mathbf{0}, \boldsymbol{\Omega})$ as

$$\mathbf{y} = \mathbf{M}\boldsymbol{\mu} + \boldsymbol{\epsilon}, \quad \boldsymbol{\epsilon} \sim N(\mathbf{0}, \boldsymbol{\Sigma})$$

for

$$\mathbf{y} = \begin{bmatrix} \boldsymbol{\pi} \\ \mathbf{q} \end{bmatrix}, \quad \mathbf{M} = \begin{bmatrix} \mathbf{I} \\ \mathbf{P} \end{bmatrix}, \quad \boldsymbol{\Sigma} = \begin{bmatrix} \mathbf{Q} & \mathbf{0} \\ \mathbf{0} & \boldsymbol{\Omega} \end{bmatrix}.$$

The estimation problem can be stated as the following weighted least-squares problem:

$$\min_{\boldsymbol{\mu}} \ (\mathbf{y} - \mathbf{M}\boldsymbol{\mu})^{\mathsf{T}}\boldsymbol{\Sigma}^{-1}(\mathbf{y} - \mathbf{M}\boldsymbol{\mu}).$$

The optimality conditions for this problem yield

$$2\mathbf{M}^{\mathsf{T}}\boldsymbol{\Sigma}^{-1}\mathbf{M}\boldsymbol{\mu} - 2\mathbf{M}^{\mathsf{T}}\boldsymbol{\Sigma}^{-1}\mathbf{y} = \mathbf{0}.$$

Hence we obtain

$$\begin{aligned}
\hat{\boldsymbol{\mu}} &= (\mathbf{M}^{\mathsf{T}}\boldsymbol{\Sigma}^{-1}\mathbf{M})^{-1}\mathbf{M}^{\mathsf{T}}\boldsymbol{\Sigma}^{-1}\mathbf{y} \\
&= (\mathbf{Q}^{-1} + \mathbf{P}^{\mathsf{T}}\boldsymbol{\Omega}^{-1}\mathbf{P})^{-1}(\mathbf{Q}^{-1}\boldsymbol{\pi} + \mathbf{P}^{\mathsf{T}}\boldsymbol{\Omega}^{-1}\mathbf{q}) \\
&= \boldsymbol{\pi} + \mathbf{Q}\mathbf{P}^{\mathsf{T}}(\mathbf{P}\mathbf{Q}\mathbf{P}^{\mathsf{T}} + \boldsymbol{\Omega})^{-1}(\mathbf{q} - \mathbf{P}\boldsymbol{\pi}).
\end{aligned}$$

We end this section with a couple of general remarks.

First we note that the Black–Litterman model can be thought of as an "inverse optimization problem": If one views the market equilibrium as the optimum solution of a portfolio optimization problem, what data would produce this outcome? In particular, given investor views, what choice of $\boldsymbol{\mu}$ would best fit the market equilibrium solution? This leads to the least-squares problem formulated above, whose solution $\hat{\boldsymbol{\mu}}$ has the expression (7.3) that we just computed. This "inverse optimization" philosophy was proposed by Bertsimas et al. (2012). It has the added flexibility of allowing investor views on volatility and market dynamics.

We also note that the analysis in the Black–Litterman model relies heavily on the assumption that the error term $\boldsymbol{\epsilon} \sim N(\mathbf{0}, \mathbf{Q})$ is normally distributed. When this is not the case, the Black–Litterman framework is still meaningful but the analysis is more complex and a closed-form solution like (7.3) does not usually exist. However, a numerical solution may be possible using the nonlinear programming algorithms discussed in Chapters 18 and 20 (Kocuk and Cornuéjols, 2017).

7.2 Shrinkage Estimation

Another approach to improve the quality of estimated expected returns is based on *shrinkage estimators*. These types of estimators are rooted in the classical finding of Stein (1956) that biased estimators may, in a formal fashion, be superior to the unbiased sample mean. As we detail below, the central idea is that the estimation can be improved by *shrinking* the sample mean towards a target. The non-technical article of Efron and Morris (1977) gives an enlightening discussion of an application of this approach to the estimation of baseball batting averages.

To formalize ideas, consider the problem of estimating the mean of an N-dimensional multivariate normal variable $\mathbf{r} \sim N(\boldsymbol{\mu}, \mathbf{V})$ from a set of observations $\mathbf{r}_1, \ldots, \mathbf{r}_T$. For a given estimate $\hat{\boldsymbol{\mu}}$, consider the quadratic loss function

$$L(\hat{\boldsymbol{\mu}}, \boldsymbol{\mu}) := (\boldsymbol{\mu} - \hat{\boldsymbol{\mu}})^\mathsf{T} \mathbf{V}^{-1} (\boldsymbol{\mu} - \hat{\boldsymbol{\mu}}). \tag{7.5}$$

For a given loss function, the *risk* of an estimator is $\mathbb{E}(L(\hat{\boldsymbol{\mu}}, \boldsymbol{\mu}))$, where the expectation is taken over the space of samples $\mathbf{r}_1, \ldots, \mathbf{r}_T$. An estimator is *inadmissible* if there exists another estimator with lower risk.

For the quadratic loss function (7.5) and $N = 1, 2$, it is known that the optimal estimator is the sample mean $\bar{\mathbf{r}} := (1/T)(\mathbf{r}_1 + \cdots + \mathbf{r}_T)$. By contrast, for $N > 2$, the *James–Stein* shrinkage estimator

$$\hat{\boldsymbol{\mu}}_{JS} := (1 - w)\bar{\mathbf{r}} + w\,\mu_0\,\mathbf{1}$$

has lower risk than the sample mean $\bar{\mathbf{r}}$ for

$$w = \min\left(1, \frac{N - 2}{T(\bar{\mathbf{r}} - \mu_0\mathbf{1})^\mathsf{T}\mathbf{V}(\bar{\mathbf{r}} - \mu_0\mathbf{1})}\right).$$

Here T is the number of observations, and μ_0 is an arbitrary number. The vector $\mu_0\mathbf{1}$ and the weight w are referred to as the *shrinkage target* and *shrinkage factor* respectively. Although some choices of μ_0 are better than others, what is surprising is that in theory μ_0 could be any fixed number. This fact is called the *Stein paradox*.

The James–Stein shrinkage estimator can be seen as a combination of two estimators:

(1) an estimator with little or no structure (like the sample mean);
(2) an estimator with a lot of structure (the shrinkage target).

The exact combination of these two estimators is determined by a certain *shrinkage intensity*. As we discuss below, this same shrinkage approach has been successfully applied to obtain improved estimators of covariance matrices and beta exposures.

The following shrinkage estimator proposed by Jorion (1986) is fairly popular in the financial literature. The estimator was derived via an empirical Bayesian approach. As a shrinkage target, use the vector $\mu_0 \mathbf{1}$ for

$$\mu_0 := \frac{\bar{\mathbf{r}}^\mathsf{T}\mathbf{V}^{-1}\mathbf{1}}{\mathbf{1}^\mathsf{T}\mathbf{V}^{-1}\mathbf{1}},$$

and as a shrinkage intensity, use

$$w := \frac{N+2}{N+2+T(\bar{\mathbf{r}}-\mu_0\mathbf{1})^\mathsf{T}\mathbf{V}^{-1}(\bar{\mathbf{r}}-\mu_0\mathbf{1})}.$$

Shrinkage can also be applied to other estimation problems. For instance, Ledoit and Wolf (2003, 2004) propose shrinkage approaches for covariance estimation in the same spirit as the James–Stein shrinkage estimator: shrink the sample covariance matrix $\bar{\mathbf{V}}$ (an unstructured estimator) towards a highly structured target estimator $\mathbf{V}_0$:

$$\hat{\mathbf{V}}_{LW} := (1-w)\bar{\mathbf{V}} + w\mathbf{V}_0.$$

The shrinkage target estimator $\mathbf{V}_0$ could be a single-factor estimator, an estimator of the covariance matrix with constant correlation, a diagonal matrix, or a multiple of the identity matrix.

Shrinkage is also routinely used for estimating benchmark exposures in a stock universe. Let r_i, $i = 1, \ldots, N$, denote the excess returns of stocks in the investment universe and r_B denote the excess return of the benchmark. Recall that the beta of stock i captures the benchmark portion of the return on stock i via the linear model

$$r_i = \beta_i r_B + \theta_i.$$

The value of β_i is the benchmark exposure of stock i. Given historical realizations of r_i, for $i = 1, \ldots, N$, and r_B, we can obtain estimates $\hat{\beta}_i$ of β_i for $i = 1, \ldots, N$ via ordinary least-squares linear regression. The forecasts given by these natural estimators tend to overestimate the betas of stocks with high benchmark exposure and underestimate the betas of the stocks with low benchmark exposure. Improved forecasts can be obtained by shrinking the betas obtained from the least-squares procedures towards one (the benchmark beta):

$$\hat{\boldsymbol{\beta}} = (1-w)\bar{\boldsymbol{\beta}} + w\mathbf{1},$$

where $\bar{\boldsymbol{\beta}}$ denotes the vector of beta estimates from the least-squares procedure.

A common rule of thumb in the above shrinkage estimators of covariance matrix and vector of betas is to use a shrinkage intensity $w = 1/2$ for estimates based on 60-month long historical data. A thorough discussion on the appropriate choice of shrinkage intensity can be found in Ledoit and Wolf (2003, 2004) and Blume (1975). Portfolio optimization can be viewed as a stochastic optimization problem (see Chapter 10). Shrinkage is relevant in this more general context as well (Davarnia and Cornuéjols, 2017).

The exercises at the end of this chapter suggest some computational experiments that illustrate the effectiveness of shrinkage estimators.

7.3 Resampled Efficiency

A different approach to address the sensitivity of mean–variance optimization to estimation error is to apply the *bootstrap technique* from statistics. The bootstrap technique is a method to estimate standard errors and confidence intervals of statistics of a dataset via random resampling from the dataset with replacement.

The application of bootstrapping to mean–variance optimization was initially explored by Jorion (1992) and later further developed and marketed by Michaud and Michaud (2008). The basic idea is to consider the joint problem of parameter estimation and portfolio construction as a statistical procedure: the efficient portfolios can be seen as a statistic on a set of financial data used for estimation. The *resampled efficiency* technique proposed by Michaud and Michaud proceeds by applying bootstrapping to this statistical process. Suppose there is a procedure that estimates the vector of expected returns and covariance from historical data. Use the available data to produce these estimates and compute efficient portfolios. Repeat this same process either by sampling from these estimates, or by bootstrapping the available data to obtain new estimates of expected returns and covariances. All these estimates are statistically equivalent. For each of them, we can generate the corresponding set of efficient portfolios. The collection of all of these portfolios forms some sort of *equivalence region*. We would like to take some average of the equivalence region so that the effects of estimation error are mitigated. However, it is not obvious how to average since the equivalence region contains portfolios with low and high variance. We do not want to mix "apples and oranges". Michaud and Michaud's suggestion is to average portfolios that are in some equivalent risk-return bucket. To that end, we propose the following procedure: For each efficient frontier, save m evenly distributed efficient portfolios. Rank them 1 to m. Then take averages of same-rank portfolios from all efficient frontiers.

This resampling procedure can be more precisely described as follows (see Algorithm 7.1). Suppose we have a procedure to produce estimates $\hat{\boldsymbol{\mu}}$ and $\hat{\mathbf{V}}$ from a history of T periods of historical data.

Algorithm 7.1 Resampling procedure

1: **for** $i = 1, \ldots, S$ **do**
2:     simulate a new history of T periods by resampling the original history
3:     use the simulated history to generate new estimates $\hat{\boldsymbol{\mu}}_i$ and $\hat{\mathbf{V}}_i$
4:     use $\hat{\boldsymbol{\mu}}_i$ and $\hat{\mathbf{V}}_i$ to generate m equally spaced efficient portfolios
        $\mathbf{x}_{1,i}, \ldots, \mathbf{x}_{m,i}$
5: **end for**

To generate the *resampled efficient portfolios*, take averages of equally ranked efficient portfolios generated above:

$$\mathbf{x}_{j,\text{resampled}} := \frac{1}{S} \sum_{i=1}^{S} \mathbf{x}_{j,i}.$$

The *resampled efficient frontier* is the expected return versus standard deviation chart of the resampled efficient portfolios with the original estimates $\hat{\boldsymbol{\mu}}$ and $\hat{\mathbf{V}}$.

There are a number of limitations to resampling (Scherer, 2002). The entire process is only a heuristic; there is no sound theory to support why the process should mitigate the effects of estimation error. The methodology does have the feature of generating portfolios that look well diversified, and this is generally well received. However, this feature can be attributed to the role of variability in the averaging process. The process is intense computationally, as it multiplies the work involved in conventional mean–variance optimization. Furthermore, the procedure does not provide any clear mechanism to facilitate the incorporation of views as in the Black–Litterman model.

7.4   Robust Optimization

Robust optimization is a fairly recent development that considers uncertainty in some parameters directly in the optimization problem. The general idea of robust optimization is to generate a solution that is *good* for *all* possible realizations of the uncertain parameters. Consider a minimization problem with inequality constraints

$$\begin{aligned} \min_{\mathbf{x}} \quad & f(\mathbf{x}, \mathbf{p}) \\ \text{s.t.} \quad & g_i(\mathbf{x}, \mathbf{p}) \leq 0, \ i = 1, \ldots, m. \end{aligned} \tag{7.6}$$

Here the vector $\mathbf{p}$ stands for some parameters that define the objective and constraints functions.

Consider first the case when the uncertain parameters occur in the constraints only. Assume that the set of parameters $\mathbf{p}$ is uncertain but it is known to be in some uncertainty set $\mathcal{U}$. In this case a *robust* version of (7.6) is one where the optimization is performed over points that are feasible for all possible realizations of the uncertain parameters $\mathbf{p} \in \mathcal{U}$; that is,

$$\begin{aligned} \min_{\mathbf{x}} \quad & f(\mathbf{x}) \\ \text{s.t.} \quad & \max_{\mathbf{p} \in \mathcal{U}} g_i(\mathbf{x}, \mathbf{p}) \leq 0, \ i = 1, \ldots, m. \end{aligned}$$

On the other hand, consider the case when the uncertain parameters occur in the objective only. In this case a robust version of (7.6) is one that finds the solution that would be best, given the worst possible realization of the uncertain parameters $\mathbf{p} \in \mathcal{U}$; that is,

$$\begin{aligned} \min_{\mathbf{x}} \max_{\mathbf{p} \in \mathcal{U}} \quad & f(\mathbf{x}, \mathbf{p}) \\ \text{s.t.} \quad & g_i(\mathbf{x}) \leq 0, \ i = 1, \ldots, m. \end{aligned}$$

If uncertain parameters occur in both the objective and constraints, then the robust version is as follows:

$$\min_{\mathbf{x}} \max_{\mathbf{p} \in \mathcal{U}} \quad f(\mathbf{x}, \mathbf{p})$$
$$\text{s.t.} \quad \max_{\mathbf{p} \in \mathcal{U}} \; g_i(\mathbf{x}, \mathbf{p}) \le 0, \; i = 1, \dots, m.$$

As we detail in Chapter 19, for suitable types of uncertainty sets the above robust versions can be rewritten as an optimization problem that is manageable albeit via more involved optimization machinery.

7.5 Other Diversification Approaches

The challenges associated with expected return estimation and the input sensitivity of mean–variance models have given rise to quantitative portfolio construction approaches that eschew expected return estimation and focus on managing risk only. We next discuss some popular approaches of this kind that have led to the development of a variety of investment products in the asset management industry.

Assume $\mathbf{V}$ is the covariance matrix of asset returns in some investment universe and $\boldsymbol{\sigma}$ is the vector of volatilities (standard deviations) of the asset returns. In particular, the diagonal entries of $\mathbf{V}$ are the squares of the entries of $\boldsymbol{\sigma}$.

The *minimum-risk* portfolio is the portfolio in the efficient frontier of minimum variance. In the absence of constraints, this portfolio is the solution to the following quadratic programming model:

$$\min_{\mathbf{x}} \quad \mathbf{x}^{\mathsf{T}} \mathbf{V} \mathbf{x}$$
$$\text{s.t.} \quad \mathbf{1}^{\mathsf{T}} \mathbf{x} = 1.$$

That is,

$$\mathbf{x}^* = \frac{1}{\mathbf{1}^{\mathsf{T}} \mathbf{V}^{-1} \mathbf{1}} \mathbf{V}^{-1} \mathbf{1}.$$

For the special case $\mathbf{V} = \mathbf{I}$ (the $N \times N$ identity matrix) the minimum-risk portfolio is the so-called *equally weighted portfolio*

$$x_i^* = \frac{1}{N}, \quad i = 1, \dots, N,$$

where N is the number of assets in the universe.

On the other hand, if $\mathbf{V}$ is diagonal, that is, $\mathbf{V} = \text{diag}(\boldsymbol{\sigma})^2$, then the portfolio components are proportional to the inverse of the squares of the volatilities:

$$x_i^* = \frac{1/\sigma_i^2}{\sum_{i=1}^{N} 1/\sigma_i^2}, \quad i = 1, \dots, N.$$

In particular, this portfolio is the *value-weighted portfolio* if the capitalization of asset i is used as a proxy for $1/\sigma_i^2$.

We next discuss two more recent diversification approaches, namely *risk parity* and *maximum diversification*. To that end, we first discuss the related concept of *risk contribution*. Observe that the risk (standard deviation) of a portfolio $\mathbf{x} = (x_1, \ldots, x_N)$ is given by

$$\sigma_P(\mathbf{x}) = \sqrt{\mathbf{x}^\mathsf{T}\mathbf{V}\mathbf{x}}.$$

If we compute the partial derivative of this portfolio with respect to x_i, we obtain the *marginal contribution to risk* of asset i:

$$MCR_i(\mathbf{x}) = \frac{\partial \sigma_P(\mathbf{x})}{\partial x_i} = \frac{(\mathbf{V}\mathbf{x})_i}{\sqrt{\mathbf{x}^\mathsf{T}\mathbf{V}\mathbf{x}}}, \quad i = 1, \ldots, N.$$

The *contribution to risk* of asset i is:

$$CR_i(\mathbf{x}) = x_i \cdot MCR_i(\mathbf{x}) = \frac{x_i \cdot (\mathbf{V}\mathbf{x})_i}{\sqrt{\mathbf{x}^\mathsf{T}\mathbf{V}\mathbf{x}}}, \quad i = 1, \ldots, N.$$

Observe that

$$\sum_{i=1}^{N} CR_i(\mathbf{x}) = \sqrt{\mathbf{x}^\mathsf{T}\mathbf{V}\mathbf{x}} = \sigma_P(\mathbf{x}).$$

Consequently, we say that $\mathbf{x}$ is a *risk-parity* portfolio if all the assets in the portfolio have the same contribution to risk; that is, if

$$CR_i(\mathbf{x}) = \frac{\sigma_P(\mathbf{x})}{N}, \quad i = 1, \ldots, N.$$

Again, in the special case when $\mathbf{V} = \mathrm{diag}(\boldsymbol{\sigma})^2$, the fully invested risk-parity portfolio is

$$x_i^* = \frac{1/\sigma_i}{\sum_{i=1}^{N} 1/\sigma_i}, \quad i = 1, \ldots, N.$$

For a general covariance matrix $\mathbf{V}$ and portfolio constraints, it may not be possible to attain perfect risk parity. In this case, we can instead minimize some kind of measure of *deviation from risk parity*. Here are some choices for examples of these kinds of measures proposed in the literature:

$$DRP_1(\mathbf{x}) = \sum_{i=1}^{N} \sum_{j=1}^{N} (x_i \cdot (\mathbf{V}\mathbf{x})_i - x_j \cdot (\mathbf{V}\mathbf{x})_j)^2$$

$$DRP_2(\mathbf{x}) = \sum_{i=1}^{N} \left(\frac{x_i \cdot (\mathbf{V}\mathbf{x})_i}{\mathbf{x}^\mathsf{T}\mathbf{V}\mathbf{x}} - \frac{1}{N} \right)^2$$

$$DRP_3(\mathbf{x}) = \sum_{i=1}^{N} \left| \frac{x_i \cdot (\mathbf{V}\mathbf{x})_i}{\mathbf{x}^\mathsf{T}\mathbf{V}\mathbf{x}} - \frac{1}{N} \right|.$$

The optimization problem associated with minimizing any of these deviation measures is in general quite a bit more challenging than other mean–variance models, as these problems are not convex. The development of efficient numerical algorithms to solve these kinds of optimization problems is a topic of current research.

Another approach to diversification is *maximum diversification* (Choueifaty and Coignard, 2008). More precisely, maximize the diversification ratio

$$\frac{\sigma^\mathsf{T} x}{\sqrt{x^\mathsf{T} V x}},$$

where σ is the vector of asset volatilities. A motivation for this approach can be given as follows: Observe that the diversification ratio is proportional to the Sharpe ratio if μ is proportional to σ. Hence maximizing diversification is equivalent to maximizing the Sharpe ratio under the assumption that the expected returns of the assets are proportional to their volatilities.

In the absence of other constraints, the fully invested maximum diversification portfolio is the solution to the optimization problem

$$\min_{x} \quad \frac{\sigma^\mathsf{T} x}{\sqrt{x^\mathsf{T} V x}} \tag{7.7}$$
$$\text{s.t.} \quad 1^\mathsf{T} x = 1.$$

In the special case when $V = \mathrm{diag}(\sigma)^2$, the solution to (7.7) coincides with the fully invested risk-parity portfolio:

$$x_i^* = \frac{1/\sigma_i}{\sum_{i=1}^N 1/\sigma_i}, \quad i = 1, \dots, N.$$

7.6   Exercises

Exercise 7.1   The purpose of this exercise is to provide a derivation of the Black–Litterman posterior formula.

(a) Consider the case when views are certain; that is, when $\Omega = 0$. In this case if we stack the equations for the prior and for the views we get

$$\begin{bmatrix} \pi \\ q \end{bmatrix} = \begin{bmatrix} \mu + \epsilon_\pi \\ P\mu \end{bmatrix}, \quad \epsilon_\pi \sim N(0, Q).$$

The estimation problem can then be stated as the following constrained weighted least-squares problem:

$$\min_{\mu} \quad (\pi - \mu)^\mathsf{T} Q^{-1}(\pi - \mu)$$
$$\text{s.t.} \quad P\mu = q.$$

(i) Write down the optimality conditions for this constrained problem.
(ii) Show that after solving the optimality conditions we obtain

$$\hat{\mu} = \pi + QP^\mathsf{T}(PQP^\mathsf{T})^{-1}(q - P\pi).$$

(b) Now consider the case $\boldsymbol{\Omega} = \begin{bmatrix} \boldsymbol{\Omega}_{11} & \mathbf{0} \\ \mathbf{0} & \mathbf{0} \end{bmatrix}$, where $\boldsymbol{\Omega}_{11}$ is non-singular. This corresponds to the case when the views can be split in two blocks and the second block of views are certain:

$$\mathbf{P}\mu = \begin{bmatrix} \mathbf{P}_1\mu \\ \mathbf{P}_2\mu \end{bmatrix} = \begin{bmatrix} \mathbf{q}_1 + \boldsymbol{\epsilon}_1 \\ \mathbf{q}_2 \end{bmatrix}, \ \boldsymbol{\epsilon}_1 \sim N(\mathbf{0}, \boldsymbol{\Omega}_{11}).$$

In this case if we stack the equations for the prior and for the views we get

$$\begin{bmatrix} \boldsymbol{\pi} \\ \mathbf{q}_1 \\ \mathbf{q}_2 \end{bmatrix} = \begin{bmatrix} \boldsymbol{\mu} + \boldsymbol{\epsilon}_\pi \\ \mathbf{P}_1\mu + \boldsymbol{\epsilon}_1 \\ \mathbf{P}_2\mu \end{bmatrix}, \ \begin{bmatrix} \boldsymbol{\epsilon}_\pi \\ \boldsymbol{\epsilon}_1 \end{bmatrix} \sim N(\mathbf{0}, \boldsymbol{\Sigma}), \ \boldsymbol{\Sigma} = \begin{bmatrix} \mathbf{Q} & \mathbf{0} \\ \mathbf{0} & \boldsymbol{\Omega}_{11} \end{bmatrix}.$$

The estimation problem can then be stated as the following constrained weighted least-squares problem:

$$\min_{\mu} \ \begin{bmatrix} \boldsymbol{\pi} - \boldsymbol{\mu} \\ \mathbf{q}_1 - \mathbf{P}_1\mu \end{bmatrix}^{\mathsf{T}} \boldsymbol{\Sigma}^{-1} \begin{bmatrix} \boldsymbol{\pi} - \boldsymbol{\mu} \\ \mathbf{q}_1 - \mathbf{P}_1\mu \end{bmatrix}$$

$$\text{s.t.} \quad \mathbf{P}_2\mu = \mathbf{q}_2.$$

(i) Write down the optimality conditions for this constrained problem.

(ii) Show that after solving the optimality conditions we obtain

$$\hat{\mu} = \boldsymbol{\pi} + \mathbf{Q}\mathbf{P}^{\mathsf{T}}(\mathbf{P}\mathbf{Q}\mathbf{P}^{\mathsf{T}} + \boldsymbol{\Omega})^{-1}(\mathbf{q} - \mathbf{P}\boldsymbol{\pi}).$$

(c) *Reduce the case when $\boldsymbol{\Omega}$ is a general covariance matrix to the case discussed in step (b). To that end, use the following fact from matrix algebra: if $\boldsymbol{\Omega}$ is symmetric and positive semidefinite, then there exists an orthogonal matrix $\mathbf{U}$ and a diagonal matrix $\boldsymbol{\Lambda}$ with non-negative entries such that $\boldsymbol{\Omega} = \mathbf{U}\boldsymbol{\Lambda}\mathbf{U}^{\mathsf{T}}$. Use $\mathbf{U}$ to make a *change of variables* so as to write the views as in step (b) and conclude that the expression (7.3) for the posterior $\hat{\mu}$ holds.

Exercise 7.2   The purpose of this exercise is to explore the effect of estimation error on the computation of efficient portfolios by comparing the "true", "estimated", and "actual" efficient frontiers. To that end, assume the expected return and covariance matrix in the Excel spreadsheet "Exercise 7.2 & 7.3 Eight Assets" are the "true" values for the expected returns and covariances for a set of eight assets. These are *monthly* expected returns and covariances.

Next, using these "true" values and assuming a multivariate normal distribution for the returns, generate monthly returns for ten years. You may find the MATLAB multivariate normal random number generator `mvnrnd` useful for this purpose.

(a) Compute the sample mean and the sample covariance matrix of the returns you generated.

(b) Compute at least ten long-only efficient portfolios along the efficient frontier based on the estimates found in part (a). Choose efficient portfolios whose expected returns range from that of the long-only minimum-variance portfolio to that of the long-only portfolio with maximum expected returns. Save these efficient portfolios.

(c) Now compute the "actual" expected returns and standard deviations for the portfolios found in step (b). These are the values of true expected returns and standard deviation of these portfolios.

(d) On the same figure plot the "estimated" efficient frontier found in (b), the "actual" frontier from step (c), and the "true" frontier (the one we would get if we used the true parameters).

(e) Repeat the above steps (generate a ten-year history, estimate, compute efficient portfolios) a few times. What do you observe?

Exercise 7.3   The Excel spreadsheet "Exercise 7.2 & 7.3 Eight Assets" provides monthly expected returns and covariance matrix for eight asset classes.

(a) Find the long-only portfolio with maximum Sharpe ratio, assuming a zero risk-free interest rate.

(b) Assume your initial portfolio is equally divided among the eight asset classes. Repeat step (a) but under the additional restriction that the two-sided turnover is at most 60%.

(c) Assume the benchmark is an equally divided portfolio and the risk-free interest rate is zero. Find the vector of equilibrium returns π.

(d) Suppose an investor has the following two views:

   View 1: the return on Euro bonds will be 0.40%.

   View 2: the return on an equally weighted portfolio of USA and UK stocks will be 1.2%.

   Use the Black–Litterman model to merge these views with the equilibrium returns. Assume the investor has total confidence in the views.

Exercise 7.4   Consider the problem of finding the *maximum diversified* fully invested portfolio in a universe of n risky assets:

$$\max_{\mathbf{x}} \quad \frac{\sigma^T \mathbf{x}}{\sqrt{\mathbf{x}^T V \mathbf{x}}}$$

$$\text{s.t.} \quad \mathbf{1}^T \mathbf{x} = 1.$$

Here σ is the vector of assets volatilities and V is the covariance matrix.

(a) Show that the maximum diversified portfolio (i.e., the solution to the above problem) is

$$\mathbf{x}_{MD} := \frac{1}{\mathbf{1}^T V^{-1} \sigma} \cdot V^{-1} \sigma.$$

(b) Use part (a) to show that

$$\mathbf{x}_{MD} := \frac{\sigma_{MD}^2}{\sigma_A} \cdot \mathbf{V}^{-1}\boldsymbol{\sigma},$$

where σ_{MD} and σ_A are respectively the volatility of $\mathbf{x}_{MD}$ and the weighted average volatility of the assets in $\mathbf{x}_{MD}$. In other words,

$$\sigma_{MD}^2 = \mathbf{x}_{MD}^{\mathsf{T}}\mathbf{V}\mathbf{x}_{MD}, \quad \sigma_A = \boldsymbol{\sigma}^{\mathsf{T}}\mathbf{x}_{MD}.$$

(c) Suppose the covariance matrix has the following *constant-correlation* form: For some $\rho \in (0,1)$

$$\mathbf{V}_{ii} = \sigma_i^2, \ \mathbf{V}_{ij} = \rho\sigma_i\sigma_j, \ \text{for} \ i = 1,\dots,n, \ \text{and} \ j = 1,\dots,n, \ \text{with} \ i \neq j.$$

In matrix form, we can write the above constant-correlation matrix as follows:

$$\mathbf{V} = \rho\boldsymbol{\sigma}\boldsymbol{\sigma}^{\mathsf{T}} + (1 - \rho)\mathrm{Diag}(\boldsymbol{\sigma})^2,$$

where $\boldsymbol{\sigma}$ is the vector with components σ_i, $i = 1,\dots,n$.

Show that in this case the holdings of the maximum diversified portfolio $\mathbf{x}_{MD}$ are given by

$$x_i = \frac{1}{\sum_{i=1}^{n} 1/\sigma_i} \cdot \frac{1}{\sigma_i}, \quad i = 1,\dots,n.$$

Exercise 7.5 The purpose of this exercise is to visualize how the covariance matrix gets distorted when it is estimated using a finite set of observations. The exercise also explores how a shrinkage technique of Ledoit and Wolf can mitigate this kind of distortion.

(a) Assume $n = 10$ assets have returns that follow a multivariate normal distribution with expected returns equal to zero and *true* covariance matrix equal to the $n \times n$ diagonal matrix

$$\mathbf{V} = \begin{bmatrix} 0.8 & & & & \\ & 0.85 & & & \\ & & \ddots & & \\ & & & 1.2 & \\ & & & & 1.25 \end{bmatrix}.$$

(The diagonal entries are equally spaced at 0.05 intervals.)

Generate $T = 120$ samples $\mathbf{r}_t$, $t = 1,\dots,T$, from this joint distribution. Each of these samples $\mathbf{r}_t \in \mathbb{R}^{10}$ is drawn from the ten-dimensional multivariate normal distribution $N(\mathbf{0}, \mathbf{V})$. You may find the MATLAB multivariate normal random number generator **mvnrnd** useful for this purpose.

(i) Use the T samples to estimate the sample covariance matrix $\hat{\mathbf{V}}$ as follows. Let $\bar{\mathbf{r}} := (1/T) \sum_{t=1}^{T} \mathbf{r}_t$, $\mathbf{z}_t := \mathbf{r}_t - \bar{\mathbf{r}}$, $t = 1, \ldots, T$, and

$$\hat{\mathbf{V}} := \frac{1}{T} \sum_{t=1}^{T} \mathbf{z}_t \mathbf{z}_t^\mathsf{T}.$$

Plot the eigenvalues both of the true covariance matrix $\mathbf{V}$ and of the estimated covariance $\hat{\mathbf{V}}$ on the same plot. Do you observe anything peculiar?

(ii) Using the estimated covariance $\hat{\mathbf{V}}$ find the estimated minimum-risk fully invested portfolio $\hat{\mathbf{x}}$. Compute the *estimated* minimum variance $\hat{\mathbf{x}}^\mathsf{T} \hat{\mathbf{V}} \hat{\mathbf{x}}$, the *actual* minimum variance $\hat{\mathbf{x}}^\mathsf{T} \mathbf{V} \hat{\mathbf{x}}$, and the *true* minimum variance $(\mathbf{x}^*)^\mathsf{T} \mathbf{V} \mathbf{x}^*$, where $\mathbf{x}^*$ is the true minimum-risk fully invested portfolio for $\mathbf{V}$. What do you observe?

(iii) Repeat parts (i) and (ii) several times (anywhere from a handful to a few thousand times). What do you observe?

(b) We will next apply the shrinkage technique of Ledoit and Wolf. To that end, let λ_i, $i = 1, \ldots, n$, denote the eigenvalues of the sample covariance matrix $\hat{\mathbf{V}}$ and $\bar{\lambda} := (1/n) \sum_{i=1}^{n} \lambda_i$. Define $\mathbf{C} := \bar{\lambda} \mathbf{I}$ and

$$\alpha := \min \left(\frac{1}{T} \cdot \frac{\frac{1}{T} \sum_{t=1}^{T} \operatorname{trace}((\mathbf{z}_t \mathbf{z}_t^\mathsf{T} - \hat{\mathbf{V}})^2)}{\operatorname{trace}((\hat{\mathbf{V}} - \mathbf{C})^2)}, 1 \right).$$

Finally consider the shrunken matrix

$$\bar{\mathbf{V}} := (1 - \alpha)\hat{\mathbf{V}} + \alpha \mathbf{C}.$$

(i) Plot the eigenvalues of the true covariance matrix $\mathbf{V}$, of the sample covariance $\hat{\mathbf{V}}$, and of the shrunken covariance $\bar{\mathbf{V}}$ on the same plot. What do you observe now?

(ii) Using the shrunken covariance $\bar{\mathbf{V}}$ find the estimated minimum-risk fully invested portfolio $\bar{\mathbf{x}}$. Compute the *estimated* minimum variance $\bar{\mathbf{x}}^\mathsf{T} \bar{\mathbf{V}} \bar{\mathbf{x}}$, the *actual* minimum variance $\bar{\mathbf{x}}^\mathsf{T} \mathbf{V} \bar{\mathbf{x}}$, and the *true* minimum variance $(\mathbf{x}^*)^\mathsf{T} \mathbf{V} \mathbf{x}^*$, where $\mathbf{x}^*$ is the true minimum-risk fully invested portfolio for $\mathbf{V}$. What do you observe? Are the results any different from part (a)(ii)?

(iii) Repeat parts (i) and (ii) several times (anywhere from a handful to a few thousand times). What do you observe? Are the results any different from part (a)(iii)?

8 Mixed Integer Programming: Theory and Algorithms

8.1 Mixed Integer Programming

The types of optimization models that we have discussed so far, namely linear and quadratic programming, allow variables to take a continuum of values. In particular, the numerical solutions to these kinds of models may have fractional values. For instance, the solution to a portfolio construction model could suggest a plan to purchase 3205.76 shares of stock XYZ. In many cases it is natural to round this value and to interpret it as a suggestion to purchase 3205 or even 3200 shares of stock XYZ. However, if a variable in an optimization model is associated with choosing among two or more alternatives, for example, as in the capital budgeting problem described below, then a model that suggests taking fractions of each of the alternatives would be of limited value. Instead, a *binary* decision, namely "to choose" or "not to choose", needs to be made for each alternative.

In general, an *integer variable* in an optimization model is a variable that is restricted to take integer values only. A *mixed integer program* is an optimization problem with the constraint that some of the variables must take integer values. In particular a *mixed integer linear program* is a problem of the form

$$\begin{aligned}
\min_{\mathbf{x}} \quad & \mathbf{c}^\mathsf{T}\mathbf{x} \\
\text{s.t.} \quad & \mathbf{A}\mathbf{x} = \mathbf{b} \\
& \mathbf{D}\mathbf{x} \geq \mathbf{d} \\
& x_j \in \mathbb{Z}, \; j \in J
\end{aligned} \tag{8.1}$$

for some vectors $\mathbf{c} \in \mathbb{R}^n$, $\mathbf{b} \in \mathbb{R}^m$, $\mathbf{d} \in \mathbb{R}^p$, matrices $\mathbf{A} \in \mathbb{R}^{m \times n}$, $\mathbf{D} \in \mathbb{R}^{p \times n}$, and subset $J \subseteq \{1, \dots, n\}$ of the variables. When all variables are restricted to be integer, that is, when $J = \{1, \dots, n\}$, the problem (8.1) is called a *pure integer linear program*.

An important case occurs when a model includes *binary* variables; that is, variables that are restricted to take the value 0 or 1. When all the variables in a mixed integer program are of this kind, it is called a *binary program*. As the examples below show, binary variables enable the modeling of important realistic features such as logical constraints, cardinality and threshold constraints, and others. However, this improvement in modeling power comes with a tradeoff in computational cost. The presence of a significant number of integer variables in an optimization problem can make it extremely difficult or impossible to solve unless there is a specific exploitable structure.

Example 8.1 (Capital budgeting)   Suppose we have a capital of 19 million dollars for long-term investment and have identified four investment opportunities with the following investment requirements and net present values (in million dollars):

	Investment 1	Investment 2	Investment 3	Investment 4
Required investment	7	10	6	3
Net present value	9	11	7	4

What investments should we choose to maximize our total net present value? Each investment is a "take it or leave it" opportunity: the investment must be funded entirely or not at all.

This problem can be formulated as the following binary linear programming model.

Binary linear programming model for capital budgeting
Variables:
$$x_i = \begin{cases} 1 & \text{if investment } i \text{ is undertaken} \\ 0 & \text{otherwise} \end{cases} \quad \text{for } i = 1, \ldots, 4.$$

Objective, in millions of dollars:
$$\max 9x_1 + 11x_2 + 7x_3 + 4x_4$$

Constraints:
$$7x_1 + 10x_2 + 6x_3 + 3x_4 \le 19 \qquad \text{(budget constraint)}$$
$$x_i \in \{0,1\} \quad \text{for } i = 1, \ldots, 4 \quad \text{(binary variables)}.$$

The optimal solution to the *linear programming relaxation* of this model, obtained by *relaxing* the binary constraints $x_i \in \{0,1\}$, for $i = 1, \ldots, 4$, to $0 \le x_i \le 1$, for $i = 1, \ldots, 4$, is

$$\mathbf{x}^* = \begin{bmatrix} 1 \\ 0.3 \\ 1 \\ 1 \end{bmatrix}.$$

This is not a feasible solution as x_2^* is not binary. If we round x_2^* to 0 we get a feasible solution. However, a better (and in fact the optimal) solution is

$$\mathbf{x}^* = \begin{bmatrix} 0 \\ 1 \\ 1 \\ 1 \end{bmatrix}.$$

This could be counterintuitive as Investment 1 has the best "bang for the buck"; that is, has the highest ratio of net present value to investment requirement.

The presence of binary variables also readily enables the modeling of logical restrictions. For example, the logical restriction

> If Investment 2 is made then Investment 4 must also be made

can be modeled via the constraint

$$x_2 \leq x_4.$$

Similarly, the logical constraint

> If Investment 1 is made then Investment 3 must not be made

can be modeled via the constraint

$$x_1 + x_3 \leq 1.$$

Example 8.2 (Clustering)   Clustering is a popular technique in data analysis. It is concerned with partitioning a collection of objects into subsets or "clusters" so that the objects within each cluster are more closely related with each other than with objects assigned to different clusters. Suppose we wish to partition a collection of N objects into $K < N$ clusters based on some kind of similarity measure:

$$\rho_{ij} = \text{ similarily measure between objects } i, j.$$

To give a financial flavor to this example, assume the objects to be clustered are N stocks and the similarity measure ρ_{ij} is the correlation between the returns of stocks i and j.

We next describe a possible approach to the above clustering problem via binary programming. This approach is closely related to the popular K-median problem. Before diving into the binary programming formulation, we describe some of the main ideas. Assume the objects are indexed $1, \ldots, N$ and are to be partitioned into the K clusters $C_1, \ldots, C_K$. A key idea is to designate an element j_ℓ in each cluster C_ℓ as the *centroid* of cluster C_ℓ. This choice suggests the following natural measure of the similarity within cluster C_ℓ:

$$\sum_{i \in C_\ell} \rho_{i,j_\ell}$$

and in turn it gives the following overall measure of the quality of the clusters $C_1, \ldots, C_K$:

$$\sum_{\ell=1}^{K} \sum_{i \in C_\ell} \rho_{i,j_\ell}.$$

The following crucial observation is key in our formulation. The centroid j_ℓ *represents* the elements in cluster C_ℓ. Indeed, each cluster contains precisely the objects assigned to its centroid, and the clusters are completely determined by the choice of the centroids. These ideas are formalized in the following binary linear programming model.

Binary linear programming model for clustering
Variables:

$$y_j = \begin{cases} 1 & \text{if } j \text{ is a centroid} \\ 0 & \text{otherwise} \end{cases} \quad \text{for } j = 1, \ldots, N.$$

$$x_{ij} = \begin{cases} 1 & \text{if } i \text{ is represented by } j \\ 0 & \text{otherwise} \end{cases} \quad \text{for } i, j = 1, \ldots, N.$$

Objective:

$$\max \sum_{j=1}^{N} \sum_{i=1}^{N} \rho_{ij} x_{ij}.$$

Constraints:

$$\sum_{j=1}^{N} y_j = K \qquad \qquad \text{(choose } K \text{ centroids)}$$

$$\sum_{j=1}^{N} x_{ij} = 1, \qquad \text{for } i = 1, \ldots, N \quad \text{(each object must be represented by one centroid)}$$

$$x_{ij} \le y_j, \qquad \text{for } i, j = 1, \ldots, N \quad (i \text{ is represented by } j \text{ only if } j \text{ is a centroid)}$$

$$x_{ij}, y_j \in \{0, 1\} \qquad \text{for } i, j = 1, \ldots, N \quad \text{(binary variables)}.$$

Another correct formulation is obtained if we replace the third set of N^2 constraints

$$x_{ij} \le y_j, \quad \text{for } i, j = 1, \ldots, N$$

with the set of N constraints

$$\sum_{i=1}^{N} x_{ij} \le N y_j, \quad \text{for } j = 1, \ldots, N.$$

8.2   Numerical Mixed Integer Programming Solvers

Excel Solver

The steps required for solving a mixed integer (or binary) linear program in Excel `Solver` are nearly identical to those for solving linear programs. The only new step is to state that some variables are integer (or binary).

Figure 8.1 displays a printout of an Excel spreadsheet implementation of the binary linear programming model for Example 8.1 as well as the dialog box obtained when we run the Excel add-in `Solver`. The spreadsheet model contains the three components of the binary program. The decision variables are in the range B7:E7. The left-hand side of the budget constraint is in cell B9 and the objective function is in cell B10. The Excel formulas in cells B9 and B10 are SUMPRODUCT(B4:E4,B7:E7) and SUMPRODUCT(B5:E5,B7:E7) respectively. In addition to these components, notice the constraint

$$\$B\$7:\$E\$7 = \texttt{binary}$$

in the **Solver** dialog box.

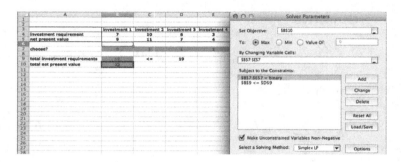

Figure 8.1 Spreadsheet implementation and the Solver dialog box for the capital budgeting model

MATLAB CVX

Figure 8.2 displays a CVX script for the same problem.

```
File  Edit  Text  Go  Cell  Tools  Debug  Desktop  Window

         1.0   +   +  1.1  ×

1        % Matlab CVX code for capital budgeting model
2 -      n = 4 ;
3 -      invest_required = [7;10;6;3] ;
4 -      npv = [9;11;7;4] ;
5 -      cvx_begin
6 -          cvx_solver gurobi
7 -          variable x(n) binary ;
8 -          maximize (npv'*x)
9 -          invest_required'*x <= 19 ;
10 -     cvx_end
```

Figure 8.2 MATLAB CVX code for capital budgeting model

Using either of these solvers we obtain the optimal solution:

$$\mathbf{x}^* = \begin{bmatrix} 0 \\ 1 \\ 1 \\ 1 \end{bmatrix}.$$

8.3 Relaxations and Duality

A *relaxation* of an optimization model

$$\min_{\mathbf{x}} \quad f(\mathbf{x})$$
$$\text{s.t.} \quad \mathbf{x} \in \mathcal{X}$$

is another optimization model

$$\min_{\mathbf{x}} \quad \tilde{f}(\mathbf{x})$$
$$\text{s.t.} \quad \mathbf{x} \in \tilde{\mathcal{X}}$$

that satisfies $\mathcal{X} \subseteq \tilde{\mathcal{X}}$ and $\tilde{f}(x) \leq f(x)$ for $x \in \mathcal{X}$. In other words, a relaxation is a less stringent optimization model obtained by "relaxing" some of the original constraints and "relaxing" its objective function. Relaxation plays a central role in a variety of algorithms for solving mixed integer programs. We next describe two widely used types of relaxations for mixed integer programming, namely *linear programming* and *Lagrangian* relaxations.

8.3.1 Linear Programming Relaxation

The *linear programming relaxation* of the mixed integer linear program (8.1) is the linear program obtained by dropping the integrality constraints; that is,

$$\min_{\mathbf{x}} \quad \mathbf{c}^\mathsf{T}\mathbf{x}$$
$$\text{s.t.} \quad \mathbf{A}\mathbf{x} = \mathbf{b} \qquad (8.2)$$
$$\mathbf{D}\mathbf{x} \geq \mathbf{d}.$$

Similarly, the linear programming relaxation of a mixed binary linear program

$$\min_{\mathbf{x}} \quad \mathbf{c}^\mathsf{T}\mathbf{x}$$
$$\text{s.t.} \quad \mathbf{A}\mathbf{x} = \mathbf{b}$$
$$\mathbf{D}\mathbf{x} \geq \mathbf{d} \qquad (8.3)$$
$$x_j \in \{0,1\}, \ j \in J$$

is the linear program

$$\min_{\mathbf{x}} \quad \mathbf{c}^\mathsf{T}\mathbf{x}$$
$$\text{s.t.} \quad \mathbf{A}\mathbf{x} = \mathbf{b}$$
$$\mathbf{D}\mathbf{x} \geq \mathbf{d} \qquad (8.4)$$
$$0 \leq x_j \leq 1, \ j \in J.$$

The proof of the following proposition is straightforward and we leave it as an exercise.

Proposition 8.3 *Consider the mixed integer program* (8.1) *and its linear programming relaxation* (8.2). *Then the following facts hold.*

(a) *The optimal value of the relaxation* (8.2) *is less than or equal to the optimal value of the mixed integer linear program* (8.1).

(b) *If the relaxation (8.2) is infeasible, then so is the mixed integer linear program (8.1).*

(c) *If the optimal solution $\mathbf{x}^*$ of the relaxation (8.2) satisfies $x_j^* \in \mathbb{Z}$ for $j \in J$ then $\mathbf{x}^*$ is also an optimal solution to the mixed integer linear program (8.1)*

The analogous facts also hold for the mixed binary linear program (8.3) and its linear relaxation (8.4). Proposition 8.3 suggests a possible avenue for solving (8.1): solve the (more tractable) linear programming relaxation (8.2). If this solution satisfies the relevant integrality constraints, then we have solved (8.1). If not, the lower bound obtained by solving (8.2) provides valuable information. For instance, if we can find a feasible solution to (8.1), then the quality of this solution can be assessed by comparing it with the lower bound obtained from solving (8.2). The sections below elaborate on this idea. In particular, Section 8.4 sketches algorithms that solve mixed linear integer programs by systematically solving a sequence of linear programming relaxations.

Proposition 8.3 also leads to the following somewhat counterintuitive conclusion about integer programming formulations. As noted in Example 8.2, there could be several correct and thus equivalent integer linear programming formulations to a given problem. Among them, it is generally better to have a formulation with a "tight" linear programming relaxation. Typically, a formulation with more constraints has a tighter linear programming relaxation and hence might be easier to solve.

8.3.2 Lagrangian Relaxation

The Lagrangian framework discussed in previous chapters can be extended to obtain relaxations of an optimization model. The intuitive idea is to obtain a relaxation of a model by shifting a set of "difficult" constraints to the objective. To be more precise, consider an optimization problem of the form

$$
\begin{aligned}
\min_{\mathbf{x}} \quad & \mathbf{c}^\mathsf{T}\mathbf{x} \\
\text{s.t.} \quad & \mathbf{A}\mathbf{x} = \mathbf{b} \\
& \mathbf{x} \in \mathcal{X},
\end{aligned} \tag{8.5}
$$

where the combined set of constraints $\mathbf{A}\mathbf{x} = \mathbf{b}$, $\mathbf{x} \in \mathcal{X}$ is "difficult" but the set of constraints $\mathbf{x} \in \mathcal{X}$ is "easy". (We will discuss a concrete example of this situation in Section 8.3.3.) A relaxation for (8.5) can be obtained as follows. Assume $\mathbf{u}$ is a vector of suitable dimension and consider the following problem without the difficult constraints:

$$
\begin{aligned}
L(\mathbf{u}) := \min_{\mathbf{x}} \quad & \mathbf{c}^\mathsf{T}\mathbf{x} + \mathbf{u}^\mathsf{T}(\mathbf{b} - \mathbf{A}\mathbf{x}) \\
\text{s.t.} \quad & \mathbf{x} \in \mathcal{X}.
\end{aligned} \tag{8.6}
$$

The problem (8.6) is a *Lagrangian relaxation* of (8.5). The following proposition is in the same spirit as Proposition 8.3. Again, its proof is straightforward and we leave it as an exercise.

Proposition 8.4 *Consider the optimization problem (8.5) and its Lagrangian relaxation (8.6) for some vector* $\mathbf{u}$. *Then the following facts hold.*

(a) *The optimal value* $L(\mathbf{u})$ *of the relaxation (8.6) is less than or equal to the optimal value of (8.5).*

(b) *If the optimal solution* $\mathbf{x}^*$ *of the relaxation (8.6) satisfies* $\mathbf{A}\mathbf{x}^* = \mathbf{b}$ *then* $\mathbf{x}^*$ *is also an optimal solution to (8.5).*

The *Lagrangian dual* of (8.5) is the problem of finding the best Lagrangian relaxation

$$\max_{\mathbf{u}} \; L(\mathbf{u}),$$

where $L(\mathbf{u})$ is the optimal value of (8.6). We note that the function $\mathbf{u} \mapsto L(\mathbf{u})$ is concave; that is, $\mathbf{u} \mapsto -L(\mathbf{u})$ is convex. Thus the Lagrangian dual is a convex optimization problem.

The Lagrangian relaxation and Lagrangian dual also extend to problems where the set of difficult constraints involves both equalities and inequalities. We simply need to be a bit careful about the sign of the multipliers for the inequality constraints. Consider the optimization problem

$$
\begin{aligned}
\min_{\mathbf{x}} \quad & \mathbf{c}^{\mathsf{T}}\mathbf{x} \\
\text{s.t.} \quad & \mathbf{A}\mathbf{x} = \mathbf{b} \\
& \mathbf{D}\mathbf{x} \geq \mathbf{d} \\
& \mathbf{x} \in \mathcal{X}.
\end{aligned}
\tag{8.7}
$$

Given vectors $\mathbf{u}, \mathbf{v}$ of suitable dimension with $\mathbf{v} \geq \mathbf{0}$ we obtain the following Lagrangian relaxation:

$$
\begin{aligned}
L(\mathbf{u}, \mathbf{v}) := \min_{\mathbf{x}} \quad & \mathbf{c}^{\mathsf{T}}\mathbf{x} + \mathbf{u}^{\mathsf{T}}(\mathbf{b} - \mathbf{A}\mathbf{x}) + \mathbf{v}^{\mathsf{T}}(\mathbf{d} - \mathbf{D}\mathbf{x}) \\
\text{s.t.} \quad & \mathbf{x} \in \mathcal{X}.
\end{aligned}
\tag{8.8}
$$

We have the following extended version of Proposition 8.4.

Proposition 8.5 *Consider the optimization problem (8.7) and its Lagrangian relaxation (8.8) for some vectors* $\mathbf{u}, \mathbf{v}$ *with* $\mathbf{v} \geq \mathbf{0}$. *Then the following facts hold.*

(a) *The optimal value* $L(\mathbf{u}, \mathbf{v})$ *of the relaxation (8.8) is less than or equal to the optimal value of (8.7).*

(b) *If the optimal solution* $\mathbf{x}^*$ *of the relaxation (8.8) satisfies* $\mathbf{A}\mathbf{x}^* = \mathbf{b}, \mathbf{D}\mathbf{x}^* \geq \mathbf{d}$, *and* $\mathbf{v}^{\mathsf{T}}(\mathbf{D}\mathbf{x}^* - \mathbf{d}) = 0$, *then* $\mathbf{x}^*$ *is also an optimal solution to (8.7).*

The Lagrangian dual of (8.7) is

$$
\begin{aligned}
\max_{\mathbf{u}, \mathbf{v}} \quad & L(\mathbf{u}, \mathbf{v}) \\
\text{s.t.} \quad & \mathbf{v} \geq \mathbf{0}.
\end{aligned}
$$

8.3.3   A Heuristic based on Lagrangian Relaxation for Clustering

We next describe a particularly successful application of Lagrangian relaxation for the clustering problem introduced in Example 8.2, namely,

$$
\begin{aligned}
Z := \max \quad & \sum_{i=1}^{N} \sum_{j=1}^{N} \rho_{ij} x_{ij} \\
\text{s.t.} \quad & \sum_{j=1}^{N} y_j = K \\
& \sum_{j=1}^{N} x_{ij} = 1 \qquad \text{for } i = 1, \ldots, N \\
& x_{ij} \leq y_j \qquad \text{for } i, j = 1, \ldots, N \\
& x_{ij}, y_j \in \{0, 1\} \quad \text{for } i, j = 1, \ldots, N.
\end{aligned}
\tag{8.9}
$$

The above model can be solved by general-purpose solvers such as Excel `Solver` or CVX, or even commercial solvers like `Gurobi` or `CPLEX`, only for relatively small values of N. One of the main difficulties is that the model involves $N^2 + N$ binary variables and $N^2 + N + 1$ constraints. For a modest value of N like $N = 100$, the model becomes unmanageable if tackled by a standard solver. At the same time, in practical clustering problem instances N can easily range in the hundreds or thousands. A heuristic based on Lagrangian relaxation developed by Cornuéjols et al. (1977) can compute approximate solutions to (8.9) for virtually unlimited values of N. We next describe the main ideas behind this heuristic. Consider the following Lagrangian relaxation to (8.9): given a vector of multipliers $\mathbf{u} = \begin{bmatrix} u_1 & \cdots & u_N \end{bmatrix}^{\mathsf{T}}$ let

$$
\begin{aligned}
L(\mathbf{u}) := \max_{\mathbf{x}, \mathbf{y}} \quad & \sum_{i=1}^{N} \sum_{j=1}^{N} \rho_{ij} x_{ij} + \sum_{i=1}^{N} u_i \left(1 - \sum_{j=1}^{N} x_{ij} \right) \\
\text{s.t.} \quad & \sum_{j=1}^{N} y_j = K \\
& x_{ij} \leq y_j, \ i, j = 1, \ldots, N \\
& x_{ij}, y_j = 0 \text{ or } 1, \ i, j = 1, \ldots, N.
\end{aligned}
\tag{8.10}
$$

This Lagrangian relaxation has moved the "difficult" constraints $\sum_{j=1}^{N} x_{ij} = 1$, with $i = 1, \ldots, N$, to the objective via the multipliers $\mathbf{u}$ and has kept the remaining constraints as the "easy" ones. This Lagrangian relaxation satisfies the following key properties:

Property 1: $L(\mathbf{u}) \geq Z$, where Z is the optimal value of (8.9). This is an immediate consequence of Proposition 8.3.

Property 2: For a given $\mathbf{u}$, (8.10) is easy to solve. To see this, first notice that

$$L(\mathbf{u}) := \max_{\mathbf{x},\mathbf{y}} \sum_{i=1}^{N}\sum_{j=1}^{N}(\rho_{ij}-u_i)x_{ij} + \sum_{i=1}^{N}u_i$$

$$\text{s.t.} \quad \sum_{j=1}^{N}y_j = K$$

$$x_{ij} \le y_j, \ i,j = 1,\ldots,N$$

$$x_{ij}, y_j \in \{0,1\}, \ i,j = 1,\ldots,N.$$

Given $\mathbf{y}$, this shows that x_{ij} should be set to its upper bound y_j or to its lower bound 0, depending on whether the objective coefficient $\rho_{ij} - u_i$ of x_{ij} is positive or negative. Therefore $L(\mathbf{u})$ can be rewritten as

$$L(\mathbf{u}) = \max_{\mathbf{y}} \quad \sum_{j=1}^{N}C_j y_j + \sum_{i=1}^{N}u_i$$

$$\text{s.t.} \quad \sum_{j=1}^{N}y_j = K \tag{8.11}$$

$$y_j \in \{0,1\}, \ j = 1,\ldots,N$$

for $C_j := \sum_{i=1}^{N}\max(0,\rho_{ij}-u_i)$. Finally, observe that the solution to (8.11) is readily computable: Sort the C_j in decreasing order, say $C_{j_1} \ge C_{j_2} \ge \cdots \ge C_{j_N}$. The optimal solution to (8.11) is obtained by setting $\bar{y}_{j_1} = \cdots = \bar{y}_{j_K} = 1$ and the remaining $\bar{y}_j$s to zero. We get $L(\mathbf{u}) = \sum_{t=1}^{K}C_{j_t} + \sum_{i=1}^{N}u_i$.

Property 3: Based on the optimal solution $\bar{\mathbf{y}}$ of $L(\mathbf{u})$ obtained in Property 2, one can get a heuristic (*ad hoc*) solution $(\bar{\mathbf{x}},\bar{\mathbf{y}})$ for (8.9) and an assessment of how good it is.

- Each $\mathbf{u}$ gives the upper bound $L(\mathbf{u}) \ge Z$ and the following heuristic feasible solution $\bar{\mathbf{x}}$ to (8.9). Let $\bar{\mathbf{y}}$ solve (8.11). Next, for each $i = 1,\ldots,N$, assign i to the most similar centroid among the K centroids such that $\bar{y}_j = 1$. That is, let $j(i) = \text{argmax}_{j:\bar{y}_j=1}\rho_{ij}$ and let $\bar{\mathbf{x}}$ be as follows:

$$\bar{x}_{ij} = \begin{cases} 1 & \text{if } j = j(i) \\ 0 & \text{otherwise.} \end{cases}$$

- If $\sum_{i,j}\rho_{ij}\bar{x}_{ij}$ and $L(\mathbf{u})$ are close to each other, then we have a near-optimal solution. To see this, observe that $\sum_{i,j}\rho_{ij}\bar{x}_{ij} \le Z \le L(\mathbf{u})$. Thus if $\sum_{i,j}\rho_{ij}\bar{x}_{ij}$ and $L(\mathbf{u})$ are close to each other, they must be close to the optimal value Z as well.

- To get the best upper bound $L(\mathbf{u})$ together with a heuristic solution of the above kind, solve

$$\min_{\mathbf{u}} L(\mathbf{u}).$$

This turns out to be a manageable convex optimization problem. In particular, it is amenable to a *subgradient* algorithm that we describe in Chapter 20.

8.4   Algorithms for Solving Mixed Integer Programs

The modeling power of mixed integer programming comes with some cost. Mixed integer programming belongs to the class of *NP-hard* computational problems (Conforti et al., 2014). In layman's terms, this means that, unlike convex optimization problems, which can be solved with fast and reliable numerical algorithms, the same cannot be expected for mixed integer programs. The algorithm that we describe next can in principle solve any mixed integer linear programs in finitely many steps. However, the NP-hardness of integer programming implies that for some problem instances the computational cost incurred by these algorithms could be insurmountable even for any foreseeable amount of computational power.

The two most popular generic methods for solving mixed integer linear programs are *cutting planes* and *branch and bound*. Both of these methods rely extensively on linear programming relaxations. A cutting plane is a new linear constraint to the linear programming relaxation that "cuts off" non-integer solutions without cutting off any feasible solution of the original mixed integer linear program. The method of cutting planes was proposed by Dantzig et al. (1954) in the context of the traveling-salesman problem, and by Gomory (1958, 1960) for pure integer linear programs and mixed integer linear programs respectively. The method is based on solving a sequence of increasingly tighter linear programming relaxations by adding cutting planes until a solution to the mixed integer linear program is found. On the other hand, Land and Doig (1960) proposed a "branch-and-bound" method to solve mixed integer linear programs. Branch and bound is an enumerative procedure based on dividing the original problem into a number of smaller problems (branching) and evaluating their quality based on their linear programming relaxations (bounding). Branch and bound was the most effective technique for solving mixed integer linear programs for multiple decades. However, in the 1990s, cutting planes made a resurgence. Current state-of-the-art integer programming solvers combine cutting planes and branch and bound into an overall procedure called "branch and cut", a term coined in Padberg and Rinaldi (1987).

8.4.1 Branch-and-Bound Method

The gist of the branch-and-bound method can be easily grasped via a couple of examples. Consider the following integer linear program (see Figure 8.3):

$$
\begin{aligned}
\max \quad & x_1 + x_2 \\
\text{s.t.} \quad & -x_1 + x_2 \leq 2 \\
& 8x_1 + 2x_2 \leq 19 \\
& x_1, x_2 \geq 0 \\
& x_1, x_2 \in \mathbb{Z}.
\end{aligned}
\tag{8.12}
$$

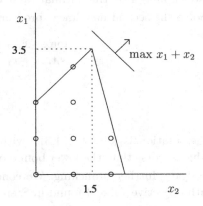

Figure 8.3 A two-variable integer program

Step 1. Solve the linear programming relaxation of (8.12), namely

$$
\begin{aligned}
\max \quad & x_1 + x_2 \\
\text{s.t.} \quad & -x_1 + x_2 \leq 2 \\
& 8x_1 + 2x_2 \leq 19 \\
& x_1, x_2 \geq 0.
\end{aligned}
$$

The solution is $\bar{\mathbf{x}} = \begin{bmatrix} 1.5 & 3.5 \end{bmatrix}^{\mathsf{T}}$ with objective value 5. Thus 5 is an upper bound on the optimal value of (8.12). The vector $\bar{\mathbf{x}} = \begin{bmatrix} 1.5 & 3.5 \end{bmatrix}^{\mathsf{T}}$ is not a feasible solution to (8.12) since the entries of $\bar{\mathbf{x}}$ are fractional. How can we exclude this fractional solution while preserving the feasible integral solutions? One way is to *branch*: create two new linear programs, one with the additional constraint $x_1 \leq 1$, and the other with the additional constraint $x_1 \geq 2$. Clearly, any solution to the integer program must be feasible to one or the other of these two problems. We will solve both of these linear programs.

Step 2. Solve the first of the two new linear programs:

$$\begin{array}{ll} \max & x_1 + x_2 \\ \text{s.t.} & -x_1 + x_2 \leq 2 \\ & 8x_1 + 2x_2 \leq 19 \\ & x_1 \leq 1 \\ & x_1, x_2 \geq 0. \end{array}$$

The solution is $\bar{\mathbf{x}} = \begin{bmatrix} 1 & 3 \end{bmatrix}^{\mathsf{T}}$ with objective value 4. This is a feasible integral solution to (8.12). So now we have the upper bound 5 and the lower bound 4 on the optimal value of (8.12).

Step 3. Solve the second new linear program:

$$\begin{array}{ll} \max & x_1 + x_2 \\ \text{s.t.} & -x_1 + x_2 \leq 2 \\ & 8x_1 + 2x_2 \leq 19 \\ & x_1 \geq 2 \\ & x_1, x_2 \geq 0. \end{array}$$

The solution is $\bar{\mathbf{x}} = \begin{bmatrix} 2 & 1.5 \end{bmatrix}^{\mathsf{T}}$ with objective value 3.5. Because this value is worse than the lower bound of 4 that we already have, we do not need any further branching. We conclude that the vector $\bar{\mathbf{x}} = \begin{bmatrix} 1 & 3 \end{bmatrix}$ with objective value 4 found in Step 2 is an optimal solution to (8.12).

The solution of the above integer program by branch and bound required the solution of three linear programs. These problems can be arranged in a *branch and-bound tree*, see Figure 8.4. Each *node* of the tree corresponds to one of the problems that were solved.

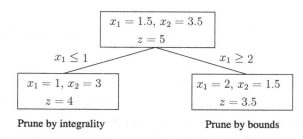

Figure 8.4 Branch-and-bound tree for (8.12)

We can stop the enumeration at a node of the branch-and-bound tree for three different reasons (when they occur, the node is said to be *pruned*).

Pruning by integrality occurs when the corresponding linear program has an optimal solution that is integral. This occurred in Step 2 in the above example.

Pruning by bounds occurs when the objective value of the linear program at that node is worse than the value of the best feasible solution found so far. This occurred in Step 3 in the above example.

Pruning by infeasibility occurs when the linear program at that node is infeasible. This did not occur in any of the steps in the above example.

We next illustrate the branch-and-bound method is a slightly modified instance that leads to a larger branch-and-bound tree. Consider the integer linear program:

$$\begin{aligned} \max \quad & 3x_1 + x_2 \\ \text{s.t.} \quad & -x_1 + x_2 \le 2 \\ & 8x_1 + 2x_2 \le 19 \\ & x_1, x_2 \ge 0 \\ & x_1, x_2 \in \mathbb{Z}. \end{aligned} \qquad (8.13)$$

Figure 8.5 depicts the branch-and-bound tree for this problem.

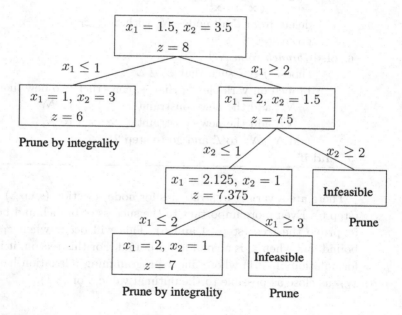

Figure 8.5 Branch-and-bound tree for (8.13)

Algorithm 8.1 sketches the branch-and-bound method for a general mixed integer linear program of the form (8.1). The branch-and-bound method keeps a list of linear programming problems obtained by relaxing the integrality requirements on the variables and imposing constraints such as $x_j \le u_j$ or $x_j \ge l_j$. Each such linear program corresponds to a *node* of the branch-and-bound tree. It will be convenient to let N_i denote both a node and its corresponding linear program in the branch-and-bound tree. Let $\mathbf{x}^i$ and z_i denote respectively the optimal solution and optimal value of the linear program N_i with the convention $z_i = \infty$

if N_i is infeasible. Let N_0 denote the root node of the branch-and-bound tree: it corresponds to the linear programming relaxation (8.2) of (8.1). Throughout the algorithm we let $\mathcal{L}$ denote the list of nodes that must still be solved. These are the nodes that have not been pruned nor branched on. Throughout the algorithm $\mathbf{x}^*$ denotes the best feasible solution found so far and z_U its objective value. The value z_U is also the best upper bound on the optimal value z of (8.1) so far. Initially, the bound z_U can be derived from a heuristic solution to (8.1), or it can be set to $+\infty$ if no heuristic solution is available.

Algorithm 8.1 Branch-and-bound method

1: $\mathcal{L} := \{N_0\}$, $z_U := +\infty$, $\mathbf{x}^* := \emptyset$ *(initialization)*
2: **if** $\mathcal{L} = \emptyset$ **then** HALT and **return** the vector $\mathbf{x}^*$ **end if** *(termination)*
3: choose and delete a node N_i from $\mathcal{L}$ and solve it *(select next node to solve)*
4: **if** $z_i \geq z_U$ **then** go to step 2 **end if** *(prune N_i)*
5: **if** $\mathbf{x}^i$ is feasible for (8.1) **then** *(update upper bound and prune)*
    $z_U := z_i$; $\mathbf{x}^* := \mathbf{x}^i$
    delete from $\mathcal{L}$ all nodes N_k with $z_k \geq z_U$
    go to step 2
6: **else** *(branch from N_i)*
    choose $j \in J$ such that $\bar{x}_j^i \notin \mathbb{Z}$
    branch on variable x_j, that is, construct two new linear programs
      N_i^1: add the new constraint $x_j \leq \lfloor \bar{x}_j^i \rfloor$ to N_i
      N_i^2: add the new constraint $x_j \geq \lceil \bar{x}_j^i \rceil$ to N_i
    add N_i^1, N_i^2 to $\mathcal{L}$ and go to step 2
7: **end if**

There are a variety of strategies for node selection (step 3) and for branching (step 6). Even more important to the success of branch and bound is the ability to prune the tree (steps 4 and 5). This will occur when z_U is a good upper bound and when z_i is a good lower bound. For this reason, it is crucial to have a formulation of (8.1) whose linear programming relaxation has an optimal value z_{LP} as close as possible to the optimal value z of (8.1).

8.4.2　Cutting-Plane Method

A *valid inequality* for a mixed integer linear program is a linear inequality that is satisfied by all feasible solutions. A *cutting plane* of a mixed integer linear program is a valid inequality that cuts off some solutions to its linear programming relaxation.

As we noted in Proposition 8.3, if an optimal solution of the linear programming relaxation satisfies the integrality constraints of a mixed integer linear program, then it is an optimal solution to the mixed integer linear program. The gist of cutting-plane methods is the observation that, when the latter does not occur, the linear programming relaxation can be strengthened by adding a cutting plane that cuts off its optimal solution.

Gomory (1960) proposed the following approach for solving mixed integer linear programs. Assume the variables in the problem are non-negative and satisfy the equality constraint

$$\sum_{j \in J} a_j x_j + \sum_{j \notin J} a_j x_j = b. \tag{8.14}$$

Assume that b is not an integer and let f_0 be its fractional part, i.e. $b = \lfloor b \rfloor + f_0$, where $0 < f_0 < 1$. For $j \in J$, let $a_j = \lfloor a_j \rfloor + f_j$, where $0 \le f_j < 1$. Replacing in (8.14) and moving sums of integer products to the right, we get

$$\sum_{j \in J:\, f_j \le f_0} f_j x_j + \sum_{j \in J:\, f_j > f_0} (f_j - 1) x_j + \sum_{j \notin J} a_j x_j = k + f_0,$$

where k is some integer. Using the fact that $k \le -1$ or $k \ge 0$, we must have either

$$-\sum_{j \in J:\, f_j \le f_0} \frac{f_j}{1 - f_0} x_j + \sum_{j \in J:\, f_j > f_0} \frac{1 - f_j}{1 - f_0} x_j - \sum_{j \notin J} \frac{a_j}{1 - f_0} x_j \ge 1$$

or

$$\sum_{j \in J:\, f_j \le f_0} \frac{f_j}{f_0} x_j - \sum_{j \in J:\, f_j > f_0} \frac{1 - f_j}{f_0} x_j + \sum_{j \notin J} \frac{a_j}{f_0} x_j \ge 1.$$

This is of the form $\sum_j c_j x_j \ge 1$ or $\sum_j d_j x_j \ge 1$, which implies $\sum_j \max(c_j, d_j) x_j \ge 1$ because the variables x_j are non-negative.

Which is the larger of the two coefficients c_j and d_j in our case? The answer is easy since one coefficient is positive and the other is negative for each variable x_j. Therefore, we get

$$\sum_{j \in J:\, f_j \le f_0} \frac{f_j}{f_0} x_j + \sum_{j \in J:\, f_j > f_0} \frac{1 - f_j}{1 - f_0} x_j + \sum_{j \notin J:\, a_j > 0} \frac{a_j}{f_0} x_j - \sum_{\notin J:\, a_j < 0} \frac{a_j}{1 - f_0} x_j \ge 1.$$
$$\tag{8.15}$$

Inequality (8.15) is valid for all $\mathbf{x} \ge 0$ that satisfy (8.14) with x_j integer for $j \in J$. It is called a *Gomory mixed integer cut*.

We illustrate the use of Gomory cuts on problem (8.12). To that end, we first add slack variables x_3 and x_4 to turn the inequality constraints into equalities:

$$\begin{aligned}
\max \quad & x_1 + x_2 \\
\text{s.t.} \quad & -x_1 + x_2 + x_3 = 2 \\
& 8x_1 + 2x_2 + x_4 = 19 \\
& x_1, x_2, x_3, x_4 \ge 0 \\
& x_1, x_2 \in \mathbb{Z}.
\end{aligned}$$

Solving the linear programming relaxation by the simplex method we get the optimal basis $B = \{1,2\}$ and so the constraints of the linear programming relaxation can be written as

$$x_1 - 0.2x_3 + 0.1x_4 = 1.5$$
$$x_2 + 0.8x_3 + 0.1x_4 = 3.5$$
$$x_1, x_2, x_3, x_4 \geq 0.$$

The corresponding basic solution is $x_3 = x_4 = 0$, $x_1 = 1.5$, $x_2 = 3.5$ with objective value $z = 5$. This solution is not integer. Let us generate the Gomory cut corresponding to the equation

$$x_1 - 0.2x_3 + 0.1x_4 = 1.5.$$

We have $f_0 = 0.5$, $f_1 = f_2 = 0$, $a_3 = -0.2$ and $a_4 = 0.1$. Applying formula (8.15), we get the Gomory cut

$$\frac{0.2}{1 - 0.5}x_3 + \frac{0.1}{0.5}x_4 \geq 1, \quad \text{i.e.,} \quad 2x_3 + x_4 \geq 5.$$

Since $x_3 = 2 + x_1 - x_2$ and $x_4 = 19 - x_1 - 2x_2$, we can express the above Gomory cut in terms of x_1, x_2:

$$3x_1 + 2x_2 \leq 9.$$

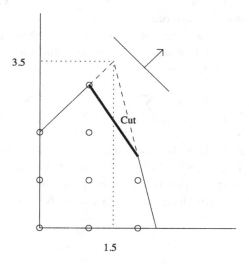

Figure 8.6 Formulation strengthened by a cut

Adding this cut to the linear programming relaxation, we get the following strengthened linear programming relaxation (see Figure 8.6):

$$\max \quad x_1 + x_2$$
$$\text{s.t.} \quad -x_1 + x_2 \leq 2$$
$$8x_1 + 2x_2 \leq 19$$
$$3x_1 + 2x_2 \leq 9$$
$$x_1, x_2 \geq 0.$$

The optimal solution to this linear program is $x_1 = 1$, $x_2 = 3$ with objective value $z = 4$. Since x_1 and x_2 are integer, this is the optimal solution to (8.12).

8.5   Exercises

Exercise 8.1   As the leader of an oil exploration drilling venture, you must determine the best selection of four out of eight possible sites. Label the sites $s_1, s_2, \ldots, s_8$ and the expected profits associated with each as $p_1, p_2, \ldots, p_8$.

(a) If site s_3 is explored, then sites s_1 and s_2 must also be explored. Furthermore, regional development restrictions are such that

(b) exploring sites s_6 *and* s_7 will prevent you from exploring site s_8;

(c) exploring at least one of the sites s_3 *or* s_4 will prevent you from exploring site s_5.

The eight expected profits are $p_i = i$ for $i = 1, \ldots, 8$. Formulate an integer program to determine the best exploration scheme and solve numerically with Solver.

Exercise 8.2   Consider the following projects for possible investments. For each project, you are given the NPV as well as the cash outflows required during each year (in millions of dollars).

	NPV	Year 1	Year 2	Year 3	Year 4
Project 1	30	12	4	4	0
Project 2	30	0	12	4	4
Project 3	20	3	4	4	4
Project 4	15	10	0	0	0
Project 5	15	0	11	0	0
Project 6	15	0	0	12	0
Project 7	15	0	0	0	13
Project 8	24	8	8	0	0
Project 9	18	0	0	10	4
Project 10	18	0	0	4	10

No partial investment is allowed in any of these projects. The firm has 18 million dollars available for investment each year.

(a) Formulate an integer linear program to determine the best investment plan.

(b) Formulate the following conditions as linear constraints.

(i) Exactly one of the projects 4, 5, 6, 7 must be invested in.

(ii) If Project 1 is invested in, then Project 2 cannot be invested in.

(iii) If Project 3 is invested in, then Project 4 must also be invested in.

(iv) If Project 8 is invested in, then either Project 9 or Project 10 or both must also be invested in.

(v) If either Project 1 or Project 2 is invested in, then neither Project 9 nor Project 10 can be invested in.

Exercise 8.3   Consider the problem

$$\begin{aligned}
\max \quad & 20x_1 + 10x_2 + 10x_3 \\
\text{s.t.} \quad & 2x_1 + 20x_2 + 4x_3 \leq 15 \\
& 6x_1 + 20x_2 + 4x_3 = 20 \\
& x_1, x_2, x_3 \geq 0 \\
& x_1, x_2, x_3 \in \mathbb{Z}.
\end{aligned}$$

Solve its linear programming relaxation. Then, show that it is impossible to obtain a feasible integral solution by rounding the values of the variables.

Exercise 8.4

(a) Compare the feasible solutions of the following three integer linear programs

$$\begin{aligned}
\max \quad & 14x_1 + 8x_2 + 6x_3 + 6x_4 \\
\text{s.t.} \quad & 28x_1 + 15x_2 + 13x_3 + 12x_4 \leq 39 \\
& x_1, x_2, x_3, x_4 \in \{0, 1\},
\end{aligned}$$
(i)

$$\begin{aligned}
\max \quad & 14x_1 + 8x_2 + 6x_3 + 6x_4 \\
\text{s.t.} \quad & 2x_1 + x_2 + x_3 + x_4 \leq 2 \\
& x_1, x_2, x_3, x_4 \in \{0, 1\},
\end{aligned}$$
(ii)

$$\begin{aligned}
\max \quad & 14x_1 + 8x_2 + 6x_3 + 6x_4 \\
\text{s.t.} \quad & 2x_1 + x_2 + x_3 + x_4 \leq 2 \\
& x_1 + x_2 \leq 1 \\
& x_1 + x_3 \leq 1 \\
& x_1 + x_4 \leq 1 \\
& x_1, x_2, x_3, x_4 \in \{0, 1\}.
\end{aligned}$$
(iii)

(b) Compare the relaxations of the above integer linear programs obtained by replacing $x_1, x_2, x_3, x_4 \in \{0, 1\}$ by $0 \leq x_j \leq 1$ for $j = 1, \ldots, 4$. Which is the best formulation among (i), (ii), (iii) for obtaining a tight bound from the linear programming relaxation?

Exercise 8.5   Prove Proposition 8.3.

Exercise 8.6   Prove Proposition 8.4.

Exercise 8.7 Prove that the function $\mathbf{u} \mapsto L(\mathbf{u})$ defined in (8.6) is a concave function.

Exercise 8.8 Prove Proposition 8.5.

Exercise 8.9 Let z_{LP} denote the value of the linear programming relaxation (8.2) and let z_{LD} be the Lagrangian dual of the following Lagrangian relaxation of (8.1):

$$L(\mathbf{u}) := \min_{\mathbf{x}} \quad \mathbf{c}^\mathsf{T}\mathbf{x} + \mathbf{u}^\mathsf{T}(\mathbf{b} - \mathbf{A}\mathbf{x})$$
$$\text{s.t.} \quad \mathbf{D}\mathbf{x} \geq \mathbf{d}$$
$$x_j \in \mathbb{Z}, \ j \in J.$$

Prove that $z_{LP} \leq z_{LD}$.

Exercise 8.10 Use the branch-and-bound method to solve the binary linear programming model:

$$\begin{aligned} \max \quad & 8x_1 + 11x_2 + 6x_3 + 4x_4 \\ \text{s.t.} \quad & 6.7x_1 + 10x_2 + 5.5x_3 + 3.4x_4 \leq 19 \\ & 8x_1 + 2x_2 \leq 19 \\ & x_1, x_2, x_3, x_4 \in \{0,1\}. \end{aligned}$$

Compare the number of nodes in the branch-and-bound tree with the following naive brute-force enumeration approach: check each of the $2^4 = 16$ possible values of $\mathbf{x} = \begin{bmatrix} x_1 & x_2 & x_3 & x_4 \end{bmatrix}^\mathsf{T}$ with $x_i \in \{0,1\}$, for $i = 1, \ldots, 4$, and pick the best feasible solution among them.

Exercise 8.11 Solve the integer linear programs of Exercise 8.4 using your favorite solver. In each case, report the number of nodes in the enumeration tree. Is it related to the tightness of the linear programming relaxation studied in Exercise 8.4(b)?

Exercise 8.12 Modify the branch-and-bound method (Algorithm 8.1) so that it stops as soon as it has found a feasible solution that is guaranteed to be within 5% of the optimum.

Exercise 8.13 Consider the integer program

$$\begin{aligned} \max \quad & 10x_1 + 13x_2 \\ \text{s.t.} \quad & 10x_1 + 14x_2 \leq 43 \\ & x_1, x_2 \geq 0 \\ & x_1, x_2 \in \mathbb{Z}. \end{aligned}$$

(a) Introduce a slack variable and solve the linear programming relaxation by the simplex method.

Hint: You should find the following optimal tableau:

$$\begin{aligned} \max \quad & x_2 + x_3 \\ \text{s.t.} \quad & x_1 + 1.4x_2 + 0.1x_3 = 4.3 \\ & x_1, x_2, x_3 \geq 0 \end{aligned}$$

with basic solution $x_1 = 4.3$, $x_2 = x_3 = 0$.

(b) Generate a Gomory mixed integer (GMI) cut that cuts off this solution.

(c) Multiply both sides of the equation $x_1 + 1.4x_2 + 0.1x_3 = 4.3$ by the constant $k = 2$ and generate the corresponding GMI cut. Repeat for $k = 3, 4$ and 5. Compare the five GMI cuts that you found.

(d) Add the GMI cut generated for $k = 3$ to the linear programming relaxation. Solve the resulting linear program by the simplex method. What is the optimum solution of the integer program?

9   Mixed Integer Programming Models: Portfolios with Combinatorial Constraints

This chapter presents several applications of integer and mixed integer programming, namely combinatorial auctions, the lockbox problem, constructing an index fund, and portfolio optimization with cardinality and threshold constraints. All of these applications involve combinatorial features that can be modeled via binary variables.

9.1   Combinatorial Auctions

A *combinatorial auction* is an auction that involves the concurrent sale of multiple items. Examples include Federal Communications Commission (FCC) spectrum auctions, electricity markets, pollution right auctions, and auctions for airport landing slots. In these kinds of auctions, bidders have preferences for sets of items usually called *bundles*. The value that a bidder has for a bundle may not necessarily be equal to the sum of the values that the bidder has for individual items in the bundle. To take the bidders' preferences into consideration, combinatorial auctions allow bidders to submit bids on combinations of items.

Specifically, let M be the set of items that the auctioneer has to sell and N the set of bidders. A *bid* is a pair $(S, b_j(S))$ where $S \subseteq M$, for $j \in N$, and $b_j(S)$ is the price that bidder j is willing to pay for the bundle S. The *combinatorial auction problem* or *winner selection problem* is the problem of identifying which bids should be accepted to maximize the auctioneer's revenue. This problem can be formulated as a binary linear program.

Binary linear programming model for the combinatorial auction problem
Variables:

$$x(S, j) = \begin{cases} 1 & \text{if bundle } S \text{ is allocated to bidder } j \\ 0 & \text{otherwise} \end{cases}$$

for $S \subseteq M, j \in N$.

Objective:

$$\max \sum_{S \subseteq M} \sum_{j \in N} b_j(S) x(S, j).$$

Constraints:

$$\sum_{S\subseteq M: i\in S}\sum_{j\in N} x(S,j) \le 1 \qquad \text{for } i \in M$$
$$\text{(allocated bundles do not overlap).}$$
$$x(S,j) \in \{0,1\} \quad \text{for } S \subseteq M,\ j \in N$$
$$\text{(binary variables).}$$

In some combinatorial auctions, bidders are awarded at most one of the bundle that they bid on, even when these bundles are disjoint. This is easy to model by adding the constraint

$$\sum_{S\subseteq M} x(S,j) \le 1 \qquad \text{for } j \in N \text{ (each bidder receives at most one bundle).}$$

For example, if there are four items for sale and the following bids have been received: $B_1 = (\{1\}, 6)$, $B_2 = (\{2\}, 3)$, $B_3 = (\{3,4\}, 12)$, $B_4 = (\{1,3\}, 12)$, $B_5 = (\{2,4\}, 8)$, $B_6 = (\{1,3,4\}, 16)$, the winners can be determined by the following integer program:

$$\begin{aligned}
\max \quad & 6x_1 + 3x_2 + 12x_3 + 12x_4 + 8x_5 + 16x_6 \\
\text{s.t.} \quad & x_1 + x_4 + x_6 \le 1 && \text{(item 1 is allocated at most once)} \\
& x_2 + x_5 \le 1 && \text{(item 2 is allocated at most once)} \\
& x_3 + x_4 + x_6 \le 1 && \text{(item 3 is allocated at most once)} \\
& x_3 + x_5 + x_6 \le 1 && \text{(item 4 is allocated at most once)} \\
& x_j \in \{0,1\} \ \text{ for } j = 1, \dots, 6.
\end{aligned}$$

If bids B_4 and B_5 come from the same bidder who wants at most one of these two bundles, it suffices to add the constraint

$$x_4 + x_5 \le 1.$$

If there are multiple units u_i of each item $i \in M$, then a bid can be more broadly defined as a pair $(\lambda, b_j(\lambda))$ where λ is an M-vector with entries $\lambda_i \in \{0, 1, \dots, u_i\}$, $i \in M$, that indicates the desired number of units λ_i of each item $i \in M$. Let Λ denote the set of all these M-vectors. The previous model is replaced by

$$\max \quad \sum_{\lambda\in\Lambda}\sum_{j\in N} b_j(\lambda)x(\lambda,j)$$
$$\text{s.t.} \quad \sum_{\lambda\in\Lambda}\sum_{j\in N} \lambda_i x(\lambda,j) \le u_i \ \text{ for } i \in M$$
$$x(\lambda,j) \in \{0,1\} \quad \text{for } \lambda \in \Lambda,\ j \in N.$$

There are further variations of the above formulations that incorporate additional features such as constraints on the kinds of bids the auctioneer accepts and constraints on the kinds of bundles that can be allocated to bidders (De Vries and Vohra, 2003). The gist of these models is essentially the same as those discussed above.

9.2   The Lockbox Problem

Consider a national firm that receives checks from all over the United States. Due to the vagaries of the Postal Service, as well as the banking system, there is a variable delay from when the check is postmarked (and hence the customer has met her obligation) and when the check clears (and when the firm can use the money). For instance, a check mailed in Pittsburgh sent to a Pittsburgh address might clear in just two days. A similar check sent to Los Angeles (L.A.) might take four days to clear. It is in the interest of the firm to have the check clear as quickly as possible since then the firm can use the money. In order to speed up this clearing, firms open offices (called lockboxes) in different cities to handle the checks.

Example 9.1   Suppose we receive payments from four regions (West, Midwest, East, and South). The average daily value from each region is as follows: $300,000 from the West, $120,000 from the Midwest, $360,000 from the East, and $180,000 from the South. We are considering opening lockboxes in L.A., Cincinnati, Boston, and/or Houston. Operating a lockbox costs $90,000 per year. The average days from mailing to clearing is given in Table 9.1. Which lockboxes should we open?

From	L.A.	Cincinnati	Boston	Houston
West	2	4	6	6
Midwest	4	2	5	5
East	6	5	2	5
South	7	5	6	3

Table 9.1 Clearing times

First we must calculate the losses due to lost interest for each possible assignment. For example, if the West sends to Boston, then on average there will be $1,800,000 ($= 6 \times \$300,000$) in process on any given day. Assuming an investment rate of 10%, this corresponds to a yearly loss of $180,000. We can calculate the losses for the other possibilities in a similar fashion to get Table 9.2.

From	L.A.	Cincinnati	Boston	Houston
West	60	120	180	180
Midwest	48	24	60	60
East	216	180	72	180
South	126	90	108	54

Table 9.2 Lost interest ('000)

The intuition for the formulation of the lockbox problem is similar to that of the formulation of the K-median model for clustering discussed in Chapter 8. We use a set of binary variables to model the lockboxes to open and another set of binary variables to model what lockbox serves each region.

Binary linear programming model for the lockbox problem
Variables:
$$y_j = \begin{cases} 1 & \text{if lockbox } j \text{ is opened} \\ 0 & \text{otherwise} \end{cases} \quad \text{for } j = 1, \ldots, 4.$$
$$x_{ij} = \begin{cases} 1 & \text{if region } i \text{ is served by lockbox } j \\ 0 & \text{otherwise} \end{cases} \quad \text{for } i, j = 1, \ldots, 4.$$
Objective: minimize total yearly costs

$$\min \sum_{i,j=1}^{4} c_{ij} x_{ij} + 90 \sum_{j=1}^{4} y_j,$$

where c_{ij} is the (i, j) entry in Table 9.2.
Constraints:

$$\sum_{j=1}^{N} x_{ij} = 1, \qquad \text{for } i = 1, \ldots, 4$$

(each region must be assigned to one lockbox)

$$x_{ij} \leq y_j, \qquad \text{for } i, j = 1, \ldots, 4$$

(region i is assigned to lockbox j only if j is opened

$$x_{ij}, y_j \in \{0, 1\} \qquad \text{for } i, j = 1, \ldots, 4$$

(binary variables).

As we observed for the binary programming formulation for clustering dis
cussed in Example 8.2, the above formulation would also be correct if we replaced
the $4^2 = 16$ constraints

$$x_{ij} \leq y_j, \ i, j = 1, \ldots, 4,$$

with the four constraints

$$\sum_{i=1}^{4} x_{ij} \leq 4y_j, \ j = 1, \ldots, 4.$$

However, we should note that the solution to the linear programming relaxation
of the first formulation (with more constraints) is

$$x_{11} = x_{21} = x_{33} = x_{43} = y_1 = y_3 = 1, \quad \text{and all other variables zero,}$$

which has binary entries and hence is an optimal solution to the binary linear
programming model. Therefore the firm should open two lockboxes, one in the
Eastern region and one in the West.

By contrast, the solution to the linear programming relaxation of the second
formulation (with fewer constraints) is

$$x_{11} = x_{22} = x_{33} = x_{44} = 1, \quad y_1 = y_2 = y_3 = y_4 = 0.25,$$

and all other variables zero,

which does not give any useful information about the binary linear programming
model.

This example highlights how different equivalent integer programming formulations can have very different properties with respect to their associated linear
program.

9.3   Constructing an Index Fund

An old and recurring debate about investing lies in the merits of active versus passive management of a portfolio. Active portfolio management tries to achieve superior performance by using technical and fundamental analysis. On the other hand, passive portfolio management relies entirely on diversification to achieve a desired performance. There are two types of passive management strategies: "buy and hold" or "indexing". In the first one, assets are selected on the basis of some fundamental criteria and there is no active selling or buying of these stocks afterwards (see the chapters on dedication (Chapter 3) and portfolio optimization (Chapter 6)). In the second approach, absolutely no attempt is made to identify mispriced securities. The goal is to choose a portfolio that mirrors the movements of a broad market population or a market index. Such a portfolio is called an index fund. Given a target population of n stocks, one selects K stocks (and their weights in the index fund) to represent the target population as closely as possible.

In the last 30 years, an increasing number of investors, both large and small, have established index funds. Simply defined, an index fund is a portfolio designed to track the movement of the market as a whole or some selected broad market segment. The rising popularity of index funds can be justified both theoretically and empirically.

Market efficiency: If the market is efficient, no superior risk-adjusted returns can be achieved by stock picking strategies since the prices reflect all the information available in the marketplace. Additionally, since the market portfolio provides the best possible return per unit of risk, to the extent that it captures the efficiency of the market via diversification, one may argue that the best theoretical approach to fund management is to invest in an index fund.

Empirical performance: Considerable empirical literature provides strong evidence that, on average, money managers have consistently under-performed the major indices. In addition, studies show that, in most cases, top performing funds for a year are no longer amongst the top performers in the following years, leaving room for the intervention of luck as an explanation for good performance.

Transaction cost: Actively managed funds incur transaction costs, which reduce the overall performance of these funds. In addition, active management implies significant research costs. Finally, fund managers may have costly compensation packages that can be avoided to a large extent with index funds.

Here we take the point of view of a fund manager who wants to construct an index fund. Strategies for forming index funds involve choosing a broad market index as a proxy for an entire market, e.g. the Standard & Poor's list of 500 stocks (S&P 500). A pure indexing approach consists in purchasing all the issues

in the index, with the same exact weights as in the index. In most instances
this approach is impractical (many small positions) and expensive (rebalancing
costs may be incurred frequently). An index fund with K stocks, where K is
substantially smaller than the size n of the target population, seems desirable.
The clustering approach introduced in Chapter 8 can be used to aggregate the
stocks in a broad market into a smaller more manageable index fund. This
approach will not necessarily yield mean/variance-efficient portfolios but will
produce a portfolio that closely replicates the underlying market population.

We describe a two-step heuristic approach for constructing an index fund.
First, select K stocks to be included in the portfolio. Second, determine weights
for these stocks so that the portfolio is as close as possible to the benchmark.
The motivation for this two-step approach is that each of the stocks selected in
the portfolio is a proxy for a portion of stocks in the index.

The first step, that is, the selection of the K stocks to be included in the
portfolio, can be formulated as the binary linear programming formulation for
clustering for Example 8.2 in Chapter 8. Recall that the model is based on the
following data:

$$\rho_{ij} = \text{similarity between stock } i \text{ and stock } j.$$

An example of this is the correlation between the returns of stocks i and j. But
one could choose other similarity measures ρ_{ij}.

Recall the binary linear programming formulation for the clustering problem

$$\max \quad \sum_{i=1}^{n}\sum_{j=1}^{n} \rho_{ij} x_{ij}$$

$$\text{s.t.} \quad \sum_{j=1}^{n} y_j = K$$

$$\sum_{j=1}^{n} x_{ij} = 1 \qquad \text{for } i = 1, \ldots, n$$

$$x_{ij} \le y_j \qquad \text{for } i, j = 1, \ldots, n$$

$$x_{ij}, y_j \in \{0, 1\} \quad \text{for } i, j = 1, \ldots, n.$$

As discussed in Chapter 8, the variables y_j describe which stocks j are in the
portfolio ($y_j = 1$ if j is selected in the portfolio, 0 otherwise). For each stock
$i = 1, \ldots, n$, the variable x_{ij} indicates which stock j in the portfolio is most
similar to i ($x_{ij} = 1$ if j is the most similar stock in the portfolio, 0 otherwise).

Once the set of K stocks has been selected, a simple approach to the second
step of portfolio construction is as follows. Assume $j_1, \ldots, j_K$ are the selected
stocks and $C_1, \ldots, C_K$ are the corresponding clusters. That is, C_ℓ is the set of
stocks represented by stock j_ℓ for $\ell = 1, \ldots, K$. Set the weight of each selected
stock j_ℓ, $\ell = 1, \ldots, K$, proportional to the total market capitalization of the

stocks in C_ℓ:

$$x_{j_\ell} := \frac{\sum\limits_{i \in C_\ell} V_i}{\sum\limits_{i=1}^{n} V_i}, \quad \ell = 1, \ldots, K,$$

where V_i is the market capitalization of stock i.

The second step can alternatively be tackled via a linear or a quadratic programming model. The variables in the model are the portfolio weights on the selected stocks. A reasonable objective is to minimize a measure associated with the quality of tracking such as active risk – this would lead to a quadratic programming problem. Alternatively, one can minimize mean absolute deviation and obtain a linear programming problem. The constraints could include bounds on beta, sector exposures, and other attributes to find weights of the selected stocks so that the portfolio is as close as possible to the benchmark.

9.4 Cardinality Constraints

In this section, we present a different approach for tracking a basket of assets, e.g., an index, with a small group of stocks. In contrast to the two-step approach in the previous section, we model the index replication problem in one step, as a mixed integer programming problem with cardinality constraints. For concreteness, consider that case when we want to track a benchmark with a portfolio containing a predetermined maximum number of stocks. Assume $\mathbf{x}_B$ is the vector of holdings in the benchmark, and $\mathbf{x}$ is the vector of holdings in the portfolio. Suppose we want to include at most K stocks in the tracking portfolio. If we have an estimate of the covariance matrix of the universe of stocks in the index, then the problem can be informally stated as follows:

$$
\begin{aligned}
\min \quad & (\mathbf{x} - \mathbf{x}_B)^\mathsf{T} \mathbf{V} (\mathbf{x} - \mathbf{x}_B) \\
\text{s.t.} \quad & \mathbf{1}^\mathsf{T} \mathbf{x} = 1 \\
& \mathbf{x} \geq 0 \\
& x_j > 0 \text{ for at most } K \text{ distinct } j = 1, \ldots, n.
\end{aligned}
\tag{9.1}
$$

In order to model the problem formally, we introduce a new set of binary variables whose role is to model the logical condition of whether each particular stock is included in the portfolio:

$$y_j = \begin{cases} 1 & \text{if } x_j > 0 \\ 0 & \text{otherwise.} \end{cases}$$

The problem (9.1) can be reformulated as the following mixed quadratic program

$$\min \quad (\mathbf{x} - \mathbf{x}_B)\mathbf{V}(\mathbf{x} - \mathbf{x}_B)$$
$$\text{s.t.} \quad \mathbf{1}^\mathsf{T}\mathbf{x} = 1$$
$$x_j \leq y_j \quad \text{for} \quad j = 1, \ldots, n$$
$$\sum_{j=1}^{n} y_j \leq K \tag{9.2}$$
$$\mathbf{x} \geq 0$$
$$y_j \in \{0, 1\} \quad \text{for} \quad j = 1, \ldots, n.$$

Observe that the *linking* constraint $x_j \leq y_j$ in (9.2) is a mathematical way o encoding the logical connection between x_j and y_j: the variable x_j is positive onl' when $y_j = 1$. Furthermore, the constraint $\sum_{j=1}^{n} y_j \leq K$ enforces the condition that at most K of the $\mathbf{x}$ variables are positive.

Consider now a more general mean–variance model with cardinality con straints where we now allow short positions:

$$\min \quad \mathbf{x}^\mathsf{T}\mathbf{V}\mathbf{x}$$
$$\text{s.t.} \quad \boldsymbol{\mu}^\mathsf{T}\mathbf{x} \geq \bar{\mu}$$
$$\mathbf{A}\mathbf{x} = \mathbf{b} \tag{9.3}$$
$$\mathbf{C}\mathbf{x} \geq \mathbf{d}$$
$$x_j \neq 0 \quad \text{for at most } K \text{ distinct} \quad j = 1, \ldots, n.$$

The above approach extends provided that there is a lower bound ℓ_j and a. upper bound u_j on the value of each holding x_j for $j = 1, \ldots, n$. In this cas the cardinality constraint can be formulated via a new set of binary variable y_j, $j = 1, \ldots, n$, together with the linking constraints

$$\ell_j y_j \leq x_j \leq u_j y_j \quad \text{for } j = 1, \ldots, n.$$

The problem (9.3) can be reformulated as the following mixed quadratic program

$$\min \quad \mathbf{x}^\mathsf{T}\mathbf{V}\mathbf{x}$$
$$\text{s.t.} \quad \boldsymbol{\mu}^\mathsf{T}\mathbf{x} \geq \bar{\mu}$$
$$\mathbf{A}\mathbf{x} = \mathbf{b}$$
$$\mathbf{C}\mathbf{x} \geq \mathbf{d}$$
$$\ell_j y_j \leq x_j \leq u_j y_j \quad \text{for } j = 1, \ldots, n \tag{9.4}$$
$$\sum_{j=1}^{n} y_j \leq K$$
$$y_j \in \{0, 1\} \qquad \qquad \text{for} \quad j = 1, \ldots, n.$$

9.5   Minimum Position Constraints

The same kind of linking constraints $\ell_j y_j \leq x_j \leq u_j y_j$ used in the mixed binary programming formulation (9.4) can be used to enforce another common practical consideration: *minimum position constraints*. Although diversification into a broad universe of assets generally has merits, there is a potential downside: some positions may become very small. Too many small positions typically generate higher research and monitoring costs. Consequently, investment managers enforce minimum position constraints. This means that if a stock j is included in the portfolio, then the holding x_j in that stock must surpass a minimum threshold $\ell_j > 0$. Consider a general mean–variance model with minimum position constraints:

$$
\begin{aligned}
\min \quad & \mathbf{x}^\mathsf{T} \mathbf{V} \mathbf{x} \\
\text{s.t.} \quad & \boldsymbol{\mu}^\mathsf{T} \mathbf{x} \geq \bar{\mu} \\
& \mathbf{A}\mathbf{x} = \mathbf{b} \\
& \mathbf{C}\mathbf{x} \geq \mathbf{d} \\
& x_j > 0 \quad \Rightarrow \quad x_j \geq \ell_j \quad \text{for } j = 1, \dots, n.
\end{aligned}
\tag{9.5}
$$

Provided there is an upper bound u_j on the value of each holding x_j, for $j = 1, \dots, n$, and proceeding as above, the problem (9.5) can be reformulated as the following mixed quadratic program:

$$
\begin{aligned}
\min \quad & \mathbf{x}^\mathsf{T} \mathbf{V} \mathbf{x} \\
\text{s.t.} \quad & \boldsymbol{\mu}^\mathsf{T} \mathbf{x} \geq \bar{\mu} \\
& \mathbf{A}\mathbf{x} = \mathbf{b} \\
& \mathbf{C}\mathbf{x} \geq \mathbf{d} \\
& \ell_j y_j \leq x_j \leq u_j y_j \quad \text{for } j = 1, \dots, n \\
& y_j \in \{0, 1\} \quad \quad \quad \text{for } j = 1, \dots, n.
\end{aligned}
\tag{9.6}
$$

9.6 Risk-Parity Portfolios and Clustering

Consider a situation where you are trying to construct a risk-parity portfolio as explained in Section 7.5. We would like to allocate equal risk to a set of assets, but several of them may be very similar. By using the risk-parity strategy directly, you end up overweighting the characteristics those assets share. Instead, you can cluster the assets first, and then allocate risk evenly to each cluster. This is more consistent with the spirit of "risk parity". This first step can be accomplished using the clustering approach suggested in Example 8.2.

9.7 Exercises

Exercise 9.1 In a combinatorial exchange, both buyers and sellers can submit combinatorial bids. Bids are like in the multiple item case, except that the values λ_i can be negative, as can the prices $p_j(\lambda)$, representing selling instead of buying. Note that a single bid can be buying some items while selling other items. Write an integer linear program that will maximize the surplus generated by the combinatorial exchange.

Exercise 9.2   You have $250,000 to invest in the following possible investments.
The cash inflows/outflows are as follows:

	Year 1	Year 2	Year 3	Year 4
Investment 1	−1.00		1.18	
Investment 2		−1.00		1.22
Investment 3			−1.00	1.10
Investment 4	−1.00	0.14	0.14	1.00
Investment 5		−1.00	0.20	1.00

For example, if you invest one dollar in Investment 1 at the beginning of
Year 1, you receive $1.18 at the beginning of Year 3. If you invest in any of these
investments, the required minimum level is $100,000 in each case. Any or all
the available funds at the beginning of a year can be placed in a money market
account that yields 3% per year. Formulate a mixed integer linear program to
maximize the amount of money available at the beginning of Year 4. Solve the
integer program using your favorite solver.

Exercise 9.3   Consider a lockbox problem where c_{ij} is the cost of assigning
region i to a lockbox in region j, for $i, j \in \{1, \ldots, n\}$. Suppose that we wish to
open exactly K lockboxes where K is a given integer, $1 \leq K \leq n$.

(a) Formulate as an integer linear program the problem of opening K lockboxes
so as to minimize the total cost of assigning each region to an open lockbox.
(b) Formulate in two different ways the constraint that regions cannot send
checks to closed lockboxes.
(c) For the following data

$$K = 2 \quad \text{and} \quad (c_{ij}) = \begin{bmatrix} 0 & 4 & 5 & 8 & 2 \\ 4 & 0 & 3 & 4 & 6 \\ 5 & 3 & 0 & 1 & 7 \\ 8 & 4 & 1 & 0 & 4 \\ 2 & 6 & 7 & 4 & 0 \end{bmatrix},$$

compare the linear programming relaxations of your two formulations in
part (b).

Exercise 9.4   You currently own a portfolio of eight stocks. Using the
Markowitz model, you computed the optimal mean–variance portfolio. The
weights of these two portfolios are shown in the following table:

Stock	A	B	C	D	E	F	G	H
Your portfolio	0.12	0.15	0.13	0.10	0.20	0.10	0.12	0.08
M/V portfolio	0.02	0.05	0.25	0.06	0.18	0.10	0.22	0.12

You would like to rebalance your portfolio in order to be closer to the mean–
variance portfolio. To avoid excessively high transaction costs, you decide to

rebalance only three stocks from your portfolio. Let x_i denote the weight of stock i in your rebalanced portfolio. The objective is to minimize the quantity

$$|x_1 - 0.02| + |x_2 - 0.05| + |x_3 - 0.25| + \cdots + |x_8 - 0.12|,$$

which measures how closely the rebalanced portfolio matches the mean–variance portfolio.

Formulate this problem as a mixed integer linear program. Note that you will need to introduce new continuous variables in order to linearize the absolute values and new binary variables in order to impose the constraint that only three stocks are traded.

9.8   Case Study

The goal of this case study is to construct a parsimonious fund that tracks a pre-specified market index.

(1) Choose a stock market index (with at least 25 stocks) to be tracked. Some possible choices are the Dow Jones Industrial Average, the S&P 100, and the Nasdaq 100. If you feel ambitious, you may choose a larger index.

Collect recent historical data over a meaningful horizon. Make sure you include more observations (ideally many more) than the number of stocks in the index. A reasonable choice is a few years (two or three) of weekly data, or a few more (six or seven) of monthly data. For larger indices, you may want to consider daily data.

Use the first 70% of your data for calibrating your model; that is, for parameter estimation, choice of stocks, choice of weights, etc. Use the remaining 30% for out-of-sample testing.

(2) Use some kind of clustering or variable selection approach to choose a small subset of stocks from the investment universe. Complete the process by assigning weights to the selected stocks using the following simple rule: Mimic the weighting method used in the index. For example, for a market-value-weighted index, assign the weight of each selected stock according to the market value of all the stocks that it represents. You can also attempt to replicate various attributes of the market index, such as exposure to particular sectors, industries, etc.

(3) Compare the performance of the constructed fund and that of the actual stock market index. To this end, test the results of your model(s) on out-of-sample data. This is more interesting if done via a rolling-time window.

(4) Construct index funds with different number of stocks. Compare their performance.

(5) Study the effect of rebalancing the index fund by periodical adjustment of the weights of the selected stocks. Try different periods for rebalancing: one week, one month, etc.

(6) Propose some alternative models and compare their results with those of the basic model. You may want to consider combinations of the following possible variations. Be as creative as you wish:

(i) Use a weighted objective function in the stock selection problem.

(ii) Use a second optimization problem to assign the weights. For example you can set the weights to minimize the tracking error (active variance) between the fund and the overall index.

(iii) Match attributes of the index such as beta, and/or exposures to factor such as industries, sectors, etc.

10 Stochastic Programming: Theory and Algorithms

Stochastic optimization is concerned with optimizing decision variables under uncertainty. As an example, Markowitz's mean–variance model can be seen as a stochastic optimization model. Stochastic optimization covers a wide class of models in a variety of disciplines. It is often associated with the terms dynamic programming, stochastic programming, and stochastic control, among others. We devote several chapters to this important and vast topic. This chapter concentrates on *single-period/two-stage* models. This provides a foundation for more general *multi-period/multi-stage* models that will be discussed in Chapters 12 through 16.

10.1 Examples of Stochastic Optimization Models

The next three examples inherently involve making decisions under uncertainty.

Example 10.1 (The newsvendor problem) A vendor purchases a particular commodity to satisfy some demand that occurs later over some time period. The demand D is random. The per-unit ordering cost, back-ordering cost, and holding costs are known to be c, p, and h, respectively. The total cost incurred by the vendor if he purchases x units and the demand turns out to be D is

$$F(x, D) = c \cdot x + p \cdot \max(D - x, 0) + h \cdot \max(x - D, 0).$$

The problem is to decide the order quantity x that minimizes the expected total cost $\mathbb{E}[F(x, D)]$.

Example 10.2 (Utility-based optimization) An investor with endowment W_0 needs to decide how to invest this initial capital over a planning horizon. The investor's preferences for her final wealth W are expressed via a concave utility function $U(W)$. Assume $\mathbf{r}$ is the vector of random returns on the assets that the investor can purchase over the planning horizon. The investor wishes to choose a portfolio $\mathbf{x} \in \mathcal{X}$ that maximizes the expected utility $\mathbb{E}[U(W)]$ of her final wealth W:

$$W = W_0(1 + \mathbf{r}^\mathsf{T}\mathbf{x}).$$

Example 10.3 (Optimal consumption and investment)   An individual may con sume some portion C_0 of her initial endowment W_0 now and invest the remaining capital $W_0 - C_0$ for consumption at a future time. Assume $\mathbf{r}$ is the vector of random returns on the assets in which she can invest her remaining capital $W_0 - C_0$. Investing in a portfolio $\mathbf{x}$ will thus produce a random wealth $W = (W_0 - C_0)(1 + \mathbf{r}^\mathsf{T}\mathbf{x})$. What should her consumption C_0 and investment decision $\mathbf{x} \in \mathcal{X}$ be to maximize her total expected utility

$$U_0(C_0) + \mathbb{E}[U_1(W)].$$

10.2   Two-Stage Stochastic Optimization

Consider the following generic type of optimization problem under uncertainty. At time 0 we need to make a set of decisions $\mathbf{x}$ subject to some constraint set $\mathcal{X}$. Between time 0 and time 1 a random outcome ω is revealed. Our goal is to choose $\mathbf{x}$ to minimize the expectation of some objective function $F(\mathbf{x}, \omega)$ that depends both on $\mathbf{x}$ as well as on the random outcome ω. This generic *stochastic optimization* problem has the following formal formulation:

$$\min_{\mathbf{x}} \quad \mathbb{E}(F(\mathbf{x}, \omega))$$
$$\mathbf{x} \in \mathcal{X}. \tag{10.1}$$

The particular form of $F(\mathbf{x}, \omega)$ may define various types of problems, as we saw in Examples 10.1, 10.2, and 10.3. The function $F(\mathbf{x}, \omega)$ could be more involved. In a problem with *recourse* the function $F(\mathbf{x}, \omega)$ depends on decisions that can be made *after* the uncertainty ω is revealed.

Stochastic optimization with recourse is a refinement of the generic formulation (10.1). In this class of problems a first set of decisions $\mathbf{x}$ must be made *here and now* at time 0. Between time 0 and time 1 a random outcome ω occurs. Then at time 1 we have the opportunity to make a new round of *wait-and-see* decisions $\mathbf{y}(\omega)$ after the random ω is revealed. This leads to a *two-stage stochastic optimization with recourse* problem formally stated as follows:

$$\min_{\mathbf{x}} \quad f(\mathbf{x}) + \mathbb{E}[Q(\mathbf{x}, \omega)]$$
$$\mathbf{x} \in \mathcal{X}. \tag{10.2}$$

The recourse term $Q(\mathbf{x}, \omega)$ depends on the initial set of decisions $\mathbf{x}$ and on the random outcome ω, and it is of the form

$$Q(\mathbf{x}, \omega) := \min_{\mathbf{y}(\omega)} \quad g(\mathbf{y}(\omega), \omega)$$
$$\mathbf{y}(\omega) \in \mathcal{Y}(\mathbf{x}, \omega).$$

The set of decisions $\mathbf{y}(\omega)$ are the *recourse decisions*. They are *adaptive* to the random outcome ω. This means that unlike $\mathbf{x}$ they are allowed to depend on ω.

Example 10.4 (Newsvendor problem revisited)   In this case the total cost is

$$F(x, D) = c \cdot x + p \cdot \max(D - x, 0) + h \cdot \max(x - D, 0).$$

We want to solve

$$\min_{x \geq 0} \ \mathbb{E}[F(x, D)] = \min_{x \geq 0} \ (c \cdot x + \mathbb{E}\left[p \cdot \max(D - x, 0) + h \cdot \max(x - D, 0)\right])$$
$$= \min_{x \geq 0} \ (c \cdot x + \mathbb{E}[Q(x, D)]),$$

where the recourse term $Q(x, D)$ is

$$\begin{aligned}
Q(x, D) := \min_{y, z} \quad & py + hz \\
\text{s.t.} \quad & y \geq D - x \\
& z \geq x - D \\
& y, z \geq 0.
\end{aligned}$$

Note that here the recourse decisions y and z are easy to compute once the demand D and the number of units purchased x are known, namely $y = (D-x)^+$ and $z = (x - D)^+$.

Sometimes it is preferable to consider a more general model obtained by replacing the objective $\mathbb{E}(F(\mathbf{x}, \omega))$ in (10.1) with $\varrho(F(\mathbf{x}, \omega))$ where $\varrho(\cdot)$ is a real-valued function. In particular, it is common to let $\varrho(\cdot)$ be a *risk measure* as illustrated in the following example. We formally define and discuss risk measures in more detail in Chapter 11.

Example 10.5 (Mean–variance revisited)   Let $\mathbf{r}$ denote the vector of random asset returns in an investment universe and let μ and $\mathbf{V}$ denote respectively the expected value and covariance matrix of $\mathbf{r}$. The classic mean–variance model

$$\begin{aligned}
\min_{\mathbf{x}} \quad & \tfrac{1}{2}\gamma \cdot \mathbf{x}^{\mathsf{T}}\mathbf{V}\mathbf{x} - \mu^{\mathsf{T}}\mathbf{x} \\
& \mathbf{x} \in \mathcal{X}
\end{aligned}$$

can be written as the stochastic optimization problem

$$\begin{aligned}
\min_{\mathbf{x}} \quad & \varrho(\mathbf{r}^{\mathsf{T}}\mathbf{x}) \\
& \mathbf{x} \in \mathcal{X}
\end{aligned}$$

for the risk measure ϱ defined by

$$\varrho(Z) = \tfrac{1}{2}\gamma \cdot \sigma^2(Z) - \mathbb{E}(Z).$$

10.3   Linear Two-Stage Stochastic Programming

A *linear two-stage stochastic program* is a problem of the form

$$\begin{aligned}
\min_{\mathbf{x}} \quad & \mathbf{c}^{\mathsf{T}}\mathbf{x} + \mathbb{E}[Q(\mathbf{x}, \omega)] \\
\text{s.t.} \quad & \mathbf{A}\mathbf{x} = \mathbf{b} \\
& \mathbf{x} \geq \mathbf{0},
\end{aligned}$$

where the recourse term $Q(\mathbf{x}, \omega)$ is the value of another linear program:

$$Q(\mathbf{x}, \omega) := \min_{\mathbf{y}} \quad \mathbf{q}(\omega)^\mathsf{T} \mathbf{y}(\omega)$$
$$\text{s.t.} \quad \mathbf{T}(\omega)\mathbf{x} + \mathbf{W}(\omega)\mathbf{y}(\omega) = \mathbf{h}(\omega)$$
$$\mathbf{y}(\omega) \geq 0.$$

Here the parameters $\mathbf{q}(\omega), \mathbf{T}(\omega), \mathbf{W}(\omega), \mathbf{h}(\omega)$ are random, and ω represents a random outcome $\omega \in \Omega$ that is revealed between stage 0 and stage 1. It is customary and convenient to think of ω itself as the array of random parameters $\omega = (\mathbf{q}, \mathbf{T}, \mathbf{W}, \mathbf{h})$. The vector $\mathbf{x}$ represents the first-stage decisions. These must be made without knowing the random draw ω. The vector $\mathbf{y}(\omega)$ denotes the second-stage decisions. These may depend on the random draw ω. To ease notation, this type of problem is often written in the following form:

$$\min \quad \mathbb{E}[\mathbf{c}^\mathsf{T}\mathbf{x} + \mathbf{q}^\mathsf{T}\mathbf{y}]$$
$$\text{s.t.} \quad \mathbf{A}\mathbf{x} = \mathbf{b}$$
$$\mathbf{T}\mathbf{x} + \mathbf{W}\mathbf{y} = \mathbf{d} \tag{10.3}$$
$$\mathbf{x} \geq 0$$
$$\mathbf{y} \geq 0,$$

but we should keep in mind that the tuple of uncertain parameters $\omega = (\mathbf{q}, \mathbf{T}, \mathbf{W}, \mathbf{h})$ is revealed between time 0 and time 1 and the recourse variables $\mathbf{y}$ may depend on this outcome.

10.4 Scenario Optimization

Scenario optimization is a computational approach to stochastic optimization. The gist of this approach is to assume a discrete distribution for the random outcome. More precisely, assume the set of possible random outcomes is a discrete probability space $\Omega = \{\omega_1, \ldots, \omega_S\}$, with probability distribution $p_k = \mathbb{P}(\omega_k)$, $k = 1, \ldots, S$. The elements in Ω are the possible realizations or *scenarios*

$$\omega_k = (\mathbf{q}_k, \mathbf{T}_k, \mathbf{W}_k, \mathbf{h}_k)$$

of the stochastic components of the model.

Under this assumption, the stochastic optimization problem (10.3) can be written as the following *deterministic equivalent*:

$$\min_{\mathbf{x}, \mathbf{y}_k} \quad \mathbf{c}^\mathsf{T}\mathbf{x} + \sum_{k=1}^{S} p_k (\mathbf{q}_k^\mathsf{T} \mathbf{y}_k)$$
$$\text{s.t.} \quad \mathbf{A}\mathbf{x} = \mathbf{b} \tag{10.4}$$
$$\mathbf{T}_k \mathbf{x} + \mathbf{W}_k \mathbf{y}_k = \mathbf{h}_k \qquad \text{for } k = 1, \ldots, S$$
$$\mathbf{x} \geq 0$$
$$\mathbf{y}_k \geq 0 \qquad \text{for } k = 1, \ldots, S.$$

The deterministic equivalent problem has S copies of the second-stage decision variables and hence can be significantly larger than the original problem before we considered the uncertainty of the parameters. Fortunately, the constraint matrix has a very special sparsity structure that can be exploited as we explain in Section 10.5 below.

Example 10.6 (Newsvendor problem revisited)   Suppose the demand D in the newsvendor problem has a discrete distribution. More precisely, suppose the scenarios for the demand D are $D_1, \ldots, D_S$ and $\mathbb{P}(D = D_i) = p_i$ for $i = 1, \ldots, S$. Hence the newsvendor problem $\min\limits_{x \geq 0} \mathbb{E}[F(x, D)]$ where

$$F(x, D) = c \cdot x + p \cdot \max(D - x, 0) + h \cdot \max(x - D, 0)$$

has the following deterministic equivalent:

$$\min_{x, \mathbf{y}, \mathbf{z}} \quad c \cdot x + p \cdot \sum_{k=1}^{S} p_i y_i + h \cdot \sum_{k=1}^{S} p_i z_i$$

$$\begin{aligned}
\text{s.t.} \quad & y_i \geq D_i - x, \quad i = 1, \ldots, S \\
& z_i \geq x - D_i, \quad i = 1, \ldots, S \\
& x \geq 0 \\
& \mathbf{y}, \mathbf{z} \geq 0.
\end{aligned}$$

10.5   *The L-Shaped Method

The constraint matrix of (10.4) has the following form:

$$\begin{bmatrix} \mathbf{A} & & & \\ \mathbf{T}_1 & \mathbf{W}_1 & & \\ \vdots & & \ddots & \\ \mathbf{T}_S & & & \mathbf{W}_S \end{bmatrix}$$

Observe that the blocks $\mathbf{W}_1, \ldots, \mathbf{W}_S$ of the constraint matrix are only interrelated through the blocks $\mathbf{T}_1, \ldots, \mathbf{T}_S$ which correspond to the first-stage decisions. In other words, once the first-stage decisions $\mathbf{x}$ have been fixed, (10.4) decomposes into S independent linear programs. The *Benders decomposition method* is an algorithm that takes advantage of this type of structure. This method is also called the *L-shaped method* in the stochastic programming literature. The idea behind this method is to solve a "master problem" involving only the variables $\mathbf{x}$ and a series of independent "recourse problems" each involving a different vector of variables $\mathbf{y}_k$. The master problem and recourse problems are linear programs. The size of these linear programs is much smaller than the size of the full model (10.4). The recourse problems are solved for a given vector $\mathbf{x}$ and their solutions are used to generate inequalities that are added to the master problem. Solving the new master problem produces a new $\mathbf{x}$ and the process is repeated. More specifically, let us write (10.4) as

$$\min_{\mathbf{x}} \quad \mathbf{c}^{\mathsf{T}}\mathbf{x} + P_1(\mathbf{x}) + \cdots + P_S(\mathbf{x})$$
$$\text{s.t.} \quad \mathbf{A}\mathbf{x} = \mathbf{b} \tag{10.5}$$
$$\mathbf{x} \geq \mathbf{0}$$

where

$$P_k(\mathbf{x}) \quad = \quad \min_{\mathbf{y}_k} \quad p_k \mathbf{q}_k^{\mathsf{T}} \mathbf{y}_k$$
$$\text{s.t.} \quad \mathbf{W}_k \mathbf{y}_k = \mathbf{h}_k - \mathbf{T}_k \mathbf{x} \tag{10.6}$$
$$\mathbf{y}_k \geq \mathbf{0}$$

for $k = 1, \ldots, S$. The dual of the linear program (10.6) is:

$$P_k(\mathbf{x}) \quad = \quad \max_{\mathbf{u}_k} \quad (\mathbf{h}_k - \mathbf{T}_k\mathbf{x})^{\mathsf{T}}\mathbf{u}_k$$
$$\text{s.t.} \quad \mathbf{W}_k^{\mathsf{T}}\mathbf{u}_k \leq p_k \mathbf{q}_k. \tag{10.7}$$

For simplicity, assume (10.7) is feasible, which is the case of interest in many applications. The recourse linear program (10.6) will be solved for a sequence of vectors $\mathbf{x}^i$, for $i = 0, 1, 2, \ldots$. The initial vector $\mathbf{x}^0$ can be obtained by solving

$$\min_{\mathbf{x}} \quad \mathbf{c}^{\mathsf{T}}\mathbf{x}$$
$$\text{s.t.} \quad \mathbf{A}\mathbf{x} = \mathbf{b} \tag{10.8}$$
$$\mathbf{x} \geq \mathbf{0}.$$

For a given vector $\mathbf{x}^i$, two possibilities can occur for the recourse linear program (10.6): either (10.6) has an optimal solution or it is infeasible.

If (10.6) has an optimal solution $\mathbf{y}_k^i$, and $\mathbf{u}_k^i$ is the corresponding optimal dual solution, then (10.7) implies that

$$P_k(\mathbf{x}) \geq (\mathbf{u}_k^i)^{\mathsf{T}}(\mathbf{T}_k\mathbf{x}^i - \mathbf{T}_k\mathbf{x}) + P_k(\mathbf{x}^i).$$

This inequality, which is called an *optimality cut*, can be added to the current master linear program. Initially, the master linear program is just (10.8).

If (10.6) is infeasible, then the dual problem is unbounded. Let $\mathbf{u}_k^i$ be a direction where (10.7) is unbounded, that is, $(\mathbf{h}_k - \mathbf{T}_k\mathbf{x}^i)^{\mathsf{T}}\mathbf{u}_k^i > 0$ and $\mathbf{W}_k^{\mathsf{T}}\mathbf{u}_k^i \leq p_k \mathbf{q}_k$. Since we are only interested in first-stage decisions $\mathbf{x}$ that lead to feasible second-stage decisions $\mathbf{y}_k$, the following *feasibility cut* can be added to the current master linear program:

$$(\mathbf{u}_k^i)^{\mathsf{T}}(\mathbf{h}_k - \mathbf{T}_k\mathbf{x}) \leq 0.$$

After solving the recourse problems (10.6) for each k, we have the following upper bound on the optimal value of (10.4):

$$UB = \mathbf{c}^{\mathsf{T}}\mathbf{x}^i + P_1(\mathbf{x}^i) + \cdots + P_S(\mathbf{x}^i),$$

where we set $P_k(\mathbf{x}^i) = +\infty$ if the corresponding recourse problem is infeasible.

Adding all the optimality and feasibility cuts found so far (for $j = 0, \ldots, i$) to the master linear program, we obtain:

$$\min_{\mathbf{x}, z_1, \ldots, z_S} \quad \mathbf{c}^\mathsf{T}\mathbf{x} + \sum_{k=1}^{S} z_k$$

$$\text{s.t.} \quad \mathbf{A}\mathbf{x} = \mathbf{b}$$

$$(\mathbf{u}_k^j)^\mathsf{T}(\mathbf{T}_k\mathbf{x}^j - \mathbf{T}_k\mathbf{x}) + P_k(\mathbf{x}^j) \le z_k \quad \text{for some pairs } (j, k)$$

$$(\mathbf{u}_k^j)^\mathsf{T}(\mathbf{h}_k - \mathbf{T}_k\mathbf{x}) \le 0 \qquad \text{for the remaining pairs } (j, k)$$

$$\mathbf{x} \ge \mathbf{0}.$$

Denoting by $\mathbf{x}^{i+1}, z_1^{i+1}, \ldots, z_S^{i+1}$ an optimal solution to this linear program, we get a lower bound on the optimal value of (10.4):

$$LB = \mathbf{c}^\mathsf{T}\mathbf{x}^{i+1} + z_1^{i+1} + \cdots + z_S^{i+1}.$$

The Benders decomposition method alternately solves the recourse problems (10.6) and the master linear program with new optimality and feasibility cuts added at each iteration until the gap between the upper bound UB and the lower bound LB falls below a given threshold. It can shown that $UB - LB$ converges to zero and indeed reaches zero after finitely many iterations. For details see Birge and Louveaux (1997, chapter 5).

10.6   Exercises

Exercise 10.1   Consider the utility-based portfolio optimization described in Example 10.2. Suppose $\mathbf{r}$ has a multivariate normal distribution and the investor has logarithmic utility function

$$U(W) = \log(W).$$

Suppose $\mathcal{X}$ is a convex set. Prove that the utility maximization problem

$$\max \quad \mathbb{E}(U(W))$$
$$\text{s.t.} \quad W = W_0 \cdot (1 + \mathbf{r}^\mathsf{T}\mathbf{x})$$
$$\mathbf{x} \in \mathcal{X}$$

is equivalent to the mean–variance problem

$$\min \quad \tfrac{1}{2}\tilde{\gamma} \cdot \mathbf{x}^\mathsf{T}\mathbf{V}\mathbf{x} - \boldsymbol{\mu}^\mathsf{T}\mathbf{x}$$
$$\mathbf{x} \in \mathcal{X}$$

for some suitable level of risk aversion $\tilde{\gamma}$.

Exercise 10.2   Repeat the above exercise when the investor has power utility function

$$U(W) = \frac{1}{1 - \gamma} \cdot W^{1-\gamma}$$

for some risk-aversion constant $\gamma > 0$, $\gamma \ne 1$.

Exercise 10.3   Consider the newsvendor problem described in Example 10.1
Suppose the demand D has a continuous cumulative distribution function Φ
that is, $\Phi(x) = \mathbb{P}(D \le x)$. Show that the solution to the newsvendor problem

$$\min_{x \ge 0} \; (c \cdot x + \mathbb{E}\left[p \cdot \max(D - x, 0) + h \cdot \max(x - D, 0)\right])$$

is

$$x^* = \Phi^{-1}\left(\frac{p - c}{p + h}\right).$$

Exercise 10.4   The purpose of this exercise is to formalize the optimality cu
described in Section 10.5. For $\mathbf{x} = \mathbf{x}^i$ assume (10.6) has an optimal solution $\mathbf{y}_k^i$
and let $\mathbf{u}_k^i$ be the corresponding optimal dual solution.

(a) Show that $P_k(\mathbf{x}^i) = (\mathbf{u}_k^i)^{\mathsf{T}}(\mathbf{h}_k - \mathbf{T}_k \mathbf{x}^i)$.
(b) Show that $P_k(\mathbf{x}) \ge (\mathbf{u}_k^i)^{\mathsf{T}}(\mathbf{h}_k - \mathbf{T}_k \mathbf{x})$ for all $\mathbf{x}$.
(c) Conclude that $P_k(\mathbf{x}) \ge (\mathbf{u}_k^i)^{\mathsf{T}}(\mathbf{T}_k \mathbf{x}^i - \mathbf{T}_k \mathbf{x}) + P_k(\mathbf{x}^i)$ for all $\mathbf{x}$.

11 Stochastic Programming Models: Risk Measures

This chapter discusses several popular risk measures. In particular, we introduce two widely used risk measures, *value at risk* and its refinement *conditional value at risk*. We show that the problem of finding a portfolio that minimizes conditional value at risk is amenable to stochastic programming techniques.

11.1 Risk Measures

In the classical Markowitz model, variance (equivalently standard deviation) is used as a measure of risk. This measure of risk is relatively easy to compute, and, as we have seen in Chapter 6, leads to a quadratic programming model when we are interested in finding efficient portfolios.

  As we illustrate below, variance has some shortcomings as a measure of risk. This has motivated the introduction of other risk measures.

Dispersion Measures

Let r denote the (random) return of an asset. The variance

$$\sigma^2 = \operatorname{var}(r) = \mathbb{E}((r - \mu)^2)$$

is a measure of *dispersion* of the distribution of r. Another dispersion measure is the *mean absolute deviation* (MAD) favored by Konno and Yamazaki (1991):

$$\mathbb{E}(|r - \mu|).$$

For the special case of normally distributed returns, the mean absolute deviation and the standard deviation are equivalent. Indeed, the following property is a straightforward exercise in probability:

Proposition 11.1 *If $r \sim N(\mu, \sigma^2)$ then $\mathbb{E}(|r - \mu|) = \sqrt{2/\pi}\,\sigma$.*

  A major difference between mean absolute deviation and standard deviation is their sensitivity to outliers. The mean absolute deviation is more robust to outliers. When the distribution of joint returns is represented via a set of scenarios, the computation of efficient portfolios for the mean absolute deviation can be formulated as a linear program. This offers an alternative with potential advantages as we will show next. Suppose the investment universe has n assets with (random) returns $r_1, r_2, \ldots, r_n$. Let $\mu_j = \mathbb{E}(r_j)$, $j = 1, \ldots, n$.

Recall the portfolio optimization problem that finds the minimum-variance portfolio among a set of portfolios $\mathcal{X}$:

$$\min_{\mathbf{x}} \quad \mathrm{var}(\mathbf{r}^\mathsf{T}\mathbf{x}) = \mathbb{E}\left(\left[(\mathbf{r} - \boldsymbol{\mu})^\mathsf{T}\mathbf{x}\right]^2 \right)$$

$$\text{s.t.} \quad \mathbf{x} \in \mathcal{X}.$$

Consider now the model obtained by using instead the mean absolute deviation as a measure of risk:

$$\min_{\mathbf{x}} \quad \mathbb{E}\left(|(\mathbf{r} - \boldsymbol{\mu})^\mathsf{T}\mathbf{x}| \right)$$

$$\text{s.t.} \quad \mathbf{x} \in \mathcal{X}. \tag{11.1}$$

Not only does the computation of efficient portfolios based on formulation (11.1) involve solving a linear program as opposed to a quadratic program, but also the linear program solves the problem directly over the set of scenarios thereby circumventing the estimation of the covariance matrix.

Mean Absolute Deviation via Scenario Optimization

Assume the possible scenarios for the vector of returns $\mathbf{r} = \begin{bmatrix} r_1 & \cdots & r_n \end{bmatrix}^\mathsf{T}$ are

$$\mathbf{r}^k = \begin{bmatrix} r_1^k & \cdots & r_n^k \end{bmatrix}^\mathsf{T}, \quad k = 1, \ldots, S,$$

and scenario k occurs with probability p_k, $k = 1, \ldots, S$. Then we can write the above mean absolute deviation model (11.1) as

$$\min_{\mathbf{x},\mathbf{w}} \quad \sum_{k=1}^{S} p_k w_k$$

$$\text{s.t.} \quad w_k = |(\mathbf{r}^k - \boldsymbol{\mu})^\mathsf{T}\mathbf{x}| \quad \text{for} \quad k = 1, \ldots, S$$

$$\mathbf{x} \in \mathcal{X}.$$

We now turn this formulation into a linear program as follows:

$$\min_{\mathbf{x},\mathbf{w}} \quad \sum_{k=1}^{S} p_k w_k$$

$$\text{s.t.} \quad w_k \geq (\mathbf{r}^k - \boldsymbol{\mu})^\mathsf{T}\mathbf{x} \quad \text{for} \quad k = 1, \ldots, S$$

$$w_k \geq -(\mathbf{r}^k - \boldsymbol{\mu})^\mathsf{T}\mathbf{x} \quad \text{for} \quad k = 1, \ldots, S$$

$$\mathbf{x} \in \mathcal{X}.$$

Note that, because $p_k > 0$ for $k = 1, \ldots, S$ and the objective is minimized, w_k in an optimal solution satisfies at equality the constraint with the larger right-hand side, that is, $w_k = |(\mathbf{r}^k - \boldsymbol{\mu})^\mathsf{T}\mathbf{x}|$.

Downside Risk Measures

Dispersion measures, such as variance and mean absolute deviation, measure the degree of uncertainty in the random return. These measures treat both positive and negative deviations from the mean as equally risky. In particular, these types of measures are blind to skewed distributions.

We will next discuss two popular downside risk measures: *value at risk* and *conditional value at risk*. Value at risk (VaR) was first introduced by a team at J.P. Morgan and made available through RiskMetrics. VaR is used by many financial institutions to track and report the market risk exposure of their trading portfolios.

VaR is a measure of the worst possible loss that a portfolio may sustain with a pre-specified likelihood. For that reason, VaR is generally measured in dollar terms, instead of percentage units. The formal definition is as follows. Assume that Y is a (random) loss function, and $\alpha \in (0,1)$ is a confidence level (typically 99%, 95%, or 90%). The α value at risk of Y is the $(1-\alpha)$ quantile of Y; that is, the value γ such that

$$\mathbb{P}(Y \geq \gamma) = 1 - \alpha.$$

We shall denote this value by $\mathrm{VaR}_\alpha(Y)$.

The value at risk has the following interpretation. Given a loss function Y and a confidence level $\alpha \in (0,1)$, the loss Y will exceed γ with probability $(1-\alpha)$. In the special case when the loss function is normally distributed, it is easy to compute VaR via well-known quantiles of the normal distribution.

Example 11.2 If $Y \sim N(\mu, \sigma^2)$ then

$$\mathrm{VaR}_{0.95}(Y) = \mu + 1.645\sigma, \quad \mathrm{VaR}_{0.99}(Y) = \mu + 2.33\sigma.$$

When Y has a discrete distribution, VaR can be computed by sorting the values of Y as detailed in the following example.

Example 11.3 Assume there are S possible scenarios for the loss Y:

$$\mathbb{P}(Y = y_k) = p_k, \quad k = 1, \ldots, S,$$

where

$$y_1 \leq y_2 \leq \cdots \leq y_S.$$

Then

$$\mathrm{VaR}_\alpha(Y) = y_K,$$

where K is the smallest index such that

$$\sum_{i=K}^{S} p_i \geq 1 - \alpha.$$

In spite of its wide popularity, VaR is known to have the following two major shortcomings (see the exercises at the end of this chapter):

- VaR is not "subadditive": The VaR of two positions combined may be greater than the sum of the VaR of each, meaning that diversification can actually increase VaR.
- VaR does not distinguish loss size beyond the VaR threshold.

These deficiencies of VaR led Artzner et al. (1999) to propose the following formal set of properties that a reasonable risk measure $\rho(Y)$ of a loss function Y should satisfy:

- Monotonicity: If $Y \geq 0$ then $\rho(Y) \geq 0$.
- Subadditivity: $\rho(Y + Z) \leq \rho(Y) + \rho(Z)$.
- Positive homogeneity: For $c > 0$, $\rho(cY) = c\rho(Y)$.
- Translational invariance: For any $c \in \mathbb{R}$, $\rho(Y + c) = \rho(Y) + c$.

A risk measure is *coherent* if it satisfies the above four properties. Neither standard deviation nor VaR are coherent. However, there is a modification of VaR that is coherent, namely the *conditional value at risk* introduced by Rockafellar and Uryasev (2000). Conditional value at risk (CVaR) is also known as expected tail loss.

CVaR can be motivated as follows. Since $\text{VaR}_\alpha(Y)$ is the most we can lose with probability α, it is equivalent to saying that with probability $(1 - \alpha)$ the loss Y will be at least $\text{VaR}_\alpha(Y)$. CVaR is the answer to the following question: What should we expect the value of that loss to be? More precisely, CVaR is defined as follows. Given a loss function Y and confidence level $\alpha \in (0, 1)$, the conditional value at risk is the expected loss Y, conditional on this loss being at least $\text{VaR}_\alpha(Y)$:

$$\mathbb{E}(Y | Y \geq \text{VaR}_\alpha(Y)).$$

We shall denote this expected value as $\text{CVaR}_\alpha(Y)$.

Again, in the special case when the loss function is normally distributed, it is easy to compute CVaR by using properties of the quantiles and expected tails of the normal distribution.

Example 11.4 If $Y \sim N(\mu, \sigma^2)$ then

$$\text{CVaR}_{0.95}(Y) = \mu + 2.06\sigma, \ \ \text{CVaR}_{0.99}(Y) = \mu + 2.67\sigma.$$

When Y has a discrete distribution, CVaR can be computed by sorting the values of Y.

Example 11.5 Assume Y takes values y_k, $k = 1, \ldots, S$, in S possible scenarios:

$$\mathbb{P}(Y = y_k) = p_k, \ k = 1, \ldots, S,$$

where

$$y_1 \leq y_2 \leq \cdots \leq y_S.$$

Then

$$\mathrm{CVaR}_\alpha(Y) = \frac{1}{1 - \alpha} \sum_{i=K}^{S} p_i y_i,$$

where K is the smallest index such that

$$\sum_{i=K}^{S} p_i = 1 - \alpha.$$

Note that here we may have to split the probability p_K.

11.2   A Key Property of CVaR

We next present a key property of CVaR that makes it possible to solve portfolio optimization problems with CVaR via convex optimization.

Proposition 11.6 *Assume Y is a loss function. Then for $\alpha \in (0, 1)$*

$$\mathrm{CVaR}_\alpha(Y) = \min_\gamma \left(\gamma + \frac{1}{1 - \alpha} \mathbb{E}[\max(Y - \gamma, 0)] \right).$$

Furthermore, the optimal solution (i.e., the minimizer) $\bar{\gamma}$ of this problem is $\mathrm{VaR}_\alpha(Y)$.

As consequence of Proposition 11.6, it follows that CVaR is subadditive. Indeed, CVaR is a coherent risk measure (see exercises at the end of the chapter for details). Another consequence of Proposition 11.6 is that CVaR can be computed as the following linear two-stage stochastic program:

$$\mathrm{CVaR}_\alpha(Y) = \min_\gamma \left[\gamma + \mathbb{E} \left(Q(\gamma, Y) \right) \right],$$

where

$$Q(\gamma, Y) := \min_z \ \frac{1}{1 - \alpha} \cdot z$$
$$\text{s.t.} \quad z \geq Y - \gamma$$
$$z \geq 0.$$

More concisely

$$\mathrm{CVaR}_\alpha(Y) = \min_{\gamma, z} \ \ \gamma + \frac{1}{1 - \alpha} \cdot \mathbb{E}(z)$$
$$\text{s.t.} \quad z \geq Y - \gamma$$
$$z \geq 0.$$

In this formulation the first-stage and second-stage decision variables are γ and z respectively. Notice that z is adapted to the random outcome Y. In the particular case when Y is discrete and takes values y_k, for $k = 1, \ldots, S$, in S possible scenarios (not necessarily sorted):

$$\mathbb{P}(Y = y_k) = p_k, \ k = 1, \ldots, S,$$

we obtain the following linear programming formulation for $\mathrm{CVaR}_\alpha(Y)$.

Variables:

$$\gamma, z_1, \ldots, z_S.$$

Linear programming formulation of CVaR:

$$
\begin{aligned}
\min_{\gamma, \mathbf{z}} \quad & \gamma + \frac{1}{1 - \alpha} \sum_{k=1}^{S} p_k z_k \\
\text{s.t.} \quad & z_k \geq y_k - \gamma, && \text{for } k = 1, \ldots, S \\
& z_k \geq 0, && \text{for } k = 1, \ldots, S.
\end{aligned}
$$

An advantage of this formulation is that it allows us to minimize the CVaR of a portfolio via linear programming as we next explain.

11.3   Portfolio Optimization with CVaR

The discussion in this section is based on Andersson et al. (2001). This study uses CVaR for measuring and controlling the credit risk of a portfolio of bonds. The loss function of interest is the loss due to credit risk; that is, the loss that the portfolio may suffer due to default or credit migration in its positions. This type of loss function is characterized by having a large likelihood of no loss and a small likelihood of a substantial loss. The loss distribution is heavily skewed. In this case, standard mean–variance analysis to characterize market risk is inadequate. VaR and CVaR are more appropriate criteria for minimizing portfolio credit risk.

Distribution of Future Values for One Single Bond

Consider a risky bond and a fixed time horizon, e.g., one year. The future value of the bond depends on the forward curve that applies to its coupon payments. The forward curve in turn depends on the current rating of the bond. The *benchmark* future value of the bond is the future value of the bond if there is no change on its credit rating. However, in the event of credit migration, the future value of the bond may differ from the benchmark value. In particular, if the credit rating deteriorates, the coupon payments will be subject to higher discount values and the future value of the bond will be lower than its benchmark value.

For a concrete illustration, suppose the one-year forward interest curves for the S&P credit ratings are as follows:

Category	Year 1	Year 2	Year 3	Year 4
AAA	0.036	0.0417	0.0473	0.0512
AA	0.0365	0.0422	0.0478	0.0517
A	0.0372	0.0432	0.0493	0.0532
BBB	0.041	0.0467	0.0525	0.0563
BB	0.0555	0.0602	0.0678	0.0727
B	0.0605	0.0702	0.0803	0.0852
CCC	0.1505	0.1502	0.1403	0.1352

Suppose the probabilities of credit rating migration for A, BBB, and B in one year are as follows:

Initial rating	Rating at year end							
	AAA	AA	A	BBB	BB	B	CCC	Default
A	0.09%	2.27%	91.05%	5.52%	0.74%	0.26%	0.01%	0.06%
BBB	0.02%	0.33%	5.95%	86.93%	5.30%	1.17%	0.12%	0.18%
B	0.00%	0.11%	0.24%	0.43%	6.48%	83.47%	4.07%	5.20%

Assuming a 50% recovery rate in default, the possible future values of a five-year, 6% BBB bond with face value 100 are as follows:

Year-end rating	Future value	Probability
AAA	109.352908	0.0002
AA	109.1723709	0.0033
A	108.6429921	0.0595
BBB	107.5309439	0.8693
BB	102.0063855	0.053
B	98.08591318	0.0117
CCC	83.6257912	0.0012
Default	50	0.0018

For example, for BBB rated bonds, the future value 107.5309439 was obtained as follows:

$$107.5309439 = 6 \cdot \left(1 + \frac{1}{1.041} + \frac{1}{1.0467^2} + \frac{1}{1.0525^3} + \frac{1}{1.0563^4}\right) + 100 \cdot \frac{1}{1.0563^4}.$$

Credit Risk Optimization for a Portfolio of Bonds

Now suppose we construct a portfolio of risky bonds. Assume there are n risky bonds and let x_j be the percentage of portfolio invested in bond j. Then the loss function of our portfolio is

$$Y(\mathbf{x}) := (\mathbf{b} - \boldsymbol{\omega})^\mathsf{T}\mathbf{x} = \sum_{j=1}^{n}(b_j - \omega_j)x_j,$$

where each b_j is the future bond value of bond j with no credit migration, and ω_j is the (random) possible future bond value of bond j with credit migration. Suppose we want to select the portfolio in the constraint set $\mathcal{X}$ with minimum CVaR_α. In other words, we want to solve

$$\min_{\mathbf{x}} \quad \text{CVaR}_\alpha(Y(\mathbf{x}))$$
$$\mathbf{x} \in \mathcal{X}.$$

Suppose that the possible scenarios for the vector of future bond values $\boldsymbol{\omega} = \begin{bmatrix} \omega_1 & \cdots & \omega_n \end{bmatrix}^\mathsf{T}$ are

$$\boldsymbol{\omega}^k = \begin{bmatrix} \omega_1^k & \cdots & \omega_n^k \end{bmatrix}^\mathsf{T}, \quad k = 1, \ldots, S.$$

Then by Proposition 11.6, this problem has the following formulation:

$$\min_{\gamma,\mathbf{x},\mathbf{z}} \quad \gamma + \frac{1}{1-\alpha}\sum_{k=1}^{S} p_k z_k$$

$$\text{s.t.} \quad z_k \geq (\mathbf{b} - \boldsymbol{\omega}^k)^\mathsf{T}\mathbf{x} - \gamma, \ k = 1, \ldots, S \qquad (11.2)$$

$$z_k \geq 0, \ k = 1, \ldots, S$$
$$\mathbf{x} \in \mathcal{X}$$
$$\gamma \quad \text{free.}$$

If the constraint set $\mathcal{X}$ is defined by linear constraints, then (11.2) is a linear program.

Scenario Generation in the Credit-Risk Example

When there is a single bond, the probability distribution of the possible future values of the bond depends on the probability of credit migration and the bond value in each of these scenarios. For instance, for the S&P ratings, the scenarios correspond to the ratings AAA, AA, A, BBB, BB, B, CCC, and default. The likelihood of each of these scenarios is given by the migration matrix, which estimates the probability of migrating from one rating to the others over a specified time period.

The discrete distribution readily yields the set of possible scenarios for the bond. Scenarios can also be generated via *normal sampling* as Figure 11.1 suggests (assuming we are working with a BB bond).

More precisely, normal sampling goes as follows:

- compute Z-scores associated with the probabilities of each of the scenarios,

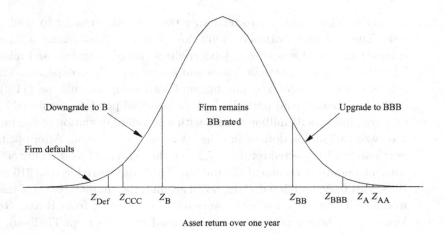

Figure 11.1

- draw samples from a standard normal distribution,
- use the Z-scores to determine the sampled scenario.

Some interesting challenges arise in the scenario generation when we need to work with multiple bonds. Under the simple assumption that the credit migrations are statistically independent, we can generate scenarios via discrete sampling or independent normal sampling. Notice that, although discrete, the joint probability distribution for a set of ten or more bonds is extremely large. Hence it is generally impractical to exhaustively generate the entire set of scenarios.

In the case when credit migrations are correlated, the scenario generation problem becomes more interesting. In this case a possible solution is to use correlated normal sampling. That is, draw samples from a correlated joint multivariate random variable. Then map each of the components in the random sample to a possible credit rating of the bonds. Some statistical packages, like the statistics toolbox in MATLAB, readily provide routines to sample from correlated multivariate normal variables. However, it is easy to generate correlated normal sampling from independent normal sampling. More precisely, to sample from a general n-dimensional normal distribution $N(\boldsymbol{\mu}, \mathbf{V})$, proceed as follows:

- Let $\mathbf{L}\mathbf{L}^{\mathsf{T}} = \mathbf{V}$ be the Cholesky factorization of the covariance matrix $\mathbf{V}$.
- Sample n standard independent normals $x_i \sim N(0, 1)$.
- Put $\mathbf{y} = \boldsymbol{\mu} + \mathbf{L}\mathbf{x}$.
- The resulting variable $\mathbf{y}$ has the desired distribution $\mathbf{y} \sim N(\boldsymbol{\mu}, \mathbf{V})$.

Solution of a Real-World Bond Example

Andersson et al. (2001) considered a portfolio of 197 bonds from 29 different countries with a market value of \$8.8 billion and duration of approximately five years. Their goal was to rebalance the portfolio in order to minimize credit risk. The one-year portfolio credit loss was generated using a Monte Carlo simulation: 20,000 scenarios of joint credit states of obligators and related losses. The distribution of portfolio losses had a long fat tail, as expected. The author rebalanced the portfolio by minimizing CVaR using formulation (11.2). For $\alpha =$ 99%, the original bond portfolio had an expected portfolio return of 7.26%. The expected loss was 95 million dollars with a standard deviation of 232 million. The VaR was 1.03 billion dollars and the CVaR was 1.32 billion. After optimizing the portfolio (with expected return of 7.26%), the expected loss was only 5000 dollars with a standard deviation of 152 million. The VaR was reduced to 210 million and the CVaR to 263 million dollars. So all around, the characteristics of the portfolio were much improved. Positions were reduced in bonds from Brazil, Russia, and Venezuela, whereas positions were increased in bonds from Thailand, Malaysia and Chile. Positions in bonds from Colombia, Poland, and Mexico remained high and each accounted for about 5% of the optimized portfolio.

11.4   Notes

As early as the 1970s and 1980s, some major financial institutions developed internal systems for risk management. The best known of these systems was RiskMetrics developed in the late 1980s at J.P. Morgan when chairman Dennis Weatherstone requested his staff provide a "4:15pm" daily one-page report measuring and explaining the risks and potential losses over the next 24 hours across the bank's entire portfolio. The RiskMetrics system featured and popularized the use of value at risk as a risk measure. The interest in a rigorous treatment of risk measures led a set of prominent scholars to develop a formal theory of *coherent measures of risk* in a landmark paper (Artzner et al., 1999). Conditional value at risk is one of the most popular coherent measures of risk.

11.5   Exercises

Exercise 11.1   Construct a counterexample to show that VaR is not necessarily subadditive. In other words, construct two loss functions Y, Z and a confidence level $\alpha \in (0, 1)$ so that

$$\text{VaR}_\alpha(Y + Z) > \text{VaR}_\alpha(Y) + \text{VaR}_\alpha(Z).$$

Exercise 11.2   Show that a coherent risk measure ρ satisfies: if $X \geq Y$ then $\rho(X) \geq \rho(Y)$.

Hint: Use the monotonicity and subadditivity of coherent risk measures.

Exercise 11.3   The purpose of this exercise is to prove Proposition 11.6 under some additional assumptions. Assume Y is a continuous loss function with density $f(y)$ and $\alpha \in (0, 1)$. Let $g(\gamma)$ be defined as

$$g(\gamma) := \gamma + \frac{1}{1-\alpha} \int_{-\infty}^{\infty} \max(y - \gamma, 0)\, f(y)\, dy.$$

(a) Show that

$$g'(\gamma) = 1 - \frac{1}{1-\alpha} \int_{\gamma}^{\infty} f(y)\, dy.$$

(b) Let $\bar{\gamma}$ be the minimizer of the optimization problem

$$\min_{\gamma} g(\gamma).$$

Use part (a) to show that

$$\mathrm{VaR}_{\alpha}(Y) = \bar{\gamma}.$$

(c) Let $\bar{\gamma}$ be as in part (b). Use parts (a) and (b) to show that

$$\mathrm{CVaR}_{\alpha}(Y) = g(\bar{\gamma}).$$

Exercise 11.4 Use Proposition 11.6 to prove that CVaR is subadditive; that is,

$$\mathrm{CVaR}_{\alpha}(Y + W) \le \mathrm{CVaR}_{\alpha}(Y) + \mathrm{CVaR}_{\alpha}(W)$$

for any two loss functions Y, W and $\alpha \in (0, 1)$.

Exercise 11.5

(a) Suppose a loss function Y has normal distribution with mean μ and variance σ^2; that is, $Y \sim N(\mu, \sigma^2)$. For $\alpha \in (0, 1)$ determine both $\mathrm{VaR}_{\alpha}(Y)$ and $\mathrm{CVaR}_{\alpha}(Y)$ in terms of the standard normal cumulative function

$$\Phi(x) = \int_{-\infty}^{x} \frac{1}{\sqrt{2\pi}} e^{-t^2/2} dt.$$

In particular, show that

$$\mathrm{VaR}_{0.95}(Y) = \mu + 1.645\sigma, \quad \mathrm{CVaR}_{0.95}(Y) = \mu + 2.06\sigma.$$

(b) Suppose a loss function Y has lognormal distribution with log mean μ and log variance σ^2, that is, $\log(Y) \sim N(\mu, \sigma^2)$. For $\alpha \in (0, 1)$ determine both $\mathrm{VaR}_{\alpha}(Y)$ and $\mathrm{CVaR}_{\alpha}(Y)$ in terms of the standard normal cumulative function

$$\Phi(x) = \int_{-\infty}^{x} \frac{1}{\sqrt{2\pi}} e^{-t^2/2} dt.$$

Exercise 11.6 The Excel spreadsheet "Exercise 11.6 Twelve Portfolios" provides scenarios (based on historical data) for the joint annual returns of 12 industry portfolios in the US market.

(a) Any given portfolio attains its "worst" return in some scenario. For example, a "NoDur" portfolio attained its worst return in 1931 whereas a "Hlth" portfolio attained its worst return in scenario 1929.

Find a long-only portfolio that maximizes the worst possible return.

(b) Find the long-only portfolio that maximizes the expected return while ensuring that its worst possible return is no lower than 2% below what you found in part (a).

Exercise 11.7　The Excel spreadsheet "Exercise 11.7 Three Bonds" provides (hypothetical) discrete distribution for the future value of three different bonds. In each case, the first value is the "benchmark" future value in the case of no credit quality change. For simplicity assume the three bonds have the same current value, say $100.

The spreadsheet also has the joint distribution for the future values of the three bonds $(64 = 4 * 4 * 4$ possible scenarios) assuming that the probability distributions are independent.

(a) For a given bond portfolio $\mathbf{x} = \begin{bmatrix} x_1 & x_2 & x_3 \end{bmatrix}^\mathsf{T}$, use the loss function discussed in Section 11.3,

$$Y = (\mathbf{b} - \boldsymbol{\omega})^\mathsf{T}\mathbf{x} = \sum_{j=1}^{3}(b_j - \omega_j)x_j,$$

where b_j is the benchmark future value of bond j, and ω_j is the random future value of bond j for $j = 1, 2, 3$. Determine the $\mathrm{VaR}_{0.95}$ and $\mathrm{CVaR}_{0.9}$ values of the portfolio $\mathbf{x} = \begin{bmatrix} 0.4 & 0.1 & 0.5 \end{bmatrix}^\mathsf{T}$.

(b) Set up a CVaR optimization model to find a portfolio $\mathbf{x} = \begin{bmatrix} x_1 & x_2 & x_3 \end{bmatrix}$ with $x_j \geq 0$, $x_1 + x_2 + x_3 = 1$, with the same benchmark future value as that of the portfolio $\begin{bmatrix} 0.4 & 0.1 & 0.5 \end{bmatrix}^\mathsf{T}$, and with minimum $\mathrm{CVaR}_{0.95}$. What is the $\mathrm{VaR}_{0.95}$ value of the optimal portfolio? What is the optimal portfolio?

Exercise 11.8　The Excel spreadsheet "Exercise 11.8 Six Bonds" provides a (hypothetical) discrete distribution for the annual return of six different bonds.

The return of each individual bond only has two possible values. The first value is the "benchmark" return in the case of no credit quality downgrade. The second value is the return in the case of credit downgrade.

The Excel file also contains the joint distribution for the returns of the six bonds $(64 = 2^6$ possible scenarios) assuming that the probability distributions are independent.

Suppose you currently have $100 and need to fulfill an obligation of $115 in a year. You intend to invest the $100 in the six bonds to try to meet the $115 obligation. You realize that because of the bonds' credit risks you may not be able to meet this financial goal in some scenarios. In such a case you will need extra money to cover the *shortfall*; that is, the difference between the obligation and whatever you can cover. For instance, if the bond portfolio in a year is worth $105 < $115, then the shortfall would be $10 = $115 - $105. On the other hand, if the bond portfolio in a year is $120 > $115, then the shortfall would be zero.

(a) Formulate a linear programming model to determine how much should be invested in each bond so that the expected value of the shortfall a year from now is minimized. Assume the portfolio must be long-only.

Formulate your linear programming model in Excel and solve it.

What is the composition of the optimal portfolio, that is, the amount of money in bond i, for $i = 1, \ldots, 6$?

What is the expected value of the shortfall?

What is the value of the worst (largest) possible shortfall?

(b) Define the *loss function* of your bond portfolio as the value of the liability minus the value of your portfolio.

  (i) Compute the numerical value of $CVaR_{0.95}$ of this loss function if your portfolio is equally divided among the six bonds.

  (ii) Find the long-only portfolio of these six bonds that minimizes $CVaR_{0.95}$. What is the optimal $CVaR_{0.95}$ value of this portfolio?

Part III

Multi-Period Models

12 Multi-Period Models: Simple Examples

The next few chapters will be devoted to *multi-period models*. Unlike the single-period models we have discussed so far, multi-period models incorporate the dynamic nature inherent when decisions are made at different stages. The decisions to be made at each particular stage can adapt to information collected in previous stages. In this chapter we discuss the following fundamental multi-period models: the Kelly criterion for repeated gambles, dynamic portfolio optimization with myopic strategies, and optimal scheduling of trades to control execution costs. The strong assumptions made in these models allow us to solve them with relatively simple techniques. Subsequent chapters will introduce more involved techniques for multi-period models, namely *dynamic programming* and *stochastic programming*. We will rely on these techniques to tackle more elaborate financial optimization models.

12.1 The Kelly Criterion

The Kelly criterion is a classical formula derived to maximize the average rate of growth of a gambler's fortune in a sequence of bets published in the landmark paper by Kelly (1956). The formula has appeal among some investment professionals. In particular, there are claims that many successful investors, including Edward Thorp, Warren Buffett, and Bill Gross, use Kelly-like methods. The popular book *Fortune's Formula* (Poundstone, 2005) gives a non-technical and engaging description of the Kelly criterion and its role in gambling and investing.

The Kelly formula can be explained as follows. Suppose a gambler can enter a bet with two possible outcomes: lose the entire amount bet or win the amount bet. Assume the probability of winning is p. Suppose a gambler starts with some initial wealth W_0 and can take this gamble repeatedly. What fraction of her current wealth should she bet each time?

To answer this question, let W_n be the gambler's wealth after n gambles. The rate of growth of the gambler's fortune is

$$g = \frac{1}{n} \log \frac{W_n}{W_0}.$$

Suppose the gambler bets a fraction f of her current wealth each time. Then

$$W_n = (1 + f)^k (1 - f)^{n-k} W_0,$$

where k is the number of wins (out of the n gambles).

Therefore we get

$$g = \frac{k}{n} \log(1 + f) + \frac{n - k}{n} \log(1 - f).$$

Taking expectations, we get

$$\mathbb{E}(g) = p \log(1 + f) + (1 - p) \log(1 - f).$$

This function of f attains its maximum at $f = 2p - 1$. That is, by betting the fraction $2p - 1$ each time, the gambler maximizes the *expected* growth rate g.

The above reasoning and formula can be extended to gambles where the payoff does not necessarily match the amount bet, and to gambles with a non-binary outcome. We can indeed see the Kelly criterion as a special case of the dynamic portfolio optimization model discussed next.

12.2   Dynamic Portfolio Optimization

Our first dynamic portfolio optimization model concerns the expected utility of final wealth, assuming the portfolio can be rebalanced at intermediate steps. More specifically, suppose that an investor starts at $t = 0$ with an initial endowment W_0. At times $t = 0, \ldots, T - 1$ the investor invests her wealth W_t in a portfolio of one risk-free asset and any number of risky assets. The investor's goal is to maximize the utility of terminal wealth $U(W_T)$ at time T for a suitable utility function $U(W)$. We next describe a formal model for this dynamic portfolio optimization problem. To that end, we introduce the following convenient notation:

- R_{t+1} = gross random returns of the assets in period $[t, t + 1]$,
    $R_{f,t+1}$ = gross risk-free return in period $[t, t + 1]$,
    $R_{p,t+1}$ = gross random return of the investor's portfolio in period $[t, t + 1]$.
- Decision variables: $\mathbf{x}_t$ = holdings (percentages) in the risky assets at time t.
- Inter-temporal constraints (also known as law of motion):

$$W_{t+1} = W_t \cdot R_{p,t+1}$$

$$= W_t \cdot \left(R_{f,t+1} + (R_{t+1} - R_{f,t+1}\mathbf{1})^\mathsf{T} \mathbf{x}_t \right), \quad t = 0, \ldots, T - 1.$$

- Objective: max $\mathbb{E}[U(W_T)]$.

12.2.1 Optimality of Myopic Policies

The solution of a multi-period model typically requires dynamic or stochastic programming techniques that we will cover in later chapters. However, under suitable assumptions the above model is sufficiently simple that we can solve it directly. Observe that the final accumulated wealth at stage T is

$$W_T = W_0 \cdot \prod_{t=1}^{T} R_{p,t}.$$

We will primarily consider the class of constant relative risk-aversion utilities given by the *power utility* $U(W) = W^{1-\gamma}/(1 - \gamma)$ with risk aversion $\gamma > 0$, $\gamma \neq 1$ and the *logarithmic utility* $U(W) = \log(W)$. Observe that $\log(W) = \lim_{\gamma \to 1} W^{1-\gamma}/(1 - \gamma)$.

For a power utility $U(W) = W^{1-\gamma}/(1 - \gamma)$ we get

$$(1 - \gamma) \cdot U(W_T) = W_T^{1-\gamma} = W_0^{1-\gamma} \cdot \prod_{t=1}^{T} R_{p,t}^{1-\gamma}.$$

For logarithmic utility $U(W) = \log(W)$, we get

$$U(W_T) = \log(W_0) + \sum_{t=1}^{T} \log(R_{p,t}).$$

From these expressions for the utility of final wealth, we can readily reach the following conclusions:

- If the risk-free return $R_{f,t} = R_f$ is the same for $t = 1, \dots, T$ and the risky returns R_t are independent for $t = 1, \dots, T$, then each $\mathbf{x}_t$ is the solution to a single-period problem. In other words, a *myopic* policy is optimal.
- If the risk-free return $R_{f,t} = R_f$ is the same for $t = 1, \dots, T$ and the risky returns R_t are independent and identically distributed (i.i.d.) for $t = 1, \dots, T$, then all $\mathbf{x}_t$ are the same.
- For $U(W) = \log(W)$, a myopic policy is optimal regardless of the distribution of the returns $R_t, R_{f,t}$, for $t = 1, \dots, T$.

The above conclusions can be related to the following two classical puzzles of finance (Kritzman, 2002):

- Half stocks all the time or all stocks half the time?
- Time diversification: Is it true that lengthening the investment horizon reduces risk?

The first puzzle can be more precisely stated as follows. Suppose there is a risky asset "stocks" with expected return μ and standard deviation σ, and a risk-free asset with return $r < \mu$. Consider the following two possible dynamic investment strategies:

- balanced strategy $(50\%, 50\%)$ in every period;
- switching strategy $(100\%, 0\%)$ half of the time and $(0\%, 100\%)$ the other half.

Suppose the risk-free return is constant, the stock returns are i.i.d. across time and an investor has a power utility. Which of the two strategies is preferable?

The second puzzle can be more precisely stated as follows. Suppose there is a risky asset "stocks" with expected return μ and standard deviation σ, and risk-free asset with return $r < \mu$. Suppose the risk-free return is constant, the stock returns are i.i.d. across time, and an investor has a power utility with risk aversion $\gamma > 0$. Is it true that if the investor's investment horizon is $T \gg 1$ then she should initially hold a higher percentage of her portfolio in stocks than if her horizon is $T = 1$? In his book, Kritzman (2002) expertly discusses the answer to these puzzles as well as a few others.

12.2.2   An Example Where a Myopic Policy Is Not Optimal

It is important to understand that the above conclusion concerning the optimality of myopic policies relies on strong assumptions on the asset returns, the investor's utility, and the fact that the investor only receives an initial endowment at time 0 and maximizes her utility of wealth at the final time T. The following simple example illustrates a case when a myopic policy is not optimal.

- Consider a three-stage (two-period) problem, i.e., $T = 2$.
- At $t = 0$ we can invest in a one-period bond or a two-period zero-coupon bond.
- At $t = 0$ we know that the risk-free interest rate is $R_{f,1} = 1.1$.
- At $t = 0$ we know that at $t = 1$ the risk-free interest rate will be as follows:

$$R_{f,2} = \begin{cases} 1.12 \text{ with prob } 1/2, \\ 1.08 \text{ with prob } 1/2. \end{cases}$$

- The one-period bond is a contract that can be entered at $t = 0$ and delivers $\$1.1$ at time $t = 1$ for each dollar invested.
- The two-period bond is a contract that can be entered at $t = 0$ and delivers $\$1.2096$ at time $t = 2$ for each dollar invested. The value of the two-period bond at time $t = 1$ depends on the risk-free interest rate at that time. Its value V at $t = 1$ per dollar invested is the 1.2096 final payout discounted at the applicable rate:

$$V = \begin{cases} \dfrac{1.2096}{1.12} = 1.08 \text{ with prob } 1/2, \\ \dfrac{1.2096}{1.08} = 1.12 \text{ with prob } 1/2. \end{cases}$$

- A myopic investor is one with investment horizon $T = 1$; a long-term investor is one with investment horizon $T = 2$.
- What would a risk-averse myopic investor do at time 0?
- What would a risk-averse long-term investor do at time 0?

12.3   Execution Costs

The efficient management of trading costs is a challenge to all institutional investors. These costs are associated with commissions, bid–ask spreads, opportunity costs of waiting, and the price impact of trading. These types of costs generally have a substantial impact on investment performance. For instance, a classical study of Pérold (1988) shows that a hypothetical "paper" portfolio constructed according to Value Line rankings outperforms the market by almost 20% during the period from 1965 to 1986. However, the actual portfolio – the Value Line Fund – outperformed the market by only 2.5% per year. The difference between these figures arises from execution costs. This "implementation shortfall" is surprisingly large and highlights the importance of execution-cost control, particularly for institutional investors whose trades often constitute a large fraction of the average trading volume of many stocks. A common and intuitive practice is to spread large execution orders over a period of time, e.g., a few hours or days. This scheduling of trades aims to find a balance between two conflicting objectives: on the one hand, fast execution generates large market impact and consequently generates large costs. These costs are related to liquidity as well as leakage of information. On the other hand, delayed execution reduces market impact but comes at the expense of greater uncertainty and opportunity risk. Suitable models of price dynamics and market impact lead to insightful conclusions on the tradeoff faced between these extremes.

One of the first formal models for trade execution was proposed by Bertsimas and Lo (1998) and is based on dynamic programming techniques. As we will see, their model shows that, under suitable conditions, the naive strategy of dividing a large order equally across the trading period minimizes expected trading cost. However, the model concentrates on expected cost only and does not take into consideration the risk (variance) of trading costs. This led Almgren and Chriss (2000) to propose a model that finds an optimal tradeoff between expected cost and risk. We next discuss the Almgren–Chriss trade execution model in detail. In its original form, this model can be presented in a relatively simple conceptual framework. Furthermore, the Almgren–Chriss model underlies several more elaborate execution models developed over the last few years.

12.3.1   Almgren–Chriss Trade Execution Model

We first introduce formal definitions associated with a trading strategy and price dynamics for the execution of a sell program for a single security. The definitions and model for a buy program are similar.

Assume we hold a block of X units of a security that needs to be liquidated by time T. Divide the time interval $[0, T]$ into N intervals of length $\tau := T/N$, and define the discrete times $t_k = k\tau$, for $k = 0, 1, \ldots, N$.

Define a *trading trajectory* as a vector

$$\mathbf{x} = \begin{bmatrix} x_0 & x_1 & \cdots & x_N \end{bmatrix}^{\mathsf{T}}.$$

Here x_k = the number of units that we plan to hold at time t_k. We have *boundary conditions* associated with our initial holding, i.e., $x_0 = X$, and liquidation at time T, i.e., $x_N = 0$.

The trading trajectory implies a *trade list*

$$\mathbf{y} = \begin{bmatrix} y_1 & \cdots & y_N \end{bmatrix}^{\mathsf{T}},$$

where $y_k = x_{k-1} - x_k$. Each y_k is the number of units sold in the time interval $[t_{k-1}, t_k]$.

An *execution trading strategy* is a rule for determining the trade size y_k given the information available at time t_{k-1}.

We also need a model for the price dynamics of the security. Assume the initial security price (at time 0) is S_0. The security price evolves according to two exogenous factors, volatility and drift, and one endogenous factor, market impact. Volatility and drift are assumed to be the result of market forces that occur independently of our trading. On the other hand, as market participants begin to detect the volume we are selling, they naturally adjust their bids downward. We distinguish two types of market impact: *temporary* and *permanent*. Temporary impact is the change in price in a single time interval due to the imbalance between supply and demand occurring as a result of our trading. Permanent impact is the equilibrium change in price due to our trading that lasts for the entire life of our liquidation.

Assume the security price evolves according to a discrete arithmetic random walk in addition to a term that accounts for permanent impact. The security price at time t_k is given by

$$S_k = S_{k-1} + \sigma \tau^{1/2} \xi_k - \tau g \left(\frac{y_k}{\tau} \right)$$

for $k = 1, \ldots, N$. Here σ represents the asset volatility, $\xi_k \sim N(0, 1)$, and the permanent impact $g(v)$ depends on the average rate of trading $v = y_k / \tau$ during the interval $[t_{k-1}, t_k]$.

We next incorporate the temporary market impact. The intuition is that a trader that liquidates y_k units during the interval $[t_{k-1}, t_k]$ may see the price decrease as a result of limited liquidity. We assume that this effect is short-lived and in particular liquidity returns after each time interval. To model this impact we incorporate a price impact function $h(v)$ that affects the actual price per share received for trade y_k:

$$\tilde{S}_k = S_{k-1} - h \left(\frac{y_k}{\tau} \right).$$

However, the effect of $h(v)$ does not appear in the next market price S_k.

Given the above trading model, we can compute the execution cost resulting from trading along a certain trajectory. The *captured value* of a trading trajectory is the total revenue obtained after liquidation. Some straightforward calculations show that this equals

$$\sum_{k=1}^{N} \tilde{S}_k y_k = S_0 X + \sum_{k=1}^{N} \left(\sigma \tau^{1/2} \xi_k - \tau g \left(\frac{y_k}{\tau} \right) \right) x_k - \sum_{k=1}^{N} y_k h \left(\frac{y_k}{\tau} \right). \qquad (12.1)$$

The *total cost of trading* or *implementation shortfall* is the difference between the initial book value of the position and the captured value:

$$C(\mathbf{x}) = S_0 X - \sum_{k=1}^{N} \tilde{S}_k y_k = \sum_{k=1}^{N} y_k h \left(\frac{y_k}{\tau} \right) - \sum_{k=1}^{N} \left(\sigma \tau^{1/2} \xi_k - \tau g \left(\frac{y_k}{\tau} \right) \right) x_k.$$

Consequently, given the above price dynamics, it follows that the expected shortfall $\mathbb{E}(\mathbf{x})$ and variance of shortfall $V(\mathbf{x})$ are respectively

$$\mathbb{E}(\mathbf{x}) := \mathbb{E}(C(\mathbf{x})) = \sum_{k=1}^{N} \left(y_k h \left(\frac{y_k}{\tau} \right) + \tau g \left(\frac{y_k}{\tau} \right) \right)$$

and

$$V(\mathbf{x}) = \sigma^2 \sum_{k=1}^{N} \tau x_k^2.$$

The units of $\mathbb{E}(\mathbf{x})$ are dollars, and the units of $V(\mathbf{x})$ are dollars squared.

For simplicity of exposition, we will make the following assumptions:

- the temporary impact function is linear $h(v) = \eta v$;
- there is no permanent impact;
- $\tau = 1$.

It is possible to extend the model and results when these assumptions are relaxed.

Under the above assumptions, the expected shortfall is

$$\mathbb{E}(\mathbf{x}) = \sum_{k=1}^{N} y_k h(y_k) = \eta \sum_{k=1}^{N} y_k^2.$$

Consider the problem of finding the trading trajectory that minimizes expected shortfall:

$$\min_{\mathbf{y}} \quad \eta \sum_{k=1}^{N} y_k^2$$

$$\text{s.t.} \quad \sum_{k=1}^{N} y_k = X.$$

It is easy to see that the solution to this problem is the equally divided trade list

$$y_k = \frac{X}{N}, \quad k = 1, \dots, N.$$

This corresponds to the *linear* trajectory

$$x_k = \frac{N-k}{N}X, \quad k = 1, \ldots, N.$$

This linear trajectory has a natural connection to the so-called volume-weighted average price (VWAP) strategy. The VWAP over the trading period $[0, T]$ defined as

$$\text{VWAP} := \frac{\sum\limits_{k=1}^{N} V_k S_k}{\sum\limits_{k=1}^{N} V_k} = \sum\limits_{k=1}^{N} u_k S_k.$$

Here V_k stands for the volume traded during the kth time interval $[t_{k-1}, t_k]$ and u_k stands for the percentage of daily volume traded during the same interval.

The VWAP strategy trades in proportion to the traded volume during a interval, i.e., it is the following strategy:

$$y_k := u_k X.$$

It is easy to see that the VWAP strategy minimizes expected shortfall when the temporary impact function is linear in the fraction of total volume traded, i.e. when $h(y_k) = \eta \cdot (y_k/V_k)$.

Observe that the linear trajectory

$$x_k = \frac{N-k}{N}X, \quad k = 1, \ldots, N$$

has expected shortfall

$$\mathbb{E}(\mathbf{x}) = \eta \frac{X^2}{N}$$

and variance

$$V(\mathbf{x}) = \frac{(N-1)(2N-1)}{6N} \sigma^2 X^2.$$

Consider the extreme urgency strategy that liquidates the entire position during the first period:

$$y_1 = X, \quad y_2 = \cdots = y_N = 0, \quad x_1 = \cdots = x_N = 0.$$

This trajectory has variance zero and expected shortfall ηX^2.

12.3.2   Efficient Frontier of Optimal Execution

The two execution strategies above suggest that we consider a tradeoff between the two objectives $\mathbb{E}(\mathbf{x})$ and $V(\mathbf{x})$. In analogy to Markowitz's mean–variance framework, Almgren and Chriss (2000) define an execution strategy to be *efficient* if no other strategy has both lower expected shortfall and lower variance.

Just like in the mean–variance context, there are several equivalent formulations for efficient execution strategies. A computationally convenient formulation is:

$$\min_{\mathbf{x}} \quad \mathbb{E}(\mathbf{x}) + \lambda V(\mathbf{x})$$
$$\text{s.t.} \quad x_0 = X \tag{12.2}$$
$$x_N = 0$$

for some risk-aversion parameter $\lambda > 0$. For convenience, put $U(\mathbf{x}) := \mathbb{E}(\mathbf{x}) + \lambda V(\mathbf{x})$. We can think of U as a "disutility" function. Using the expressions for expected value and variance of shortfall, we obtain

$$U(\mathbf{x}) = \eta \sum_{k=1}^{N} (x_k - x_{k-1})^2 + \lambda \sigma^2 \sum_{k=1}^{N} x_k^2.$$

In particular, $U(\mathbf{x}) = \mathbb{E}(\mathbf{x}) + \lambda V(\mathbf{x})$ is a convex quadratic function. The optimality conditions for (12.2) are

$$\frac{\partial U}{\partial x_k}(\mathbf{x}) = 2(\lambda \sigma^2 + 2\eta)x_k - 2\eta(x_{k-1} + x_{k+1}) = 0,$$

for $k = 1, 2, \ldots, N-1$, together with the boundary conditions

$$x_0 = X, \quad x_N = 0.$$

The latter system of equations can be written as

$$\begin{bmatrix} 2 + \lambda\sigma^2/\eta & -1 & 0 & \cdots & & 0 \\ -1 & 2 + \lambda\sigma^2/\eta & -1 & \cdots & & 0 \\ \vdots & \ddots & \ddots & \ddots & & \vdots \\ 0 & \cdots & -1 & 2 + \lambda\sigma^2/\eta & -1 \\ 0 & \cdots & & 0 & -1 & 2 + \lambda\sigma^2/\eta \end{bmatrix} \begin{bmatrix} x_1 \\ x_2 \\ \vdots \\ x_{N-1} \end{bmatrix} = \begin{bmatrix} X \\ 0 \\ \vdots \\ 0 \end{bmatrix}$$

and $x_0 = X, x_N = 0$.

For N large, the discussion below shows that the solution to this system of equations is approximately

$$x_j = \frac{\sinh(\kappa(N-j))}{\sinh(\kappa N)} \cdot X, \quad j = 0, 1, \ldots, N,$$

where κ is the *urgency* parameter,

$$\kappa := \sqrt{\frac{\lambda\sigma^2}{\eta}}.$$

Efficient Frontier in Continuous Time

We can extend the results from the previous section to the continuous-time setting. The mathematics is a bit cleaner and more elegant. This is an idealized model where we assume that the trade can be executed continuously over the time interval $[0, T]$.

In this case we need to determine a continuous trading trajectory $x(t)$, t [0, T], with boundary conditions $x(0) = X$, $x(T) = 0$. We extend the securi price dynamics to the continuous-time setting. Again, for simplicity we sha assume that there is only temporary impact. The price dynamics for the mark price is an arithmetic Brownian motion

$$S(t) = S(0) + \sigma B(t),$$

and the actual execution price received at time t is

$$\tilde{S}(t) = S(t) - \eta y(t),$$

where $y(t) := -\dot{x}(t)$ is the rate of execution at time t.

From properties of the stochastic integral, we get the following expression f the execution shortfall:

$$C(x) = XS(0) - \int_0^T \tilde{S}(t)y(t)dt = \eta \int_0^T \dot{x}(t)^2 dt - \sigma \int_0^T x(t)dB(t).$$

Therefore the expected shortfall $\mathbb{E}(x)$ and variance of shortfall $V(x)$ are follows:

$$\mathbb{E}(x) = \eta \int_0^T \dot{x}(t)^2 dt, \quad V(x) = \sigma^2 \int_0^T x(t)^2 dt.$$

An efficient execution trajectory is the solution to the following problem:

$$\min_{x(t)} \quad \int_0^T (\eta \dot{x}(t)^2 + \lambda \sigma^2 x(t)^2)dt$$
$$\text{s.t.} \quad x(0) = X$$
$$x(T) = 0.$$

This is a problem in *calculus of variations*. The *Euler equation* for this prol lem (see formula (A.1) in Section A.3) yields the following ordinary differenti equation:

$$\ddot{x}(t) = \frac{\lambda \sigma^2}{\eta} \cdot x(t),$$

with boundary conditions

$$x(0) = X, \quad x(T) = 0.$$

The solution to this differential equation is

$$x(t) = \frac{\sinh(\kappa(T - t))}{\sinh(\kappa T)} \cdot X, \quad t \in [0, T],$$

where κ is an "urgency" parameter, defined by

$$\kappa := \sqrt{\frac{\lambda \sigma^2}{\eta}}.$$

The parameter κ has the following nice interpretation. The reciprocal $\theta := 1/\kappa$ measured in units of time and can be interpreted as the "half-life" of the trade

More precisely, when $T \to \infty$, the trade is reduced by a factor of $e = 2.71828\ldots$ by time θ.

It is insightful to verify the units of the various parameters of our model. The units of σ^2, η, and λ are as follows:

$$\sigma^2 : \frac{\text{currency}^2}{\text{volume}^2 \cdot \text{time}}$$

$$\eta : \frac{\text{currency} \cdot \text{time}}{\text{volume}^2}$$

$$\lambda : \frac{1}{\text{currency}}.$$

Recall that the urgency parameter is

$$\kappa = \sqrt{\frac{\lambda \sigma^2}{\eta}}.$$

Therefore, the units of $\theta = 1/\kappa$ are indeed units of time.

Multiple-Security Portfolios

The previous execution model and results can be extended to the case when we need to liquidate a whole portfolio $X = \begin{bmatrix} X_1 & \cdots & X_m \end{bmatrix}^T$ of m securities. In this case, the *trading trajectory* is a sequence of m-dimensional vectors $\mathbf{x}_k = \begin{bmatrix} x_{1k} & \cdots & x_{mk} \end{bmatrix}^T$, *for* $k = 0, \ldots, N$. The trade list is also a sequence of m-dimensional vectors $\mathbf{y}_k = \mathbf{x}_{k-1} - \mathbf{x}_k$, $k = 1, \ldots, N$.

For simplicity we shall assume that there is only a linear temporary impact and $\tau = 1$. Hence the security prices S_k follow a multi-dimensional random walk:

$$S_k = S_{k-1} + \xi_k.$$

Here $\xi_k \sim N(0, \Sigma)$, where Σ is the covariance matrix of the m security prices. We assume Σ to be symmetric and positive definite.

The prices actually received are

$$\tilde{S}_k = S_k - H\mathbf{y}_k,$$

where H is symmetric and positive semidefinite.

Proceeding as before, we get the following expressions for expected shortfall and variance of shortfall respectively:

$$\mathbb{E}(\mathbf{x}) = \sum_{k=1}^{N} \mathbf{y}_k^T H \mathbf{y}_k = \sum_{k=1}^{N} (\mathbf{x}_k - \mathbf{x}_{k-1})^T H (\mathbf{x}_k - \mathbf{x}_{k-1})$$

and

$$V(\mathbf{x}) = \sum_{k=1}^{N} \mathbf{x}_k^T \Sigma \mathbf{x}_k.$$

Now the set of efficient trading strategies is characterized by the solutions to the quadratic program:

$$\min_{\mathbf{x}} \quad \mathbb{E}(\mathbf{x}) + \lambda V(\mathbf{x})$$
$$\text{s.t.} \quad \mathbf{x}_0 = X$$
$$\mathbf{x}_N = 0.$$

This is again a convex quadratic optimization problem. Its solution is

$$
\begin{bmatrix} \mathbf{x}_1 \\ \mathbf{x}_2 \\ \vdots \\ \mathbf{x}_{N-1} \end{bmatrix}
=
\begin{bmatrix}
2H + \lambda\Sigma & -H & 0 & \cdots & 0 \\
-H & 2H + \lambda\Sigma & -H & \cdots & 0 \\
\vdots & & \ddots & \ddots & \vdots \\
0 & \cdots & -H & 2H + \lambda\Sigma & -H \\
0 & \cdots & 0 & -H & 2H + \lambda\Sigma
\end{bmatrix}^{-1}
\begin{bmatrix} HX \\ 0 \\ \vdots \\ 0 \\ 0 \end{bmatrix}
$$

and $\mathbf{x}_0 = X$, $\mathbf{x}_N = \mathbf{0}$.

Unlike for the one-security model, the above solution may not necessarily satisfy the monotonicity constraints $\mathbf{x}_k \leq \mathbf{x}_{k-1}$, with $k = 1, \ldots, N$. This means that the above strategy may have trades that are "buys" at intermediate steps even though the execution is meant to liquidate a vector of positions. If this possibility is not desirable, we can introduce the constraints $\mathbf{x}_k \leq \mathbf{x}_{k-1}$, for $k = 1, \ldots, N$, into the optimization problem for efficient trajectories. The resulting model no longer has a closed-form solution but it is still a quadratic program. For particularly large portfolios, the size of the quadratic programs poses an interesting computational challenge.

Adaptive Strategies

The models described above in discrete and continuous time assume *static* trajectories. That is, the trajectories do not respond to changes during execution. It is conceivable that an adaptive strategy that depends on the initial portion of the trajectory could do better. In order to solve these kinds of optimization problems we need to rely on *dynamic programming* techniques. These are generally more challenging optimization problems. The next chapter introduces this powerful technique.

12.3.3 Trade Execution Models in Practice

A trade execution model used by an institutional investor typically includes other bells and whistles that we have not discussed, such as short-term alpha, spread, permanent impact, and temporary impact. These additional features can be incorporated in the model discussed in Section 12.3.1. An important empirical observation is that, instead of a linear market impact, other forms of market impact such as $1/2$ or $3/5$ powers of volume traded appear to be more appropriate. In these cases the optimal execution problem is no longer a quadratic program but it is still a convex program.

The estimation of market impact is a challenging practical problem. One difficulty is the need for data at the execution level. Furthermore, even when such data are available, market impact is not directly observable. Instead, we can only observe the total realized impact, which includes permanent and temporary impact as well as some random noise. Almgren et al. (2005) estimate the market impact using linear regression, based on a large dataset of US equity brokerage executions from Citigroup. The model is used to calibrate a version of the Almgren–Chriss model with nonlinear temporary costs and is validated with out-of-sample backtesting.

12.4   Exercises

Exercise 12.1   Show that the Kelly criterion can be seen as a special case of the dynamic portfolio optimization problem with the logarithmic utility of the final wealth and a peculiar pair of risk-free and risky assets. Conclude that if the payoff for each dollar invested is b instead of 1, then the optimal fraction to bet is

$$\frac{(b+1)p - 1}{b}.$$

Exercise 12.2   Consider the following variation of the Kelly criterion. Suppose at each betting round the gambler can bet on *two independent gambles*. For each of the two gambles the following applies: if a gambler bets one dollar, then she wins one dollar with probability p and loses the dollar she bet with probability $1 - p$.

Suppose the gambler starts with some initial wealth W_0 and repeatedly bets on the above two gambles. Determine the fractions f_1, f_2 of wealth that she should bet at each round in each of the two gambles to maximize the average growth rate of her wealth.

Exercise 12.3   Recall the example described in Section 12.2.2. You may choose to prove the statements below formally or to verify them numerically. For the latter, you may use the Excel spreadsheet "Exercise 12.3 Two Periods".

(a) Suppose a myopic investor is risk-neutral; that is, her objective is to maximize $\mathbb{E}(W_1)$ by investing her initial wealth in a long-only portfolio composed of the one-period and the two-period bonds. Show that the investor would be indifferent between the one- and the two-period bonds. In other words, any long-only portfolio is optimal for the investor.

(b) Suppose a myopic investor has power utility with risk-aversion parameter $\gamma > 0$; that is, her objective is to maximize $\mathbb{E}\left(W_1^{1-\gamma}/(1 - \gamma)\right)$. Show that the investor would prefer to hold her entire portfolio in the one-period bond.

(c) Suppose a long-term investor is risk-neutral; that is, her objective is to maximize the expected wealth $\mathbb{E}(W_2)$ at time 2. Show that the investor would choose to place her entire portfolio in the one-period bond at $t =$ and roll it over at the risk-free rate at $t = 1$.

(d) Suppose a long-term investor has power utility with risk-aversion paramet $\gamma > 1$; that is, her objective is to maximize $\mathbb{E}\left(W_2^{1-\gamma}/(1-\gamma)\right)$. Show th the investor would prefer to hold part of her entire portfolio in the two-peri bond. Furthermore, show that the higher the risk aversion γ, the higher t holding in the two-period bond.

Exercise 12.4   Prove identity (12.1) in the Almgren–Chriss model.

Exercise 12.5   Consider the following variation of the one-asset trading mod of Almgren and Chriss. Assume the security price at period k is

$$S_k = S_{k-1} + \xi_k$$

and the actual security price received at period k is

$$\tilde{S}_k = S_{k-1} - h_k(y_k),$$

where $h_k(y_k) = cy_k/v_k$, c is a constant, and v_k is the volume traded during t kth interval $[k-1, k]$.

(a) Write down the expression for the shortfall, as a function of the tradi trajectory $\begin{bmatrix} x_0 & x_1 & \cdots & x_N \end{bmatrix}^\mathsf{T}$ and/or the trading list $\begin{bmatrix} y_1 & \cdots & y_N \end{bmatrix}^\mathsf{T}$.
(b) Write down the formulation for the problem of finding the trading li $\begin{bmatrix} y_1 & \cdots & y_N \end{bmatrix}^\mathsf{T}$ that minimizes expected shortfall.
(c) Prove that the solution to the problem in part (b) is the VWAP strategy

$$y_k = \frac{v_k}{\sum_{j=1}^N v_j} X, \quad k = 1, \ldots, N.$$

Exercise 12.6   Suppose today (year $t = 0$) you have an initial endowme $W_0 = \$10,000$. Now (beginning of year 0), and at the beginning of the next years, you can allocate your wealth to two investment choices:

1. A risk-free asset "cash" that generates a 5% annual return.
2. A risky asset "stocks" with annual return that is normally distributed wi mean 10% and standard deviation 20%.

Let W_{20} denote the endowment at the end of the 20-year investment peri (i.e., beginning of year 20).

Consider the following three investment strategies:

1. Buy and hold: Invest the initial endowment 50% in cash and 50% in stoc and never rebalance.
2. Balanced: Rebalance the portfolio at the beginning of every year to a 50 cash and 50% stocks mix.
3. Switching: Alternate each year between 100% cash and 100% stocks.

The Excel spreadsheet "Exercise 12.6 Twenty Years" contains a random sample of the risk-free and risky returns over a 20-year period.

(a) Compute the total accumulated return achieved by each of the three strategies. That is,

$$\frac{W_{20}}{W_0} - 1.$$

(b) Compute the annualized return achieved by each of the three strategies. That is,

$$\left(\frac{W_{20}}{W_0}\right)^{1/20} - 1.$$

(c) Use your favorite simulation software to generate 10,000 random samples of the risk-free and risky returns over a 20-year period. Report the sample mean and sample variance of the total accumulated return and of the annualized return for each of the three strategies.

(d) Produce charts (e.g., histograms) to visualize the distribution of total accumulated and annualized returns achieved by the three strategies. Which strategy seems to have a higher expected annualized return? Which one seems to have a higher variance of annualized return?

(e) Which strategy would you prefer? Why?

13 Dynamic Programming: Theory and Algorithms

Dynamic programming is an approach to model and solve multi-period decisio problems. The fundamental principle of dynamic programming is the *Bellma equation*, a certain kind of optimality condition. As we detail in this chapter, th central idea of the Bellman equation is to break down a multi-stage problem int multiple two-stage problems. Under suitable conditions, the Bellman equatio yields a recursion that helps in characterizing the solution and in computing i Before embarking on a formal description, we illustrate the dynamic progran ming approach via some examples.

13.1 Some Examples

Example 13.1 (Matches puzzle) Suppose there are 30 matches on a table an I play the following game with a clever opponent: I begin by picking up 1, : or 3 matches. Then my opponent must pick 1, 2, or 3 matches. We continu alternating until the last match is picked up. The player who picks up the las match loses. How can I (the first player) be sure of winning?

Solution. If I can ensure that it will be my opponent's turn when 1 match remain I certainly win. Let us work backwards one step: If I can ensure that it will b my opponent's turn when 5 matches remain, I will also win. The reason for th is that no matter what he does when there are 5 matches left, I can make sur that when he has his next turn, only 1 match will remain. Hence it is clea that I win if I can force my opponent to play when 5 matches remain. We ca continue working backwards and conclude that I will ensure victory if I can forc my opponent to play when 5, 9, 13, 17, 21, 25, or 29 matches remain. Sinc the game starts with 30 matches on the table, I can ensure victory by pickin 1 match at the beginning, bringing the number down to 29.

Example 13.2 (Knapsack problem) Given a set of items, each with a certai weight and value, select the collection of items with total maximum value suc that their total weight does not exceed some fixed weight limit W.

Solution. Let $w_t > 0$ and $v_t > 0$ be the weight and value respectively of item t fc $t = 1, \ldots, n$. The knapsack problem can be formulated as an integer program an solved via the technique covered in Chapter 8.2. We next illustrate an alternative approach via dynamic programming. Consider the problem as a sequence of binary decisions $x_t \in \{0,1\}$ corresponding to "include" or "do not include" item t for $t = 1, \ldots, n$. To find the optimal selection of items, we can work "backwards" as we did in the matches puzzle. Let W_t be the remaining amount of weight available at stage $t = 1, \ldots, n$ with $W_1 = W$, and let $J_t(W_t)$ denote the value of an optimal collection of items if we started selecting items at stage t with remaining weight limit W_t. The value function $J_t(W_t)$ satisfies the following backward recursion for $t = 1, 2, \ldots, n - 1$:

$$J_t(W_t) = \begin{cases} J_{t+1}(W_t) & \text{if} \quad w_t > W_t \\ \max\{J_{t+1}(W_t), J_{t+1}(W_t - w_t) + v_t\} & \text{if} \quad w_t \leq W_t. \end{cases} \quad (13.1)$$

Our goal is to obtain the value $J_1(W)$ and the corresponding optimal collection of items. The steps that lead to $J_1(W)$ in the above recursion are tied to the optimal decisions x_t^* for $t = 1, \ldots, n - 1$. On the one hand, $x_t^* = 0$ corresponds to $J_t(W_t) = J_{t+1}(W_t)$; that is, do not select item t. On the other hand, $x_t^* = 1$ corresponds to $J_t(W_t) = J_{t+1}(W_t - w_t) + v_t$. Observe that for $0 \leq W_n \leq W$ the last-stage value function satisfies

$$J_n(W_n) = \begin{cases} 0 & \text{if} \quad w_n > W_n \\ v_n & \text{if} \quad w_n \leq W_n. \end{cases}$$

Example 13.3 (Optimal consumption problem)   Assume that now (beginning of year 0) you have an initial amount of wealth $W_0 > 0$. At the beginning of year t you choose to consume C_t dollars and invest the rest of your wealth in one-year treasury bills. You can consume at most the wealth available in year t. Consuming C_t in year t provides a utility $U(C_t)$. On the other hand, each dollar invested in one-year treasury bill yields $1 + r$ dollars cash at the beginning of the next year. Suppose you want to maximize your total utility of consumption over the next T years:

$$\max_{C_0, \ldots, C_T} \sum_{t=0}^{T} U(C_t).$$

How much should you consume each year?

Solution. The key to solving the optimal consumption problem is again to work "backwards" in time just like we did in the matches puzzle and knapsack problem. Let W_t denote the amount of wealth available at the beginning of year t and let $J_t(W_t)$ be the total utility of consumption from year t to year T if we start at year t with wealth W_t. The value function $J_t(W_t)$ satisfies the following backwards recursion for $t = 0, 1, 2, \ldots, T - 1$:

$$J_t(W_t) = \max_{0 \leq C_t \leq W_t} \{J_{t+1}((W_t - C_t) \cdot (1 + r)) + U(C_t)\} \quad (13.2)$$

and the maximizer C_t^* is the optimal consumption level at year t. Observe that for $W_T \geq 0$ the last-stage value function satisfies

$$J_T(W_T) = U(W_T)$$

attained at the optimal consumption level $C_T^* = W_T$.

13.2   Model of a Sequential System (Deterministic Case)

We next introduce the formal notation and terminology of dynamic programming. The presentation follows the approach popularized in the classical book of Bertsekas (2005). For ease of exposition, we first consider the deterministic case. That is, the context without random components.

A *sequential system* is defined by the following elements.

Stages: These are the points in time when decisions are made. We will normally consider $t = 0, 1, \ldots, T$ or $t = 1, 2, \ldots, T$.

States: The state of the system at a particular stage is the information that is relevant for subsequent decisions. We will generally denote the state at stage t as s_t, for $t = 0, 1, \ldots, T$. Sometimes it is convenient to include also a "final state" s_{T+1}.

Decisions: These are also called controls or actions that we can make at each stage and that affect the behavior of the system. We will generally denote the decisions as x_t, for $t = 0, 1, \ldots, T$.

Law of motion: This defines how the state of the system evolves. A general law of motion has the form

$$s_{t+1} = f_t(s_t, x_t), \quad t = 0, 1, \ldots, T.$$

Assume we are interested in optimizing some overall objective function

$$\sum_{t=0}^{T} g_t(s_t, x_t) + g_{T+1}(s_{T+1}), \tag{13.3}$$

where each $g_t(s_t, x_t)$, for $t = 0, 1, \ldots, T$, and $g_{T+1}(s_{T+1})$ is some cost or reward per stage. This defines a *sequential decision problem*: find x_t, for $t = 0, 1, \ldots, T$ to minimize the total cost or maximize the reward (13.3).

Both Examples 13.2 and 13.3 can be readily stated in this framework.

Dynamic programming formulation for the knapsack problem

Stages: $t = 1, 2, \ldots, n$.

State at stage t: remaining weight capacity W_t.

Decision at stage t: binary variable $x_t \in \{0, 1\}$ indicating whether to include item t or not. This decision is constrained to be $x_t = 0$ if $w_t > W_t$ as in this case the weight of item t exceeds the remaining weight capacity.

Law of motion: the remaining weight capacity at stage $t + 1$ is the one from stage t reduced by w_t if item t is included. Otherwise they are the same. More precisely,

$$W_{t+1} = W_t - w_t x_t, \quad t = 1, 2, \ldots, n - 1.$$

Objective: maximize the total value of the selected items

$$\max_{t=1,\ldots,n} \sum_{t=1}^{n} v_t x_t.$$

Dynamic programming formulation for the optimal consumption problem
Stages: $t = 0, 1, 2, \ldots, T$.
State at stage t: available wealth W_t. It is also convenient to assume that
terminal wealth $W_{T+1} = 0$.
Decision at stage t: consumption $C_t \in [0, W_t]$.
Law of motion: the wealth at stage $t+1$ is the portion of wealth from stage t
that was not consumed increased by a factor $1 + r$. More precisely,

$$W_{t+1} = (W_t - C_t)(1 + r), \ t = 0, 1, 2, \ldots, T.$$

Objective: maximize the total utility of consumption

$$\max_{C_0,\ldots,C_T} \sum_{t=0}^{T} U(C_t).$$

13.3 Bellman's Principle of Optimality

The heart of dynamic programming is a principle of optimality due to Bellman.
Its flavor was suggested by the solutions to Examples 13.1, 13.2, and 13.3.
To state the principle precisely, we need a bit of notation. Suppose we are
maximizing total reward

$$J(s_0) := \max_{x_0,\ldots,x_T} \left\{ \sum_{t=0}^{T} g_t(s_t, x_t) + g_{T+1}(s_{T+1}) \right\}.$$

Consider the "tail problem" that starts at stage t:

$$J_t(s_t) := \max_{x_t,\ldots,x_T} \left\{ \sum_{\tau=t}^{T} g_\tau(s_\tau, x_\tau) + g_{T+1}(s_{T+1}) \right\}.$$

Bellman's optimality principle can be stated as follows. The value-to-go functions
$J_t(s_t)$ satisfy the recursive relationship

$$J_t(s_t) = \max_{x_t} \left\{ g_t(s_t, x_t) + J_{t+1}(f_t(s_t, x_t)) \right\}. \tag{13.4}$$

The recursive relationship (13.4) is called the *Bellman equation*. Observe that
the recursive relationships (13.1) and (13.2) are exactly the Bellman equation
(13.4) in the particular context of Examples 13.2 and 13.3 respectively.

There is a certain jargon associated with the solution to a sequential decision
problem and Bellman's optimality principle. The function $J_t(s_t)$ is called the
value-to-go function at stage t. If the objective is to minimize a total cost,

sometimes it is called the *cost-to-go function*. The solution $\mathbf{x}_t^*(\mathbf{s}_t)$ of Bellman equation (13.4) at stage t is called an *optimal decision rule at stage t*. Notice tha this solution depends on the state $\mathbf{s}_t$ at stage t. The vector of optimal decisio rules $(\mathbf{x}_0^*(\cdot), \ldots, \mathbf{x}_T^*(\cdot))$ is called the *optimal policy*.

Bellman's optimality principle can be phrased as:

> If $(\mathbf{x}_0^*(\cdot), \ldots, \mathbf{x}_T^*(\cdot))$ is an optimal policy for the entire problem, then $(\mathbf{x}_t^*(\cdot), \ldots, \mathbf{x}_T^*(\cdot))$ is an optimal policy for the tail problem beginning at stage t.

13.4   Linear–Quadratic Regulator

We next illustrate Bellman's optimality principle with a popular model fro control engineering called the *linear–quadratic regulator*. It provides the founda tion for a model of dynamic investment with transaction costs and predictab returns that we will discuss in the next chapter. The linear–quadratic regulato is a model for the problem of steering the location $\mathbf{s}_t$ of an object towards th origin via a control input $\mathbf{u}_t$. Instead of a constraint on the location of the objec the linear–quadratic regulator imposes a penalty for deviating from the origi Assume the states and controls evolve according to the following linear law motion:

$$\mathbf{s}_{t+1} = \mathbf{A}\mathbf{s}_t + \mathbf{B}\mathbf{u}_t, \ t = 0, 1, \ldots, N-1.$$

Assume we have a quadratic cost function

$$\sum_{t=0}^{N-1} (\mathbf{s}_t^\mathsf{T} \mathbf{Q} \mathbf{s}_t + \mathbf{u}_t^\mathsf{T} \mathbf{R} \mathbf{u}_t) + \mathbf{s}_N^\mathsf{T} \mathbf{Q} \mathbf{s}_N,$$

where $\mathbf{Q}, \mathbf{R}$ are symmetric positive definite matrices of appropriate sizes.

The goal is to determine the optimal sequence of controls $\mathbf{u}_t, t = 0, 1, \ldots, N-$ that minimize the above cost when the initial position of the object is s_0:

$$J(\mathbf{s}_0) := \min_{\mathbf{u}_0, \ldots, \mathbf{u}_{N-1}} \left\{ \sum_{t=0}^{N-1} (\mathbf{s}_t^\mathsf{T} \mathbf{Q} \mathbf{s}_t + \mathbf{u}_t^\mathsf{T} \mathbf{R} \mathbf{u}_t) + \mathbf{s}_N^\mathsf{T} \mathbf{Q} \mathbf{s}_N \right\}.$$

We next apply the backwards dynamic programming principle. For the last stag N we evidently have

$$J_N(\mathbf{s}_N) = \mathbf{s}_N^\mathsf{T} \mathbf{Q} \mathbf{s}_N.$$

For stage $N - 1$ we have the Bellman equation

$$J_{N-1}(\mathbf{s}_{N-1}) = \min_{\mathbf{u}_{N-1}} \left\{ \mathbf{s}_{N-1}^{\mathsf{T}} \mathbf{Q} \mathbf{s}_{N-1} + \mathbf{u}_{N-1}^{\mathsf{T}} \mathbf{R} \mathbf{u}_{N-1} + J_N(\mathbf{s}_N) \right\}$$

$$= \min_{\mathbf{u}_{N-1}} \left\{ \mathbf{s}_{N-1}^{\mathsf{T}} \mathbf{Q} \mathbf{s}_{N-1} + \mathbf{u}_{N-1}^{\mathsf{T}} \mathbf{R} \mathbf{u}_{N-1} \right.$$
$$\left. + (\mathbf{A}\mathbf{s}_{N-1} + \mathbf{B}\mathbf{u}_{N-1})^{\mathsf{T}} \mathbf{Q} (\mathbf{A}\mathbf{s}_{N-1} + \mathbf{B}\mathbf{u}_{N-1}) \right\}$$

$$= \min_{\mathbf{u}_{N-1}} \left\{ \mathbf{s}_{N-1}^{\mathsf{T}} \mathbf{Q} \mathbf{s}_{N-1} + \mathbf{s}_{N-1}^{\mathsf{T}} \mathbf{A}^{\mathsf{T}} \mathbf{Q} \mathbf{A} \mathbf{s}_{N-1} + 2\mathbf{s}_{N-1}^{\mathsf{T}} \mathbf{A}^{\mathsf{T}} \mathbf{Q} \mathbf{B} \mathbf{u}_{N-1} \right.$$
$$\left. + \mathbf{u}_{N-1}^{\mathsf{T}} (\mathbf{R} + \mathbf{B}^{\mathsf{T}} \mathbf{Q} \mathbf{B}) \mathbf{u}_{N-1} \right\}.$$

The latter is a convex quadratic function of $\mathbf{u}_{N-1}$. To find its minimum, we compute its gradient and equate it to zero to obtain:

$$2\mathbf{B}^{\mathsf{T}} \mathbf{Q} \mathbf{A} \mathbf{s}_{N-1} + 2(\mathbf{R} + \mathbf{B}^{\mathsf{T}} \mathbf{Q} \mathbf{B}) \mathbf{u}_{N-1} = \mathbf{0}.$$

Thus, the optimal control at stage $N - 1$ is

$$\mathbf{u}_{N-1}^* = -(\mathbf{R} + \mathbf{B}^{\mathsf{T}} \mathbf{Q} \mathbf{B})^{-1} \mathbf{B}^{\mathsf{T}} \mathbf{Q} \mathbf{A} \mathbf{s}_{N-1} = \mathbf{L}_{N-1} \mathbf{s}_{N-1},$$

where

$$\mathbf{L}_{N-1} = -(\mathbf{R} + \mathbf{B}^{\mathsf{T}} \mathbf{Q} \mathbf{B})^{-1} \mathbf{B}^{\mathsf{T}} \mathbf{Q} \mathbf{A}.$$

Plugging this value of $\mathbf{u}_{N-1}^*$ in the above expression for $J_{N-1}(\mathbf{s}_{N-1})$ we get

$$J_{N-1}(\mathbf{s}_{N-1}) = \mathbf{s}_{N-1}^{\mathsf{T}} \mathbf{Q} \mathbf{s}_{N-1} + \mathbf{s}_{N-1}^{\mathsf{T}} \mathbf{A}^{\mathsf{T}} \mathbf{Q} \mathbf{A} \mathbf{s}_{N-1}$$
$$- \mathbf{s}_{N-1}^{\mathsf{T}} \mathbf{A}^{\mathsf{T}} \mathbf{Q} \mathbf{B} (\mathbf{R} + \mathbf{B}^{\mathsf{T}} \mathbf{Q} \mathbf{B})^{-1} \mathbf{B}^{\mathsf{T}} \mathbf{Q} \mathbf{A} \mathbf{s}_{N-1}$$
$$= \mathbf{s}_{N-1}^{\mathsf{T}} \mathbf{K}_{N-1} \mathbf{s}_{N-1},$$

where

$$\mathbf{K}_{N-1} = \mathbf{Q} + \mathbf{A}^{\mathsf{T}} (\mathbf{Q} - \mathbf{Q} \mathbf{B} (\mathbf{R} + \mathbf{B}^{\mathsf{T}} \mathbf{Q} \mathbf{B})^{-1} \mathbf{B}^{\mathsf{T}} \mathbf{Q}) \mathbf{A}.$$

Next we will prove by induction that

$$J_t(\mathbf{s}_t) = \mathbf{s}_t^{\mathsf{T}} \mathbf{K}_t \mathbf{s}_t, \quad \mathbf{u}_t^* = \mathbf{L}_t \mathbf{s}_t,$$

where

$$\mathbf{K}_N = \mathbf{Q},$$
$$\mathbf{K}_t = \mathbf{Q} + \mathbf{A}^{\mathsf{T}} (\mathbf{K}_{t+1} - \mathbf{K}_{t+1} \mathbf{B} (\mathbf{R} + \mathbf{B}^{\mathsf{T}} \mathbf{K}_{t+1} \mathbf{B})^{-1} \mathbf{B}^{\mathsf{T}} \mathbf{K}_{t+1}) \mathbf{A}, \quad t = N - 1, \ldots, 0,$$

and

$$\mathbf{L}_t = -(\mathbf{R} + \mathbf{B}^{\mathsf{T}} \mathbf{K}_{t+1} \mathbf{B})^{-1} \mathbf{B}^{\mathsf{T}} \mathbf{K}_{t+1} \mathbf{A}, \quad t = N - 1, \ldots, 0.$$

We already showed that the above holds for $t = N - 1$. Assume that it holds for $t + 1$. At stage t we have the Bellman equation

$$J_t(\mathbf{s}_t)$$
$$= \min_{\mathbf{u}_t} \left\{ \mathbf{s}_t^\mathsf{T} \mathbf{Q} \mathbf{s}_t + \mathbf{u}_t^\mathsf{T} \mathbf{R} \mathbf{u}_t + J_{t+1}(\mathbf{s}_{t+1}) \right\}$$
$$= \min_{\mathbf{u}_t} \left\{ \mathbf{s}_t^\mathsf{T} \mathbf{Q} \mathbf{s}_t + \mathbf{u}_t^\mathsf{T} \mathbf{R} \mathbf{u}_t + (\mathbf{A} \mathbf{s}_t + \mathbf{B} \mathbf{u}_t)^\mathsf{T} \mathbf{K}_{t+1}(\mathbf{A} \mathbf{s}_t + \mathbf{B} \mathbf{u}_t) \right\}$$
$$= \min_{\mathbf{u}_t} \left\{ \mathbf{s}_t^\mathsf{T} \mathbf{Q} \mathbf{s}_t + \mathbf{s}_t^\mathsf{T} \mathbf{A}^\mathsf{T} \mathbf{K}_{t+1} \mathbf{A} \mathbf{s}_t + 2\mathbf{s}_t^\mathsf{T} \mathbf{A}^\mathsf{T} \mathbf{K}_{t+1} \mathbf{B} \mathbf{u}_t + \mathbf{u}_t^\mathsf{T} (\mathbf{R} + \mathbf{B}^\mathsf{T} \mathbf{K}_{t+1} \mathbf{B}) \mathbf{u}_t \right\}$$

The latter is a convex quadratic function of $\mathbf{u}_t$. To find its minimum, we compute its gradient and equate it to zero to obtain:

$$2\mathbf{B}^\mathsf{T} \mathbf{K}_{t+1} \mathbf{A} \mathbf{s}_t + 2(\mathbf{R} + \mathbf{B}^\mathsf{T} \mathbf{K}_{t+1} \mathbf{B}) \mathbf{u}_t = \mathbf{0}.$$

Thus, the optimal control at stage t is

$$\mathbf{u}_t^* = -(\mathbf{R} + \mathbf{B}^\mathsf{T} \mathbf{K}_{t+1} \mathbf{B})^{-1} \mathbf{B}^\mathsf{T} \mathbf{K}_{t+1} \mathbf{A} \mathbf{s}_t = \mathbf{L}_t \mathbf{s}_t,$$

where

$$\mathbf{L}_t = -(\mathbf{R} + \mathbf{B}^\mathsf{T} \mathbf{K}_{t+1} \mathbf{B})^{-1} \mathbf{B}^\mathsf{T} \mathbf{K}_{t+1} \mathbf{A}.$$

Plugging this value of $\mathbf{u}_t^*$ in the above expression for $J_t(\mathbf{s}_t)$ we get

$$J_t(\mathbf{s}_t) = \mathbf{s}_t^\mathsf{T} \mathbf{Q} \mathbf{s}_t + \mathbf{s}_t^\mathsf{T} \mathbf{A}^\mathsf{T} \mathbf{K}_{t+1} \mathbf{A} \mathbf{s}_t - \mathbf{s}_t^\mathsf{T} \mathbf{A}^\mathsf{T} \mathbf{K}_{t+1} \mathbf{B} (\mathbf{R} + \mathbf{B}^\mathsf{T} \mathbf{K}_{t+1} \mathbf{B})^{-1} \mathbf{B}^\mathsf{T} \mathbf{K}_{t+1} \mathbf{A} \mathbf{s}_t$$
$$= \mathbf{s}_t^\mathsf{T} \mathbf{K}_t \mathbf{s}_t,$$

where

$$\mathbf{K}_t = \mathbf{Q} + \mathbf{A}^\mathsf{T} (\mathbf{K}_{t+1} - \mathbf{K}_{t+1} \mathbf{B} (\mathbf{R} + \mathbf{B}^\mathsf{T} \mathbf{K}_{t+1} \mathbf{B})^{-1} \mathbf{B}^\mathsf{T} \mathbf{K}_{t+1}) \mathbf{A}.$$

13.5   Sequential Decision Problem with Infinite Horizon

Infinite horizon problems are often appropriate models for problems where there is no terminal stage, such as investments for an endowment or a foundation. They are also often appropriate to model problems with very long time horizons. The infinite horizon setting tends to simplify some issues since the dependence of the value function on t can be eliminated.

Consider an infinite horizon problem whose law of motion is of the form

$$\mathbf{s}_{t+1} = f(\mathbf{x}_t, \mathbf{s}_t)$$

and whose objective function is

$$\max_{\mathbf{x}_0, \mathbf{x}_1, \dots} \sum_{t=0}^{\infty} \theta^t \cdot g(\mathbf{x}_t, \mathbf{s}_t),$$

where $\theta \in (0, 1)$ is a given discount factor. Define the value-to-go function $V(\cdot)$ as

$$V(\mathbf{s}_0) := \max_{\mathbf{x}_0, \mathbf{x}_1, \dots} \sum_{t=0}^{\infty} \theta^t \cdot g(\mathbf{x}_t, \mathbf{s}_t).$$

Observe that at any intermediate stage t we have

$$V(\mathbf{s}_t) := \max_{\mathbf{x}_t, \mathbf{x}_{t+1}, \dots} \sum_{\tau=t}^{\infty} \theta^{\tau-t} \cdot g(\mathbf{x}_\tau, \mathbf{s}_\tau).$$

Thus, in this case the Bellman equation can be written as

$$V(\mathbf{s}_t) = \max_{\mathbf{x}_t} g(\mathbf{x}_t, \mathbf{s}_t) + \theta \cdot V(\mathbf{s}_{t+1}).$$

13.6 Linear–Quadratic Regulator with Infinite Horizon

Consider now the infinite horizon version of the linear–quadratic regulator that we discussed in Section 13.4. The goal now is to determine the optimal sequence of controls $\mathbf{u}_t$, $t = 0, 1, \dots$, that minimizes the following cost:

$$V(\mathbf{s}_0) := \min_{\mathbf{u}_0, \mathbf{u}_1, \dots} \left\{ \sum_{t=0}^{\infty} (\mathbf{s}_t^{\mathsf{T}} \mathbf{Q} \mathbf{s}_t + \mathbf{u}_t^{\mathsf{T}} \mathbf{R} \mathbf{u}_t) \right\}.$$

A common technique to solve the Bellman equation (and similar differential equations) is "ansatz", which can be loosely described as "make an educated guess and later verify". In this problem, we try the following quadratic ansatz for the form of the value function:

$$V(\mathbf{s}_t) = \mathbf{s}_t^{\mathsf{T}} \mathbf{K} \mathbf{s}_t$$

for some symmetric positive definite matrix $\mathbf{K}$.

With this educated guess we now apply the Bellman equation (infinite horizon case):

$$
\begin{aligned}
V(\mathbf{s}_t) &= \min_{\mathbf{u}_t} \left[\mathbf{s}_t^{\mathsf{T}} \mathbf{Q} \mathbf{s}_t + \mathbf{u}_t^{\mathsf{T}} \mathbf{R} \mathbf{u}_t + V(\mathbf{s}_{t+1}) \right] \\
&= \min_{\mathbf{u}_t} \left[\mathbf{s}_t^{\mathsf{T}} \mathbf{Q} \mathbf{s}_t + \mathbf{u}_t^{\mathsf{T}} \mathbf{R} \mathbf{u}_t + (\mathbf{A} \mathbf{s}_t + \mathbf{B} \mathbf{u}_t)^{\mathsf{T}} \mathbf{K} (\mathbf{A} \mathbf{s}_t + \mathbf{B} \mathbf{u}_t) \right] \\
&= \min_{\mathbf{u}_t} \left[\mathbf{s}_t^{\mathsf{T}} \mathbf{Q} \mathbf{s}_t + \mathbf{s}_t^{\mathsf{T}} \mathbf{A}^{\mathsf{T}} \mathbf{K} \mathbf{A} \mathbf{s}_t + 2 \mathbf{s}_t^{\mathsf{T}} \mathbf{A}^{\mathsf{T}} \mathbf{K} \mathbf{B} \mathbf{u}_t + \mathbf{u}_t^{\mathsf{T}} (\mathbf{R} + \mathbf{B}^{\mathsf{T}} \mathbf{K} \mathbf{B}) \mathbf{u}_t \right].
\end{aligned}
$$

The latter is a convex quadratic function of $\mathbf{u}_t$. To find its minimum, we compute its gradient and equate it to zero to obtain:

$$2 \mathbf{B}^{\mathsf{T}} \mathbf{K} \mathbf{A} \mathbf{s}_t + 2(\mathbf{R} + \mathbf{B}^{\mathsf{T}} \mathbf{K} \mathbf{B}) \mathbf{u}_t = 0.$$

Thus, the optimal control at stage t is

$$\mathbf{u}_t^* = -(\mathbf{R} + \mathbf{B}^{\mathsf{T}} \mathbf{K} \mathbf{B})^{-1} \mathbf{B}^{\mathsf{T}} \mathbf{K} \mathbf{A} \mathbf{s}_t = \mathbf{L} \mathbf{s}_t,$$

where

$$\mathbf{L} = -(\mathbf{R} + \mathbf{B}^{\mathsf{T}} \mathbf{K} \mathbf{B})^{-1} \mathbf{B}^{\mathsf{T}} \mathbf{K} \mathbf{A}.$$

Plugging this value of $\mathbf{u}_t^*$ in the above Bellman equation we get

$$V(\mathbf{s}_t) = \mathbf{s}_t^\mathsf{T}\mathbf{Q}\mathbf{s}_t + \mathbf{s}_t^\mathsf{T}\mathbf{A}^\mathsf{T}(\mathbf{K} - \mathbf{K}\mathbf{B}(\mathbf{R} + \mathbf{B}^\mathsf{T}\mathbf{K}\mathbf{B})^{-1}\mathbf{B}^\mathsf{T}\mathbf{K})\mathbf{A}\mathbf{s}_t.$$

Hence for the above guess to be correct, we must have:

$$\mathbf{K} = \mathbf{Q} + \mathbf{A}^\mathsf{T}(\mathbf{K} - \mathbf{K}\mathbf{B}(\mathbf{R} + \mathbf{B}^\mathsf{T}\mathbf{K}\mathbf{B})^{-1}\mathbf{B}^\mathsf{T}\mathbf{K})\mathbf{A}.$$

This is the so-called *Ricatti equation*. Under suitable assumptions on $\mathbf{Q}, \mathbf{R}, \mathbf{A}, \mathbf{B}$, this equation is known to have a unique symmetric positive definite solution $\mathbf{K}$.

Consider the following special case: $\mathbf{A} = \mathbf{B} = \mathbf{I}$ and $\mathbf{R} = \lambda\mathbf{Q}$ with $\lambda > 0$. In this case the law of motion is

$$\mathbf{s}_{t+1} = \mathbf{s}_t + \mathbf{u}_t$$

and the Ricatti equation is

$$\mathbf{K} = \mathbf{Q} + \mathbf{K} - \mathbf{K}(\lambda\mathbf{Q} + \mathbf{K})^{-1}\mathbf{K}.$$

We thus obtain

$$\mathbf{Q} = \mathbf{K}(\lambda\mathbf{Q} + \mathbf{K})^{-1}\mathbf{K}.$$

To solve for $\mathbf{K}$, try to find a solution of the form $\mathbf{K} = a\mathbf{Q}$. Plugging this in the above equation yields

$$1 = \frac{a^2}{\lambda + a}.$$

This is a quadratic equation in a with two roots, but only one that is positive, namely

$$a = \frac{1 + \sqrt{1 + 4\lambda}}{2}.$$

Therefore we get

$$\mathbf{K} = a\mathbf{Q} = \frac{1 + \sqrt{1 + 4\lambda}}{2}\mathbf{Q},$$

and consequently

$$\mathbf{L} = -\frac{1 + \sqrt{1 + 4\lambda}}{2\lambda + 1 + \sqrt{1 + 4\lambda}}\mathbf{I}.$$

In particular, the optimal control at time t is

$$\mathbf{u}_t = -\frac{1 + \sqrt{1 + 4\lambda}}{2\lambda + 1 + \sqrt{1 + 4\lambda}}\mathbf{s}_t.$$

Note that when $\lambda = 0$, there is no direct cost associated with the control variable $\mathbf{u}_t$ and therefore it is optimal to select $\mathbf{u}_t$ to minimize the cost of $\mathbf{s}_{t+1} = \mathbf{s}_t + \mathbf{u}_t$, which is given by $\mathbf{s}_{t+1}^\mathsf{T}\mathbf{Q}\mathbf{s}_{t+1}$. Clearly, this is minimized when $\mathbf{s}_{t+1} = 0$, or equivalently, when $\mathbf{u}_t = -\mathbf{s}_t$. On the other hand, for $\lambda > 0$, the cost $\lambda\mathbf{u}_t^\mathsf{T}\mathbf{Q}\mathbf{u}$ keeps $\mathbf{u}_t$ from reaching all the way to $-\mathbf{s}_t$. Instead, $\mathbf{u}_t$ is a scalar multiple of $-\mathbf{s}_t$ where the scalar multiple is less than 1. In addition, the larger λ, the higher the cost of the control variable $\mathbf{u}_t$, and therefore the smaller this scalar multiple.

13.7   Model of Sequential System (Stochastic Case)

The above dynamic programming machinery has a straightforward extension to a more general context that includes a stochastic component in the law of motion.

A *stochastic sequential system* is an extension of the deterministic case. Like a deterministic sequential system, the main components of a stochastic sequential system are stages, states, decisions, and law of motion. The first three are exactly as before. On the other hand, the law of motion of a stochastic sequential system is of the more general form

$$s_{t+1} = f_t(s_t, x_t, \omega_t), \quad t = 0, 1, \ldots, T.$$

As before, s_t, x_t are the state and action at stage t and s_{t+1} is the state at stage $t + 1$. In addition, ω_t is some random disturbance that occurs at stage t.

Assume we are interested in optimizing some overall objective function

$$\mathbb{E}\left[\sum_{t=0}^{T} g_t(s_t, x_t, \omega_t) + g_{T+1}(s_{T+1})\right], \tag{13.5}$$

where each $g_t(s_t, x_t, \omega_t)$, $t = 0, 1, \ldots, T$, and $g_{T+1}(s_{T+1})$ is a cost or a reward per stage. This defines a *stochastic sequential decision problem*: find x_t, $t = 0, 1, \ldots, T$, to minimize or maximize the expected total cost or reward (13.5).

Bellman's optimality principle also extends in a natural fashion. Suppose we are maximizing the expected reward

$$J(s_0) := \max_{x_0, \ldots, x_T} \mathbb{E}\left[\sum_{t=0}^{T} g_t(s_t, x_t, \omega_t) + g_{T+1}(s_{T+1})\right].$$

Consider the "tail problem" that starts at stage t:

$$J_t(s_t) := \max_{x_t, \ldots, x_T} \mathbb{E}\left[\sum_{\tau=t}^{T} g_\tau(s_\tau, x_\tau, \omega_\tau) + g_{T+1}(s_{T+1})\right].$$

Bellman's optimality principle can be stated as follows. The value-to-go functions $J_t(s_t)$ satisfy the following Bellman equation:

$$J_t(s_t) = \max_{x_t} \mathbb{E}_t\left[g_t(s_t, x_t, \omega_t) + J_{t+1}(f_t(s_t, x_t, \omega_t))\right]. \tag{13.6}$$

The stochastic case also has an infinite horizon version. Consider an infinite horizon problem with a law of motion of the form

$$s_{t+1} = f(x_t, s_t, \omega_t)$$

and objective function

$$\max_{x_0, x_1, \ldots} \mathbb{E}\left[\sum_{t=0}^{\infty} \theta^t \cdot g(x_t, s_t, \omega_t)\right],$$

where $\theta \in (0,1)$ is a given discount factor.

Define the value-to-go function $V(\cdot)$ as

$$V(\mathbf{s}_0) := \max_{\mathbf{x}_0, \mathbf{x}_1, \ldots} \mathbb{E} \left[\sum_{t=0}^{\infty} \theta^t \cdot g(\mathbf{x}_t, \mathbf{s}_t, \omega_t) \right].$$

Observe that at any intermediate stage t we have

$$V(\mathbf{s}_t) := \max_{\mathbf{x}_t, \mathbf{x}_{t+1}, \ldots} \mathbb{E} \left[\sum_{\tau=t}^{\infty} \theta^{\tau-t} \cdot g(\mathbf{x}_\tau, \mathbf{s}_\tau, \omega_\tau) \right].$$

In this case the Bellman equation can be written as

$$V(\mathbf{s}_t) = \max_{\mathbf{x}_t} \mathbb{E}_t \left[g(\mathbf{x}_t, \mathbf{s}_t, \omega_t) + \theta \cdot V(\mathbf{s}_{t+1}) \right].$$

13.8   Notes

Dynamic programming was introduced by Bellman (1954, 1957), who state
the fundamental principle of optimality. Dynamic programming is pervasive
many disciplines, including finance, economics, biology, management, etc. Tl
book of Bertsekas (2005) is a popular modern reference on this topic. The bo(
by Porteus (2002) gives a treatment on dynamic programming with focus (
inventory theory.

13.9   Exercises

Exercise 13.1   Consider the following puzzle. There are 40 matches on a tabl
You begin by picking up 1, 2, 3, or 4 matches. Then your opponent must pick
2, 3, or 4 matches. The two of you continue taking turns until the last match
picked up. The player who picks up the last match loses.

(a) Can you find a strategy that guarantees your victory? If so, how?

(b) What if the initial number of matches is 39, 38, 37, or 36 instead of 40?

(c) Suppose the game starts with 40 matches and you and your opponent tal
    turns as above but the player who picks up the last match wins. Can yo
    find a strategy that guarantees your victory? If so, how?

Exercise 13.2   Consider the following capital budgeting example fro
Chapter 8:

$$\begin{aligned}
\max \quad & 9x_1 + 11x_2 + 7x_3 + 4x_4 \\
\text{s.t.} \quad & 7x_1 + 10x_2 + 6x_3 + 3x_4 \leq 19 \\
& x_i \in \{0, 1\}, \quad i = 1, \ldots, 4.
\end{aligned}$$

Observe that this is a knapsack problem. Prove that the vector $\mathbf{x}^* = \begin{bmatrix} 0 & 1 & 1 & 1 \end{bmatrix}$
is an optimal solution to this problem by showing that it satisfies the Bellma
equation.

Exercise 13.3   Consider the optimal consumption problem described in Example 13.3. Suppose the consumer has a logarithmic utility of consumption $U(C_t) = \log(C_t)$.

(a) Show that the optimal consumption and value-to-go function at stage T are respectively $C_T^*(W_T) = W_T$ and $J_T(W_T) = \log(W_T)$.

(b) Assume $r = 0$. Use the Bellman equation and induction to show that the optimal consumption and value-to-go function at stages $t = 0, 1, \ldots, T-1$ are respectively $C_t(W_t) = W_0/(T-t+1)$ and $J_t(W_t) = (T-t+1)\log(W_t)$.

(c) Assume $r > 0$. Use the Bellman equation and induction to find the optimal consumption and value-to-go function at stages $t = 0, 1, \ldots, T-1$.

Exercise 13.4   Consider the following infinite horizon variation of the previous consumption problem. Assume the consumer lives forever and her objective is to maximize the following total discounted utility of consumption

$$\sum_{t=0}^{\infty} \theta^t \cdot \log(C_t)$$

for some $\theta \in (0,1)$. Use the following educated guess ("ansatz") for the optimal value function:

$$V(W_t) = a \cdot \log(W_t) + b$$

for some constants a, b with $a > 0$. Use the Bellman equation to verify this educated guess and determine the optimal consumption rule $C_t^*(W_t)$ and optimal value function $V(W_t)$ (that is, the values of a and b). For simplicity, assume $r = 0$.

Exercise 13.5   Consider the following variation of the optimal consumption problem described in Example 13.3. At each stage t the amount of non-consumed wealth $W_t - C_t$ can be split between treasury bills and an index fund. Funds placed in treasury bills earn an annual risk-free return $r > 0$ whereas the funds placed in the index fund earn a risky return r_t with expected value $\mathbb{E}(r_t) = \mu > r$ and variance $\text{var}(r_t) = \sigma^2 > 0$. Assume the returns are i.i.d. across different periods.

Use dynamic programming to formulate the following optimal investment and consumption problem: Determine the consumption $C_t \in [0, W_t]$ and fraction of wealth $x_t \in \mathbb{R}$ invested in the index fund at stage $t = 0, 1, \ldots, T$ that maximize the total expected utility of consumption

$$\max_{\substack{C_0, \ldots, C_T \\ x_0, \ldots, x_T}} \mathbb{E}\left[\sum_{t=0}^{T} U(C_t)\right]$$

over the next T years. Proceed as follows.

(a) Write the law of motion; that is, the equation that describes the state W_{t+1} in terms of the state W_t and decisions C_t, x_t at stage t.

(b) Write the Bellman equation for the value-to-go function $J_t(W_t)$.

(c) Consider the special case of logarithmic utility of consumption $U(C_t) = \log(C_t)$. Use the Bellman equation and induction to determine the optimal consumption C_t^* and investment fraction x_t^* as well as the value-to-go function $J_t(W_t)$ at stage t for $t = 0, 1, \ldots, T$.

(d) Indicate how your model changes if the fraction of wealth x_t invested in the index fund at each stage t is subject to the constraint $x_t \in [0, 1]$. (That is, no leverage is allowed in the investment portfolio.)

14 Dynamic Programming Models: Multi-Period Portfolio Optimization

This chapter describes four types of dynamic portfolio optimization problems that are amenable to dynamic programming technology. The first two deal respectively with optimization of final wealth and its extension, optimal consumption and investment. These two classical models date back multiple decades. The last two problems are much more modern developments. One of them is a model for dynamic trading when returns are predictable and trading is costly. The other one is a model for dynamic portfolio optimization that incorporates capital gains taxes.

14.1 Utility of Terminal Wealth

Let us revisit the dynamic portfolio optimization model with initial endowment and utility of terminal wealth that we discussed in Chapter 12. However, this time we consider a more general setting where there are forecasting variables available at each stage. Such forecasting variables could be associated with a factor model. For instance, they could be macroeconomic indicators, or certain measurable parameters of a particular asset or firm.

As before, suppose that an investor starts at $t = 0$ with an initial endowment W_0. At times $t = 0, \ldots, T-1$ the investor invests her wealth W_t in a portfolio of risk-free and risky assets. The investor's goal is to maximize the expected utility of terminal wealth $U(W_T)$ at time T for some utility function $U(\cdot)$. Define the following convenient notation:

- $R_{f,t+1}$ = gross risk-free return in period $[t, t+1]$;
- $\mathbf{r}_{t+1}$ = vector of excess returns of the risky assets in period $[t, t+1]$;
- $R_{p,t+1}$ = gross random return of the investor's portfolio in period $[t, t+1]$;
- $\mathbf{z}_t$ = forecasting state variables available at stage t;
- W_t = wealth at stage t.

We have the inter-temporal budget constraint:

$$W_{t+1} = W_t \cdot R_{p,t+1} = W_t \cdot (R_{f,t+1} + \mathbf{r}_{t+1}^\mathsf{T} \mathbf{x}_t), \quad t = 0, \ldots, T-1.$$

The specific components of this sequential decision problem are as follows:

Stages: these are $t = 0, \ldots, T-1$.

State at stage t: this is $(W_t, \mathbf{z}_t)$.

Decision variables at stage t: these are the vector $\mathbf{x}_t$ of portfolio holdings (percentages) in the risky assets.

Law of motion: this is the same as the above inter-temporal constraint

$$W_{t+1} = W_t \cdot (R_{f,t+1} + \mathbf{r}_{t+1}^{\mathsf{T}} \mathbf{x}_t), \ t = 0, \ldots, T-1.$$

We next apply Bellman's optimality principle. In this case the value-to-go function is

$$J_t(W_t, \mathbf{z}_t) = \max_{\mathbf{x}_t, \ldots, \mathbf{x}_{T-1}} \mathbb{E}_t(U(W_T))$$

$$= \max_{\mathbf{x}_t, \ldots, \mathbf{x}_{T-1}} \mathbb{E}_t \left[U \left(W_t \cdot \prod_{\tau=t}^{T-1} (R_{f,\tau+1} + \mathbf{r}_{\tau+1}^{\mathsf{T}} \mathbf{x}_\tau) \right) \right].$$

At the final stage T we get

$$J_T(W_T, \mathbf{z}_T) = U(W_T).$$

For earlier stages, we have the Bellman equation

$$J_t(W_t, \mathbf{z}_t) = \max_{\mathbf{x}_t} \mathbb{E}_t \left[J_{t+1}(W_{t+1}, \mathbf{z}_{t+1}) \right]$$

$$= \max_{\mathbf{x}_t} \mathbb{E}_t \left[J_{t+1}(W_t(R_{f,t+1} + \mathbf{r}_{t+1}^{\mathsf{T}} \mathbf{x}_t), \mathbf{z}_{t+1}) \right].$$

In the special case of power utility $U(W) = W^{1-\gamma}/(1-\gamma)$, where $\gamma > 0$, we rewrite the Bellman equation as follows. Define $\psi_t(\mathbf{z}_t) := J_t(1, \mathbf{z}_t)$. Then it is easy to see that the Bellman equation is equivalent to

$$\psi_t(\mathbf{z}_t) = \max_{\mathbf{x}_t} \mathbb{E}_t \left[(R_{f,t+1} + \mathbf{r}_{t+1}^{\mathsf{T}} \mathbf{x}_t)^{1-\gamma} \cdot \psi_{t+1}(\mathbf{z}_{t+1}) \right].$$

We can draw the following interesting conclusions from here. On the one hand if $\mathbf{r}_{t+1}$ and $\mathbf{z}_{t+1}$ are independent at time t, then the term on the right-hand side above satisfies

$$\mathbb{E}_t \left[(R_{f,t+1} + \mathbf{r}_{t+1}^{\mathsf{T}} \mathbf{x}_t)^{1-\gamma} \cdot \psi_{t+1}(\mathbf{z}_{t+1}) \right]$$
$$= \mathbb{E}_t \left[(R_{f,t+1} + \mathbf{r}_{t+1}^{\mathsf{T}} \mathbf{x}_t)^{1-\gamma} \right] \cdot \mathbb{E}_t \left(\psi_{t+1}(\mathbf{z}_{t+1}) \right). \tag{14.1}$$

Thus, to find $\mathbf{x}_t$ we need to solve

$$\max_{\mathbf{x}_t} \mathbb{E}_t \left[\frac{(R_{f,t+1} + \mathbf{r}_{t+1}^{\mathsf{T}} \mathbf{x}_t)^{1-\gamma}}{1-\gamma} \right] = \max_{\mathbf{x}_t} \mathbb{E}_t \left[U(R_{p,t+1}) \right].$$

In this case the optimal policy is myopic.

On the other hand, if $\mathbf{r}_{t+1}$ and $\mathbf{z}_{t+1}$ are correlated, then (14.1) no longer holds. In this case $\mathbf{x}_t$ may include some kind of "inter-temporal hedging component". The intuition is that the correlation between $\mathbf{r}_{t+1}$ and $\mathbf{z}_{t+1}$ would induce some kind of serial dependence in our returns. In other words, the current forecasted return $\mathbf{r}_{t+1}$ conveys information about future returns. Unlike the myopic strategy, the optimal dynamic strategy incorporates this serial dependence.

14.2 Optimal Consumption and Investment

Consider an extension of the previous dynamic portfolio optimization model where the goal is to maximize an expected utility that combines two terms: consumption along the planning horizon and terminal wealth. The latter component is sometimes called *bequest*.

There are three key differences from the previous model. First, there is an additional decision variable $C_t \in [0, W_t]$ at each stage t that denotes the amount of wealth the investor consumes at stage t. Second, the objective function is

$$\max_{\substack{\mathbf{x}_0, \ldots, \mathbf{x}_{T-1} \\ C_0, \ldots, C_{T-1}}} \mathbb{E}\left(\sum_{t=0}^{T-1} U(C_t) + B(W_T) \right)$$

for some utility functions $U(C)$ and $B(W)$. Third, the new law of motion, or inter-temporal budget constraint, is

$$W_{t+1} = (W_t - C_t) \cdot R_{p,t+1} = (W_t - C_t) \cdot (R_{f,t+1} + \mathbf{r}_{t+1}^\mathsf{T} \mathbf{x}_t), \ t = 0, \ldots, T-1.$$

To simplify our discussion we consider the case when there are no forecasting variables $\mathbf{z}_t$. In particular this implies that the returns on the risky assets are independent across different time periods. At the final stage T we have the following value-to-go function

$$J_T(W_T) = B(W_T).$$

For earlier stages, we have the Bellman equation

$$J_t(W_t) = \max_{C_t, \mathbf{x}_t} \mathbb{E}_t \left[J_{t+1}(W_{t+1}) + U(C_t) \right].$$

The first-order optimality conditions yield

$$U'(C_t) = \mathbb{E}_t \left[J'_{t+1}(W_{t+1}) R_{p,t+1} \right]$$

and

$$\mathbb{E}_t \left[J'_{t+1}(W_{t+1}) \mathbf{r}_{t+1} \right] = \mathbf{0}.$$

The first one is obtained by differentiating with respect to C_t and the second one is obtained by differentiating with respect to $\mathbf{x}_t$.

If we plug the optimal $C_t, \mathbf{x}_t$ back into the Bellman equation and differentiate with respect to the state variable W_t, we obtain the following *envelope condition*:

$$U'(C_t) = J'_t(W_t).$$

In the special case of a logarithmic utility of consumption and bequest $U(C) = \log(C)$, $B(W) = \log(W)$, we can draw a more explicit conclusion about the problem. In this case the Bellman equation yields the following expressions for the value function and optimal consumption:

$$J_t(W_t) = \frac{\log(W_t)}{T - t + 1} + b_t$$

and

$$C_t^*(W_t) = \frac{W_t}{T - t + 1}.$$

The specific value b_t and the optimal portfolio $\mathbf{x}_t^*(W_t)$ depend on the joint probability distribution of $R_{f,t+1}$ and $\mathbf{r}_{t+1}$. By contrast, the optimal consumption $C_t^*(W_t)$ only depends on W_t.

14.3 Dynamic Trading with Predictable Returns and Transaction Costs

We next discuss a recent model due to Gârleanu and Pedersen (2013) for dynamic portfolio optimization when asset returns are predictable by signals and trading is costly. This problem is quite timely and especially relevant for active investors. The optimal trading policy should balance various tradeoffs. Fast trading generates more alpha and lower risk but also higher transaction costs. Slow trading does the opposite. On the other hand, there may be fast signals that require quick action and slow signals associated with longer-lasting alpha. The model that we discuss next provides an insightful solution to this problem.

Consider a universe of assets, whose returns evolve according to the following law of motion:

$$\mathbf{r}_{t+1} = \mathbf{B}\mathbf{f}_t + \mathbf{u}_{t+1}.$$

Here $\mathbf{f}_t$ is a vector of factor returns that predict asset returns, $\mathbf{B}$ is a matrix of exposures or sensitivities of the asset returns to factor returns, and $\mathbf{u}_t$ is an idiosyncratic zero-mean noise term with constant covariance matrix

$$\text{var}_t(\mathbf{u}_{t+1}) := \Sigma.$$

The vector of factor returns $\mathbf{f}_t$ is known to the investor at time t and evolves according to

$$\Delta\mathbf{f}_{t+1} = -\Phi\mathbf{f}_t + \epsilon_{t+1},$$

where $\Delta\mathbf{f}_{t+1} = \mathbf{f}_{t+1} - \mathbf{f}_t$.

Trading is costly. The transaction cost associated with trading the vector of shares $\Delta\mathbf{x}_t = \mathbf{x}_t - \mathbf{x}_{t-1}$ is

$$TC(\Delta\mathbf{x}_t) = \tfrac{1}{2}\Delta\mathbf{x}_t^\mathsf{T}\Lambda\,\Delta\mathbf{x}_t$$

for some symmetric positive definite matrix Λ.

The model objective is

$$\max_{\mathbf{x}_0,\mathbf{x}_1,\ldots} \mathbb{E}_0\left[\sum_{t=0}^{\infty}(1-\rho)^{t+1}\left(\mathbf{r}_{t+1}^\mathsf{T}\mathbf{x}_t - \frac{\gamma}{2}\mathbf{x}_t^\mathsf{T}\Sigma\mathbf{x}_t\right) - \frac{(1-\rho)^t}{2}\Delta\mathbf{x}_t^\mathsf{T}\Lambda\,\Delta\mathbf{x}_t\right].$$

Gârleanu and Pedersen (2013) apply a dynamic programming approach to characterize the optimal trading strategy. We summarize the main results below.

The state at time t is the pair $(\mathbf{x}_{t-1}, \mathbf{f}_t)$. The value-to-go function is

$$V(\mathbf{x}_{t-1}, \mathbf{f}_t) = \max_{\mathbf{x}_t, \mathbf{x}_{t+1}, \dots} \mathbb{E}_t \left[\sum_{\tau=t}^{\infty} (1-\rho)^{\tau+1-t} \left(\mathbf{r}_{\tau+1}^{\mathsf{T}} \mathbf{x}_\tau - \frac{\gamma}{2} \mathbf{x}_\tau^{\mathsf{T}} \Sigma \mathbf{x}_\tau \right) - \frac{(1-\rho)^{\tau-t}}{2} \Delta \mathbf{x}_\tau^{\mathsf{T}} \Lambda \, \Delta \mathbf{x}_\tau \right].$$

Hence the Bellman equation is

$$V(\mathbf{x}_{t-1}, \mathbf{f}_t) = \max_{\mathbf{x}_t} \left\{ -\frac{1}{2} \Delta \mathbf{x}_t^{\mathsf{T}} \Lambda \, \Delta \mathbf{x}_t \right.$$
$$\left. + (1-\rho) \left(\mathbb{E}_t(\mathbf{r}_{t+1}^{\mathsf{T}} \mathbf{x}_t) - \frac{\gamma}{2} \mathbf{x}_t^{\mathsf{T}} \Sigma \mathbf{x}_t + \mathbb{E}_t\left[V(\mathbf{x}_t, \mathbf{f}_{t+1}) \right] \right) \right\}.$$

We make an educated guess and later verify (ansatz) the following quadratic form for the value function:

$$V(\mathbf{x}_t, \mathbf{f}_{t+1}) = -\tfrac{1}{2} \mathbf{x}_t^{\mathsf{T}} \mathbf{A}_{xx} \mathbf{x}_t + \mathbf{x}_t^{\mathsf{T}} \mathbf{A}_{xf} \mathbf{f}_{t+1} + \tfrac{1}{2} \mathbf{f}_{t+1}^{\mathsf{T}} \mathbf{A}_{ff} \mathbf{f}_{t+1} + a_0.$$

Using this ansatz, it can be shown that the optimal trading policy is

$$\mathbf{x}_t = \mathbf{x}_{t-1} + \Lambda^{-1} \mathbf{A}_{xx} \left(\mathbf{aim}_t - \mathbf{x}_{t-1} \right)$$

where

$$\mathbf{aim}_t = \mathbf{A}_{xx}^{-1} \mathbf{A}_{xf} \, \mathbf{f}_t.$$

The Bellman equation also yields expressions for the matrices $\mathbf{A}_{xx}, \mathbf{A}_{xf}, \mathbf{A}_{ff}$. See the exercises at the end of the chapter. In the special case $\Lambda = \lambda \Sigma$ we obtain

$$\mathbf{x}_t = \left(1 - \frac{a}{\lambda} \right) \mathbf{x}_{t-1} + \frac{a}{\lambda} \, \mathbf{aim}_t,$$

where

$$a = \frac{-(\gamma(1-\rho) + \lambda\rho) + \sqrt{(\gamma(1-\rho) + \lambda\rho)^2 + 4\gamma\lambda(1-\rho)^2}}{2(1-\rho)}.$$

Next, we get a more explicit expression of the $\mathbf{aim}$ portfolio. To that end, first observe that the myopic solution in the absence of transaction costs is precisely the solution to the static Markowitz model at time t; that is,

$$\mathtt{Markowitz}_t = (\gamma\Sigma)^{-1} \mathbf{B} \mathbf{f}_t.$$

Again we consider the special case $\Lambda = \lambda\Sigma$. For $z := \gamma/(\gamma + a)$ we get

$$\mathbf{aim}_t = z \cdot \mathtt{Markowitz}_t + (1-z)\mathbb{E}_t(\mathbf{aim}_{t+1})$$
$$= \sum_{\tau=t}^{\infty} z(1-z)^{\tau-t} \mathbb{E}_t(\mathtt{Markowitz}_\tau).$$

Furthermore, the portfolio $\mathbf{aim}_t$ has a similar form to $\mathtt{Markowitz}_t$ provided the forecasting signals are appropriately scaled down:

$$\mathbf{aim}_t = (\gamma\Sigma)^{-1} \mathbf{B} \left(\mathbf{I} + \frac{a}{\gamma} \Phi \right)^{-1} \mathbf{f}_t.$$

The optimal strategy is characterized by two principles. First, aim in front the target. Second, trade partially towards the current aim. More precisely, the optimal updated portfolio is a linear combination of the existing portfolio and a *aim* portfolio. The latter is a weighted average of the current Markowitz portfol (the moving target) and the expected Markowitz portfolios on all future date (where the target is moving).

14.4   Dynamic Portfolio Optimization with Taxes

Taxes pose a significant friction to most investors in financial markets. The are a variety of taxes that apply in different ways to income, dividends, an capital gains. It is common to ignore taxes in traditional finance and portfol theory. This simplification is in part due to the difficulties involved in modelir the effects of taxes.

Capital taxes introduce a peculiar type of challenge in portfolio managemen Since the sale of an appreciated asset triggers a capital gain tax liability, there a tradeoff between the benefits of diversification versus the tax costs triggered b rebalancing the portfolio. In addition to the tradeoff between diversification an taxes, many individual investors also have to deal with both a tax-deferred an a taxable account. In this context an investor faces an *asset location* proble in addition to the usual asset allocation problem. Asset location refers to th problem of how the investor should locate her portfolio holdings across the ta: deferred and taxable accounts.

Basic Case: Tax Management Only

In the United States tax code, capital gains and losses are triggered when asse are sold. This feature means that the investor could manage her assets in way that reduce her tax liabilities by choosing when to realize gains or losses. In th section we describe some models for optimal tax trading.

One of the earliest and most basic models for optimal tax trading was intro duced by Constantinides (1983). In this model it is assumed that the tax rat on capital gains is independent of the length of the holding period. It is als assumed that capital losses generate tax rebates. Finally, it is assumed tha there are no transaction costs, no capital loss restrictions, and no wash-sa restrictions. A wash sale occurs when an asset is sold at a capital loss and th same or substantially identical one is also purchased within 30 days before or afte the sale. Under these assumptions the optimal tax-trading strategy is relativel simple: Realize losses as soon as they occur and defer gains indefinitely. B realizing losses, the investor gets a tax rebate. If the investor did not realize th loss as soon as it happened, the opportunity for a tax rebate could disappea Constantinides's model can be extended to account for proportional transactio costs. If there are proportional transaction costs, then the optimal tax-tradin strategy would still be to defer gains but to realize losses only beyond a certai threshold. The exact size of the threshold depends on the size of the transactio costs, the tax rate, and the asset's volatility.

In a more elaborate follow-up article Constantinides (1984) proposed a model that considers a more realistic setting where the tax rate depends on the length of the holding period. In this model the sale of assets with long-term status is taxed at a rate lower than that of assets with short-term status. In this case the optimal tax-trading strategy still calls for realizing losses as soon as they occur. In addition and somewhat surprisingly, it is also sometimes optimal to sell (and immediately repurchase) assets with an embedded long-term gain. The rationale for this action is that there is a "re-start" option associated with resetting the tax basis and having the opportunity to realize short-term losses. The value of this re-start option depends on the asset volatility and the ratio of the short-term and long-term capital tax rates. The following example provided by C. Spatt[1] illustrates this phenomenon.

Example 14.1   Consider an asset with current price $P_0 = \$20$. Suppose that at dates $t = 0, 1$ we have

$$P_{t+1} = \begin{cases} P_t + k & \text{with probability } 0.5 \\ P_t - k & \text{with probability } 0.5. \end{cases}$$

Assume an investor buys one share of this asset at date $t = 0$. Our goal is to determine the trading strategy (realize/not realize) at dates $t = 1, 2$ that minimizes expected taxes.

(a) First consider the following case. The short-term and long-term capital gain tax rates are respectively $\tau_s = 0.5$, and $\tau_\ell = 0.5y$, where $0 < y < 1$. The sale of shares held for one period can be treated as either short-term or long-term depending on what is more advantageous to the investor. The sale of shares held for two periods is treated as long-term. Assume there are no transaction costs.

In this case at dates $t = 1$ and $t = 2$ it is optimal to realize losses.

At date $t = 1$ it is optimal to realize a long-term gain if $y < 0.5$. See Exercise 14.2.

(b) Now consider the case when the capital gain tax rate is $\tau = 0.2$ for both long-term and short-term gains or losses. Assume a transaction cost of 0.5 per share traded.

In this case at date $t = 2$ it is optimal to realize a loss if $k > 1.25$.

At date $t = 1$ it is optimal to realize a loss if $k > 5$.

Since short-term and long-term rates are the same, it is optimal not to realize gains at any date.

Portfolio Choice with Taxes

We now turn our attention to the problem of dynamic portfolio choice in the presence of capital gains taxes. The model below is a simplified version of a model proposed by Dammon et al. (2001).

---

[1] Personal communication.

We consider an economy with a risky and a risk-free asset where investors live for T periods. We also assume that in this economy investors are endowed with some initial capital and their goal is to maximize some expected utility of consumption C_t at dates $t = 0, 1, \ldots, T$ and bequest W_T at date T. The return of the risk-free asset between date $t-1$ and date t is r. The price of the risky asset is serially independent and follows a binomial process. Let P_t denote the price of the risky asset at date t. Let n_t and m_t denote respectively the number of shares of the risky and risk-free assets held right after trading at date t. Throughout the model we will assume no shorting, i.e., we will impose the constraints $n_t \geq 0$ and $m_t \geq 0$.

We assume that capital gains are taxed at a rate τ, and capital losses are credited at the same rate. To compute the capital gain triggered by an asset sale, we assume that the tax basis P_t^* for the shares at date t is the weighted average price of those shares. Therefore, the tax basis P_t^* evolves according to the following law of motion:

$$P_t^* = \begin{cases} \dfrac{n_{t-1} \cdot P_{t-1}^* + (n_t - n_{t-1})^+ \cdot P_t}{n_{t-1} + (n_t - n_{t-1})^+} & \text{if } P_{t-1}^* < P_t \\ P_t & \text{if } P_{t-1}^* \geq P_t. \end{cases}$$

Right after trading at date t, the realized capital gain or loss G_t is given by

$$G_t = \begin{cases} (n_{t-1} - n_t)^+(P_t - P_{t-1}^*) & \text{if } P_{t-1}^* \leq P_t \\ n_{t-1}(P_t - P_{t-1}^*) & \text{if } P_{t-1}^* \geq P_t. \end{cases}$$

We have the following inter-temporal balance of wealth equation that relates the portfolio holdings at dates $t-1$ to the portfolio holdings at dates $t = 1, \ldots, T-1$:

$$n_t P_t + m_t + C_t = n_{t-1} P_t + m_{t-1}(1 + r) - \tau G_t.$$

Similarly, at date T we have

$$W_T = n_{T-1} P_T + m_{T-1}(1 + r) - \tau G_T.$$

The portfolio choice problem can be stated as a dynamic programming problem where the state variables at date t are $(P_t, P_{t-1}^*, n_{t-1}, m_{t-1})$, the actions at time t are (n_t, m_t, C_t) and the objective is

$$\max_{C_t, n_t, m_t} \mathbb{E}\left[\sum_{t=0}^{T} U(C_t, t) + B(W_T)\right].$$

The following example illustrates the striking effect of taxes in portfolio choice

Example 14.2   Assume that at date $t = 0$ the holdings are $n_{-1} > 0$ and $m_{-1} = 0$. In other words, our entire portfolio is invested in the risky asset. Assume $r = 0$, $P_0 = 1$, $0 < 1 - k < P_{-1}^* = 1 - \delta < 1$, and

$$P_1 = \begin{cases} P_0 + k & \text{with prob } 1/2 \\ P_0 - k & \text{with prob } 1/2. \end{cases}$$

Assume $T = 1$ and our goal is to determine the portfolio holdings at date $t =$ so as to maximize some utility of final wealth $\mathbb{E}(U(W_1))$. Since there is n consumption at date $t = 0$ we have

$$n_0 P_0 + m_0 = n_{-1} P_0 + m_{-1} - \tau(n_{-1} - n_0)^+ (P_0 - P_{-1}^*).$$

Thus

$$m_0 = (n_{-1} - n_0)(1 - \tau\delta).$$

And so the only variable in our problem is n_0 subject to the constraints $0 \leq n_0 \leq n_1$.

Furthermore, the balance of wealth equation yields

$$W_1 = n_0 P_1 + m_0 + \tau n_0 (P_0^* - P_1)^+$$
$$= \begin{cases} n_0(\tau\delta + k) + n_{-1}(1 - \tau\delta) & \text{with prob } 1/2 \\ n_0(\tau k - k) + n_{-1}(1 - \tau\delta) & \text{with prob } 1/2. \end{cases}$$

It is evident that in the absence of taxes ($\tau = 0$), the optimal holding of the risky asset is $n_0 = 0$ for any positive level of risk aversion. However, it is easily checked numerically that n_0 may vary all the way between 0 and n_{-1} for positive values of τ. In particular, for $\tau = 0.2$, $\delta = 0.1$, $k = 0.2$ we get $n_0 = 0.8352$ for the logarithmic utility function.

The numerical solution to the more general model in Dammon et al. (2001) reveals the following interesting insights. As expected, it is optimal to realize capital losses as soon as they occur. Since diversification is more valuable to young investors, it is optimal for them to sell assets with large embedded capital gains to rebalance their portfolios. On the other hand, elderly investors defer most capital gains. Because in the US tax code there is a tax forgiveness at death, it is optimal for elderly investors to increase their allocations to equity as they approach their terminal age.

If in addition to some initial endowment an investor receives income, then the following insights are again revealed by a numerical solution to the model. Young investors hold more equity, very much in line with popular financial planning advice. Because of capital gain taxes, it is optimal to use income to adjust asset allocation instead of selling assets with embedded capital gains. In years immediately prior to retirement it is optimal to reduce equity allocation, again in line with popular financial planning advice. Finally, beyond retirement it is optimal to have a gradual increase in equity holdings.

Asset Allocation and Asset Location

The availability of various kinds of tax-deferred retirements accounts such as 401K, 403(b), IRA, and Keough give investors the ability to shelter some of their assets from taxes. Since assets may be held both in a tax-deferred as well as in a taxable account, the *location* decision has important implications on portfolio choice. Dammon et al. (2004) developed a model to study this problem. Via an arbitrage argument, they show that it is optimal to allocate assets to the tax-deferred account in descending order of tax exposure until the limit of the

tax-deferred account is reached. In particular, the assets with the highest taxab
yields, such as taxable bonds, should go in the tax-deferred account. If the limit
the tax-deferred account is reached, then assets with lower taxable yields shou
be allocated to the taxable account.

The numerical solution to the model in Dammon et al. (2004) also show
that, the larger the fraction of wealth in the tax-deferred account, the higher th
fraction of total wealth allocated to assets with higher taxable yield.

14.5   Exercises

Exercise 14.1   Consider the optimal consumption and investment model di
cussed in Section 14.2.

(a) Prove the envelope condition by proceeding as follows. Let $C_t^*(W_t)$ an
$\mathbf{x}_t^*(W_t)$ denote the optimal consumption and optimal portfolio at stage
respectively. Thus

$$J_t(W_t) = \mathbb{E}_t \left[J_{t+1} \left[(W_t - C_t^*(W_t)) \cdot (1 + \mathbf{r}_{t+1}^\mathsf{T} \mathbf{x}_t^*(W_t)) \right] + U(C_t^*(W_t)) \right].$$

Use the chain rule to differentiate both sides above with respect to W_t. The
use the optimality conditions for the Bellman equation to show that th
expression for the derivative of the right-hand side simplifies to $U'(C_t^*(W_t)$
thereby giving the envelope condition

$$J_t'(W_t) = U'(C_t^*(W_t)).$$

(b) Consider the special case $U(C) = \log(C)$, $B(W) = \log(W)$. Use inductio
to prove the following expressions for the value function and optimal con
sumption:

$$J_t(W_t) = \frac{\log(W_t)}{T - t + 1} + b_t$$

and

$$C_t^*(W_t) = \frac{W_t}{T - t + 1},$$

where b_t depends on the joint distribution of $R_{f,t+1}, \mathbf{r}_{t+1}$.

Exercise 14.2   Consider the model for short-term versus long-term taxe
described in Example 14.1.

(a) Suppose the short-term and long-term capital gain tax rates are respectivel
$\tau_s = 0.5$, and $\tau_\ell = 0.5y$, where $0 < y < 1$.

Prove that indeed at dates $t = 1$ and $t = 2$ it is optimal to realize losse
and at date $t = 1$ it is optimal to realize a long-term gain if $y < 0.5$.

(b) Suppose the capital gain tax rate is $\tau = 0.2$ for both long-term and short-term gains or losses and there is a transaction cost of 0.5 per share traded. Prove that at date $t = 2$ it is optimal to realize a loss if $k > 1.25$, and at date $t = 1$ it is optimal to realize a loss if $k > 5$. Prove that it is optimal not to realize gains at any date.

Exercise 14.3   Consider the model for portfolio choice with taxes described in Example 14.2.

(a) Prove that for $\tau = 0$ then the optimal holding of the risky asset at $t = 0$ is $n_0 = 0$ for any risk-averse concave utility function $U(W)$.

(b) Check numerically that for $\tau = 0.2$, $\delta = 0.1$, $k = 0.2$, and logarithmic utility function $U(W) = \log(W)$, the optimal holding at $t = 0$ is $n_0 = 0.8352$.

Exercise 14.4

(a) Consider the following minimum-variance portfolio optimization problem with risky assets, transaction costs, and no constraints:

$$\min_{\mathbf{x}} \left\{ \frac{1}{2} \mathbf{x}^\mathsf{T} \mathbf{V} \mathbf{x} + \frac{1}{2} (\mathbf{x} - \mathbf{x}_0)^\mathsf{T} \mathbf{R}(\mathbf{x} - \mathbf{x}_0) \right\}. \tag{14.2}$$

Here $\mathbf{x}_0$ is some initial portfolio, $\mathbf{V}$ is the covariance matrix of asset returns, and $\mathbf{R}$ is a symmetric positive definite matrix that models the transaction cost incurred in changing the initial portfolio $\mathbf{x}_0$ to the new portfolio $\mathbf{x}$. Prove that the solution to (14.2) is

$$\mathbf{x}^* = (\mathbf{V} + \mathbf{R})^{-1} \mathbf{R} \mathbf{x}_0.$$

(b) Now consider a multi-period version of the previous problem. Assume the investor starts with an initial portfolio $\mathbf{x}_0$ and her objective is

$$\min_{\mathbf{x}_1, \dots, \mathbf{x}_T} \sum_{t=1}^{T} \left(\frac{1}{2} \mathbf{x}_t^\mathsf{T} \mathbf{V} \mathbf{x}_t + \frac{1}{2} (\mathbf{x}_t - \mathbf{x}_{t-1})^\mathsf{T} \mathbf{R}(\mathbf{x}_t - \mathbf{x}_{t-1}) \right).$$

Apply dynamic programming to solve this problem. Proceed as follows:
- the stages are $t = 1, \dots, T$;
- the state at stage t is $\mathbf{x}_{t-1}$, that is, the portfolio previously set at stage $t - 1$;
- the action at stage t is the vector of holdings $\mathbf{x}_t$;
- the cost at stage t is the quadratic term

$$\frac{1}{2} \mathbf{x}_t^\mathsf{T} \mathbf{V} \mathbf{x}_t + \frac{1}{2} (\mathbf{x}_t - \mathbf{x}_{t-1})^\mathsf{T} \mathbf{R}(\mathbf{x}_t - \mathbf{x}_{t-1}).$$

(i) Show that the optimal optimal decision rule $\mathbf{x}_T^*$ and the value-to-go function $J_T(\mathbf{x}_{T-1})$ at the last stage T are

$$\mathbf{x}_T^* = (\mathbf{V} + \mathbf{R})^{-1} \mathbf{R} \mathbf{x}_{T-1}$$

and

$$J_T(\mathbf{x}_{T-1}) = \frac{1}{2} \mathbf{x}_{T-1}^\mathsf{T} \mathbf{K}_T \mathbf{x}_{T-1},$$

where $\mathbf{K}_T = \mathbf{R} - \mathbf{R}(\mathbf{V} + \mathbf{R})^{-1} \mathbf{R}$.

(ii) Use the Bellman equation and induction to prove that the optimal decision rule $\mathbf{x}_t^*$ and the value-to-go function $J_t(\mathbf{x}_{t-1})$ at each stage $t = T-1$, $T-2, \ldots, 1$ are of the form

$$\mathbf{x}_t^* = \mathbf{L}_t \mathbf{x}_{t-1}$$

and

$$J_t(\mathbf{x}_{t-1}) = \frac{1}{2} \mathbf{x}_{t-1}^\mathsf{T} \mathbf{K}_t \mathbf{x}_{t-1}$$

for some suitable matrices $\mathbf{L}_t$ and $\mathbf{K}_t$.

(c) Now consider an infinite horizon version of the previous problem. Assume there is only one risky asset, the investor starts with an initial portfolio x_0 and her objective is

$$\min_{x_1, x_2, \ldots} \sum_{t=1}^{\infty} \left(\frac{q}{2} x_t^2 + \frac{r}{2} (x_t - x_{t-1})^2 \right).$$

Use the following educated guess ("ansatz") for the optimal value function:

$$V(x_{t-1}) = \frac{k}{2} x_{t-1}^2 \qquad (14.3)$$

for some constant $k > 0$. In other words, our educated guess is that the value function starting at stage t with state x_{t-1}, namely

$$V(x_{t-1}) := \min_{x_t, x_{t+1}, \ldots} \sum_{\tau=t}^{\infty} \left(\frac{q}{2} x_\tau^2 + \frac{r}{2} (x_\tau - x_{\tau-1})^2 \right),$$

is of the form (14.3) for some constant $k > 0$.

Use the Bellman equation to verify this educated guess, determine the optimal decision rule x_t^*, and find the value of k in terms of q, r.

Give expressions for both k and the optimal portfolio x_t^* that are as explicit as possible.

Exercise 14.5

(a) Consider the mean–variance portfolio optimization problem with risk-free asset and no constraints:

$$\max_{\mathbf{x}} \left\{ \boldsymbol{\mu}^\mathsf{T} \mathbf{x} - \frac{\gamma}{2} \mathbf{x}^\mathsf{T} \mathbf{V} \mathbf{x} \right\}. \qquad (14.4)$$

Here $\boldsymbol{\mu}$ is the vector of excess returns, $\mathbf{V}$ is the covariance matrix, $\mathbf{x}$ is the vector of holdings in risky assets, and $\gamma > 0$ is a risk-aversion constant.

Prove that the solution to (14.4) is

$$\mathbf{x}^* = \frac{1}{\gamma} \mathbf{V}^{-1} \boldsymbol{\mu}.$$

(b) Now consider a variation of the previous problem that includes a quadratic term for transaction costs:

$$\max_{\mathbf{x}} \left\{ \boldsymbol{\mu}^\mathsf{T} \mathbf{x} - \frac{\gamma}{2} \mathbf{x}^\mathsf{T} \mathbf{V} \mathbf{x} - \frac{\lambda}{2} (\mathbf{x} - \mathbf{x}_0)^\mathsf{T} \mathbf{V} (\mathbf{x} - \mathbf{x}_0) \right\}. \qquad (14.5)$$

Here $\mathbf{x}_0$ is some initial portfolio and $\lambda > 0$ is a transaction cost constant.

Prove that the solution to (14.5) is

$$\mathbf{x}^* = \frac{1}{\gamma + \lambda} \mathbf{V}^{-1} \boldsymbol{\mu} + \frac{\lambda}{\gamma + \lambda} \mathbf{x}_0.$$

(c) Now consider a multi-period version of the previous problem. Assume the objective is

$$\max_{\mathbf{x}_1,\ldots,\mathbf{x}_T} \mathbb{E}\left[\sum_{t=1}^{T}\left(\boldsymbol{\mu}_t^\mathsf{T}\mathbf{x}_t - \frac{\gamma}{2}\mathbf{x}_t^\mathsf{T}\mathbf{V}\mathbf{x}_t - \frac{\lambda}{2}(\mathbf{x}_t - \mathbf{x}_{t-1})^\mathsf{T}\mathbf{V}(\mathbf{x}_t - \mathbf{x}_{t-1})\right)\right].$$

This assumes that the vector of expected returns may vary with time but not the covariance matrix. Apply dynamic programming to solve this problem. Proceed as follows:

- the stages are $t = 1, \ldots, T$;
- the state at stage t is $(\mathbf{x}_{t-1}, \boldsymbol{\mu}_t)$;
- the action at stage t is the vector of holdings $\mathbf{x}_t$;
- the reward at stage t is the quadratic term

$$\boldsymbol{\mu}_t^\mathsf{T}\mathbf{x}_t - \frac{\gamma}{2}\mathbf{x}_t^\mathsf{T}\mathbf{V}\mathbf{x}_t - \frac{\lambda}{2}(\mathbf{x}_t - \mathbf{x}_{t-1})^\mathsf{T}\mathbf{V}(\mathbf{x}_t - \mathbf{x}_{t-1}).$$

(i) Show that the optimal decision rule $\mathbf{x}_T^*$ and the value-to-go function $J_T(\mathbf{x}_{T-1}, \boldsymbol{\mu}_T)$ at the last stage T are

$$\mathbf{x}_T^* = a\,\mathbf{V}^{-1}\boldsymbol{\mu}_T + b\,\mathbf{x}_{T-1},$$

and

$$J_T(\mathbf{x}_{T-1}, \boldsymbol{\mu}_T) = \frac{c}{2}\boldsymbol{\mu}_T^\mathsf{T}\mathbf{V}^{-1}\boldsymbol{\mu}_T + \frac{d}{2}\mathbf{x}_{T-1}^\mathsf{T}\mathbf{V}\mathbf{x}_{T-1} + e\,\boldsymbol{\mu}_T^\mathsf{T}\mathbf{x}_{T-1},$$

for some constants a, b, c, d, e.

(ii) *Assume $\boldsymbol{\mu}_t$ has the following (simple and somewhat unrealistic) law of motion:

$$\boldsymbol{\mu}_{t+1} = \bar{\boldsymbol{\mu}} + \rho(\boldsymbol{\mu}_t - \bar{\boldsymbol{\mu}}) + \boldsymbol{\epsilon}_t,$$

where $\bar{\mu}, \rho$ are constants, $|\rho| < 1$, and $\boldsymbol{\epsilon}_t$ is a vector of independently normally distributed random shocks each with mean 0 and variance 1. Use the Bellman equation and induction to prove that the optimal decision rule $\mathbf{x}_t^*$ and the value-to-go function $J_t(\mathbf{x}_{t-1}, \boldsymbol{\mu}_t)$ at each stage $t = T - 1, T - 2, \ldots, 1$ are

$$\mathbf{x}_t^* = a_t\mathbf{V}^{-1}\boldsymbol{\mu}_t + b_t\mathbf{x}_{t-1},$$

and

$$J_t(\mathbf{x}_{t-1}, \boldsymbol{\mu}_t) = \frac{c_t}{2}\boldsymbol{\mu}_t\mathbf{V}^{-1}\boldsymbol{\mu}_t + \frac{d_t}{2}\mathbf{x}_{t-1}^\mathsf{T}\mathbf{V}\mathbf{x}_{t-1} + e_t\boldsymbol{\mu}_t^\mathsf{T}\mathbf{x}_{t-1} + f_t,$$

for some constants $a_t, b_t, c_t, d_t, e_t, f_t$.

15 Dynamic Programming Models: the Binomial Pricing Model

One of the most common uses of dynamic programming in financial mathematics is through lattice models. In particular, the *binomial lattice model* of Cox et al (1979) has become an indispensable tool for pricing derivative securities. This chapter describes this model and the underlying dynamic programming principle for the pricing of European options and the pricing and optimal exercising American options.

15.1 Binomial Lattice Model

The binomial lattice provides a model for the price movements of a risky asset. It can be seen as a multi-period version of the single-period binomial model discussed in Section 4.3. The binomial lattice model describes the price of risky asset at some discrete times $0, 1, \ldots, N$. A basic period length, such as week, day, or second, is assumed to elapse between any two consecutive times. The model assumes that if the share price of the risky asset is S_k at time k the the share price S_{k+1} at time $k+1$ can take two values, namely $S_{k+1} = u \cdot S_k$ and $S_{k+1} = d \cdot S_k$ where $u > d > 0$ are multiplicative factors (u stands for "up" and d for "down" factors). The probabilities assigned to these two possible states a p and $1 - p$ respectively, where $0 < p < 1$. The multi-stage price structure can be represented on a lattice as illustrated in Figure 15.1.

After k time periods, the asset price can take $k+1$ different values. If the price at stage 0 is S_0, then the price S_k at stage k is $u^j d^{k-j} S_0$ if there are j up moves and $k - j$ down moves. Observe that there are $\binom{k}{j}$ possible paths to reach the node corresponding to j up moves and $t-j$ down moves after t periods. Therefore the probability that the price is $u^j d^{k-j} S_0$ in stage k is $\binom{k}{j} p^j (1 - p)^{k-j}$ because between two consecutive times the probability of an up move is p whereas that of a down move is $1 - p$.

15.2 Option Pricing

Using the above binomial lattice model for the price process of an underlying risky asset, the value of an option on this asset can be computed by dynamic programming by using backward recursion, working from the maturity date (time N

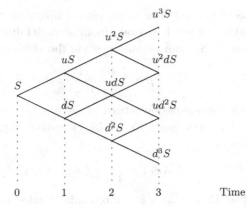

Figure 15.1 Asset price in the binomial lattice model

back to time 0 (the current time). The approach fits within the stochastic setting introduced in Section 13.7. More precisely, the stages of the dynamic program are the discrete times $k = 0, \ldots, N$. The state at stage k is the asset price S_k. Thus S_k can take the $k+1$ possible values defined by the $k+1$ nodes in the kth layer of the lattice. The state S_N at the final stage N is the *terminal state*. The law of motion for the asset price is as follows:

$$S_{k+1} = \begin{cases} uS_k & \text{with probability } p \\ dS_k & \text{with probability } q = 1 - p \end{cases}$$

for $k = 0, 1, \ldots, N$. However, the following adjustment of utmost importance must be made: for option pricing purposes, we do not use these actual probabilities p and $q = 1 - p$ but instead the risk-neutral probabilities $\tilde{p}$ and $\tilde{q} = 1 - \tilde{p}$ as explained below.

15.2.1 European Options

Consider a European option contract that matures at time N with payoff $g(S_N)$ for some function $g(\cdot)$ of the underlying asset price S_N. For instance, if the contract is a European call option maturing at time N with strike price K, the payoff at maturity is $g(S_N) = (S_N - K)^+$. Similarly, if the contract is a European put option maturing at time N with strike price K, the payoff at maturity is $g(S_N) = (K - S_N)^+$.

Let $V_k(S_k)$ denote the value of the option at stage k when the asset price is S_k. This is the value-to-go function in our dynamic program. The value of the option at stage 0 is given by $V_0(S_0)$. This is the quantity that we have to compute in order to solve the option pricing problem. At the final time N the value-to-go function is given by the payoff of the option contract. That is,

$$V_N(S_N) = g(S_N).$$

Since we are dealing with a European option, we can compute the value $V_k(\cdot)$

in terms of $V_{k+1}(\cdot)$. The single-period subproblem between stages k and $k+1$
identical to the single-period binomial model discussed in Section 4.3. Therefor
the value $V_k(S_k)$ can be obtained via the *risk-neutral* probabilities (4.4), name

$$\tilde{p} = \frac{1+r-d}{u-d} \quad \text{and} \quad \tilde{q} = \frac{u-1-r}{u-d},$$

where r is the one-period return on the risk-free asset between time k and time k
1. Thus for European options the value-to-go functions $V_k(\cdot)$ can be recursive
computed as

$$V_k(S_k) = \frac{1}{1+r}\left(\tilde{p}V_{k+1}(uS_k) + \tilde{q}V_{k+1}(dS_k)\right). \tag{15.}$$

Example 15.1   Consider a binomial lattice model with $N = 3$, $u = 2$, d
$\frac{1}{2}$, $r = 0.25$, and $S_0 = 40$ as depicted in Figure 15.2. Compute the price of
European call option with strike price $K = 50$.

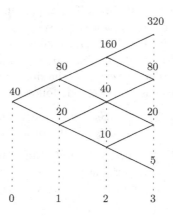

Figure 15.2 Binomial lattice with $u = 2, d = 0.5, S_0 = 40, N = 3$

For these values of u, d, r the risk-neutral probabilities are

$$\tilde{p} = \tilde{q} = \frac{0.75}{1.5} = \frac{1}{2}.$$

The option value $V_3(S_3) = (S_3 - 50)^+$ at the final stage 3 is as follows:

$$V_3(5) = V_3(20) = 0, \quad V_3(80) = 30, \quad V_3(320) = 270.$$

Next, applying (15.1) we get $V_2(S_2)$:

$$V_2(10) = 0, \quad V_2(40) = \frac{1}{1.25}\cdot\frac{30}{2} = 12, \quad V_2(160) = \frac{1}{1.25}\cdot\frac{30+270}{2} = 120.$$

Applying (15.1) again we get $V_1(S_1)$:

$$V_1(20) = \frac{1}{1.25}\cdot\frac{12}{2} = 4.8, \quad V_1(80) = \frac{1}{1.25}\cdot\frac{120+12}{2} = 52.8.$$

Finally applying (15.1) one more time, we get the option value $V_0(S_0)$:

$$V_0(40) = \frac{1}{1.25} \cdot \frac{4.8 + 52.8}{2} = 23.04.$$

Example 15.2 Consider a binomial lattice model with $N = 3$, $u = 2$, $d = \frac{1}{2}$, $r = 0.25$, and $S_0 = 40$. Compute the price of a European put option with strike price $K = 60$.

Again the risk-neutral probabilities are $\tilde{p} = \tilde{q} = \frac{1}{2}$. We can proceed as in Example 15.1. The option value $V_3(S_3) = (60 - S_3)^+$ at the final stage 3 is

$$V_3(5) = 55, \quad V_3(20) = 40, \quad V_3(80) = V_3(320) = 0.$$

Next, applying (15.1) we get $V_2(S_2)$:

$$V_2(10) = \frac{1}{1.25} \cdot \frac{55 + 40}{2} = 38, \quad V_2(40) = \frac{1}{1.25} \cdot \frac{40}{2} = 16, \quad V_2(160) = 0.$$

Applying (15.1) again we get $V_1(S_1)$:

$$V_1(20) = \frac{1}{1.25} \cdot \frac{38 + 16}{2} = 21.6, \quad V_1(80) = \frac{1}{1.25} \cdot \frac{16}{2} = 6.4.$$

Finally applying (15.1) one more time, we get the option value $V_0(S_0)$:

$$V_0(40) = \frac{1}{1.25} \cdot \frac{21.6 + 6.4}{2} = 11.2.$$

15.2.2 American Options

Consider now an American option contract that can be exercised at any time $k = 0, 1, \ldots, N$ with payoff $g(S_k)$ for some function $g(\cdot)$ of the underlying asset price S_k. The key difference between this type of *American option* contract and the above type of *European option* contract is the possibility of early exercise. Because of this additional feature in the contract, the pricing problem of an American option needs to account for the optimal exercise timing of the option. This is accomplished via an adjustment to the previous recursion in the calculation of the value function.

Once again, at the final time N the value-to-go function is given by the payoff of the option contract. That is,

$$V_N(S_N) = g(S_N).$$

The computation of the value-to-go function $V_k(\cdot)$ in terms of $V_{k+1}(\cdot)$ needs to reflect the possibility of early exercise. To that end, the recursive formula (15.1) needs to be amended as follows:

$$V_k(S_k) = \max\left\{\frac{1}{1+r}\left(\tilde{p}V_{k+1}(uS_k) + \tilde{q}V_{k+1}(dS_k)\right), g(S_k)\right\}. \tag{15.2}$$

In words, the value of the option at stage k is the maximum of the following two quantities: the first one is the discounted value of the option at stage $k+1$ or the payoff obtained if the option is exercised immediately. When the latter is larger it is optimal to exercise the option at stage k.

Example 15.3 Consider a binomial lattice model with $N = 3$, $u = 2$, $d = \frac{1}{2}$, $r = 0.25$, and $S_0 = 40$. Compute the price of an American call option with strike price $K = 50$.

Again the risk-neutral probabilities are $\tilde{p} = \tilde{q} = \frac{1}{2}$. The option value $V_3(S_3) =$ $(S_3 - 50)^+$ at the final stage 3 is

$$V_3(5) = V_3(20) = 0, \quad V_3(80) = 30, \quad V_3(320) = 270.$$

Next, applying (15.2) we get $V_2(S_2)$:

$$V_2(10) = 0, \quad V_2(40) = \max\left\{\frac{1}{1.25} \cdot \frac{30}{2}, (40 - 50)^+\right\} = 12,$$

$$V_2(160) = \max\left\{\frac{1}{1.25} \cdot \frac{30 + 270}{2}, (160 - 50)^+\right\} = 120.$$

Observe that regardless of the value of S_2, it is not optimal to exercise the option at stage 2.

Applying (15.2) again we get $V_1(S_1)$:

$$V_1(20) = \max\left\{\frac{1}{1.25} \cdot \frac{12}{2}, (20 - 50)^+\right\} = 4.8,$$

$$V_1(80) = \max\left\{\frac{1}{1.25} \cdot \frac{120 + 12}{2}, (80 - 50)^+\right\} = 52.8.$$

Observe that regardless of the value of S_1, it is not optimal to exercise the option at stage 1.

Finally applying (15.2) one more time, we get the option value $V_0(S_0)$:

$$V_0(40) = \max\left\{\frac{1}{1.25} \cdot \frac{52.8 + 4.8}{2}, (40 - 50)^+\right\} = 23.04.$$

Observe that there is no difference in price and in exercise policy between the American and the European call options.

Example 15.4 Consider a binomial lattice model with $N = 3$, $u = 2$, $d = \frac{1}{2}$, $r = 0.25$, and $S_0 = 40$. Compute the price of an American put option with strike price $K = 60$.

Once again, the risk-neutral probabilities are $\tilde{p} = \tilde{q} = \frac{1}{2}$. The option value $V_3(S_3) = (60 - S_3)^+$ at the final stage 3 is

$$V_3(5) = 55, \quad V_3(20) = 40, \quad V_3(80) = V_3(320) = 0.$$

Next, applying (15.2) we get $V_2(S_2)$

$$V_2(10) = \max\left\{\frac{1}{1.25} \cdot \frac{55 + 40}{2}, (60 - 10)^+\right\} = 50,$$

$$V_2(40) = \max\left\{\frac{1}{1.25} \cdot \frac{40}{2}, (60 - 40)^+\right\} = 20, \quad V_2(160) = 0.$$

Observe that it is optimal to exercise the option at stage 2 when $S_2 = 10$ and $S_2 = 40$.

Applying (15.2) again we get $V_1(S_1)$

$$V_1(20) = \max\left\{\frac{1}{1.25} \cdot \frac{50 + 20}{2}, (60 - 20)^+\right\} = 40,$$

$$V_1(80) = \max\left\{\frac{1}{1.25} \cdot \frac{20}{2}, (60 - 80)^+\right\} = 8.$$

Observe that it is optimal to exercise the option at stage 1 when $S_1 = 20$.

Finally applying (15.2) one more time, we get the option value $V_0(S_0)$

$$V_0(40) = \max\left\{\frac{1}{1.25} \cdot \frac{40 + 8}{2}, (60 - 40)^+\right\} = 20.$$

Observe that it is optimal to exercise the option at this stage.

Notice that the prices of the American and European calls in Example 15.1 and Example 15.3 are identical. This happens because there is nothing to gain by early exercising of the American call. By contrast, there is a substantial difference in the prices of the American and European puts in Example 15.2 and Example 15.4. This happens because sometimes it is advantageous to exercise an American put early. The results of these examples illustrate the following far more general property of American options.

Theorem 15.5   *Consider the binomial lattice model described in Section 15.1, and an American option contract on the underlying risky asset that can be exercised at any time $k = 0, 1, \ldots, N$ with payoff $g(S_k)$ for some function $g(\cdot)$. If $r \geq 0$ and the function $g(\cdot)$ is convex and satisfies $g(0) = 0$, then the value of the American option contract is the same as that of a European option contract with payoff $g(S_N)$ that can only be exercised at stage N. In other words, early exercising of the American option yields no advantage.*

Proof   By (15.2), it suffices to show that the following equation and inequality hold for $k = 0, 1, \ldots, N - 1$:

$$V_k(S_k) = \frac{1}{1+r}\left(\tilde{p}V_{k+1}(uS_k) + \tilde{q}V_{k+1}(dS_k)\right) \geq g(S_k). \tag{15.3}$$

First, observe that for $k = 0, 1, \ldots, N - 1$

$$S_k = \frac{1}{1+r}(\tilde{p}uS_k + \tilde{q}dS_k), \tag{15.4}$$

since $\tilde{p}, \tilde{q}$ are the risk-neutral probabilities.

Next, we prove (15.3) by (backward) induction on k. The assumptions on the function $g(\cdot)$, equation (15.4), and $V_N(S_N) = g(S_N)$ imply that

$$g(S_{N-1}) = g\left(\frac{r \cdot 0 + \tilde{p}uS_{N-1} + \tilde{q}dS_{N-1}}{1+r}\right)$$

$$\leq \frac{r}{1+r} \cdot g(0) + \frac{1}{1+r}(\tilde{p}g(uS_{N-1}) + \tilde{q}g(dS_{N-1}))$$

$$= \frac{1}{1+r}(\tilde{p}V_N(uS_{N-1}) + \tilde{q}V_N(dS_{N-1})).$$

Therefore (15.3) holds for $k = N-1$. Suppose (15.3) holds for $k = j+1 \leq N-$
The assumptions on $g(\cdot)$, equation (15.4), and the induction hypothesis imp
that

$$g(S_j) = g\left(\frac{r \cdot 0 + \tilde{p}uS_j + \tilde{q}dS_j}{1+r}\right)$$

$$\leq \frac{r}{1+r} \cdot g(0) + \frac{1}{1+r}(\tilde{p}g(uS_j) + \tilde{q}g(dS_j))$$

$$\leq \frac{1}{1+r}(\tilde{p}V_{j+1}(uS_j) + \tilde{q}V_{j+1}(dS_j)).$$

Hence (15.3) holds for $k = j$ as well.

15.3   Option Pricing in Continuous Time

The binomial lattice model can be seen as a discrete version of a popula
continuous-time *geometric Brownian motion model*. We next sketch some of th
main ideas and results of this continuous model and its relation to the binomi
lattice model discussed above. A full treatment of this topic is beyond the scop
of this book. We refer the reader to Shreve (2000) for a detailed exposition
this topic.

Suppose the continuous-time price S_t, with $t \in [0, T]$, of a risky asset evolv
according to the stochastic differential equation

$$\frac{dS_t}{S_t} = \mu dt + \sigma dW_t, \tag{15.5}$$

where μ and σ are constants representing the instantaneous *drift* and *volatili*
of the asset price S_t, and W_t is a Brownian motion. The stochastic differenti
equation (15.5) can be seen as a continuous-time analog of the one-period up
down price movement in the binomial lattice model. The solution to (15.5) is th
continuous-time process

$$S_t = S_0 e^{(\mu - \sigma^2/2)t + \sigma W_t}, \tag{15.6}$$

which can equivalently be written as

$$\log \frac{S_t}{S_0} = \left(\mu - \frac{\sigma^2}{2}\right)t + \sigma W_t.$$

Techniques from stochastic calculus have led to the development of pricin
models for a wide variety of options provided the underlying risky asset
modeled via a suitable stochastic differential equation. In particular, in the
seminal and ground-breaking work Black and Scholes (1973) and Merton (197
derived a pricing formula for a European option on an underlying risky asset wit
a price process modeled as a geometric Brownian motion. In particular, consid
a call option maturing at time $T > 0$ with payoff $(S_T - K)^+$. Assume the pric
of the underlying risky asset is as in (15.6) and the risk-free asset compoun
continuously at an instantaneous rate $r \geq 0$; that is, the price B_t of the risk-fre

asset is

$$B_t = B_0 e^{rt}.$$

The Black–Scholes–Merton model yields the following explicit formula for the price $V_t(S_t)$ at time $t \in [0, T]$ of a European call option with payoff $(S_T - K)^+$:

$$V_t(S_t) = \Phi(d_1)S_t - \Phi(d_2)Ke^{-r(T-t)}, \qquad (15.7)$$

where

$$\Phi(x) = \frac{1}{2\pi} \int_{-\infty}^{x} e^{-t^2/2} dt, \quad d_1 = \frac{1}{\sigma\sqrt{T-t}} \left[\log\left(\frac{S_t}{K}\right) + \left(r + \frac{\sigma^2}{2}\right)(T-t) \right],$$

$$d_2 = d_1 - \sigma\sqrt{T-t}.$$

The Black–Scholes–Merton model also yields the following formula for the price $V_t(S_t)$ at time $t \in [0, T]$ of a European put option with payoff $(K - S_T)^+$:

$$V_t(S_t) = \Phi(-d_2)Ke^{-r(T-t)} - \Phi(-d_1)S_t,$$

where $\Phi(\cdot), d_1, d_2$ are the same as above.

The binomial lattice model can be seen as a discrete approximation of the geometric Brownian motion. The following section and Exercise 15.4 at the end of the chapter elaborate on this approximation.

15.4 Specifying the Model Parameters

To specify the binomial lattice model, one needs to choose values for u, d, and p. This is done by matching the mean and volatility of the asset price to the mean and volatility of the above binomial distribution. Because the model is multiplicative (the price S of the asset being either $u \cdot S$ or $d \cdot S$ in the next stage), it is convenient to work with $\log(S_{k+1}/S_k)$.

Let S_k denote the asset price in stages $k = 0, \ldots, N$. Let μ and σ be the mean and volatility of $\ln(S_N/S_0)$. (We assume that this information about the asset is known.) Let $\Delta = 1/N$ denote the length between consecutive stages. Then for $k = 0, 1, \ldots, N-1$ the mean and volatility of $\ln(S_{k+1}/S_k)$ are $\mu\Delta$ and $\sigma\sqrt{\Delta}$ respectively. In the binomial lattice, a direct computation shows that for $k = 0, 1, \ldots, N-1$ the mean and variance of $\ln(S_{k+1}/S_k)$ are $p\ln u + (1-p)\ln d$ and $p(1-p)(\ln u - \ln d)^2$ respectively. Matching these values we get two equations:

$$p\ln u + (1-p)\ln d = \mu\Delta$$
$$p(1-p)(\ln u - \ln d)^2 = \sigma^2\Delta.$$

Note that there are three parameters but only two equations, so we can set $d = 1/u$ as in Cox et al. (1979). Then the equations simplify to

$$(2p-1)\ln u = \mu\Delta$$
$$4p(1-p)(\ln u)^2 = \sigma^2\Delta.$$

Squaring the first and adding it to the second, we get $(\ln u)^2 = \sigma^2 \Delta + (\mu \Delta)^2$. This yields

$$u = e^{\sqrt{\sigma^2 \Delta + (\mu \Delta)^2}}$$

$$d = e^{-\sqrt{\sigma^2 \Delta + (\mu \Delta)^2}}$$

$$p = \frac{1}{2} \left(1 + \frac{1}{\sqrt{1 + \sigma^2/\mu^2 \Delta}} \right).$$

When Δ is small, these values can be approximated as

$$u \approx e^{\sigma \sqrt{\Delta}}$$

$$d \approx e^{-\sigma \sqrt{\Delta}}$$

$$p \approx \frac{1}{2} \left(1 + \frac{\mu}{\sigma} \sqrt{\Delta} \right).$$

In other words, for small Δ

$$\log \frac{S_{k+1}}{S_k} \approx \begin{cases} \sigma \sqrt{\Delta} & \text{with probability } \frac{1}{2} \left(1 + \frac{\mu}{\sigma} \sqrt{\Delta} \right) \\ -\sigma \sqrt{\Delta} & \text{with probability } \frac{1}{2} \left(1 - \frac{\mu}{\sigma} \sqrt{\Delta} \right), \end{cases}$$

which is a discrete approximation of (15.5).

As an example, consider a binomial model with 52 periods of one week each. Consider also a stock with current known price S_0 and random price S_{52} a year from today. We are given the mean μ and volatility σ of $\ln(S_{52}/S_0)$, say $\mu = 10\%$ and $\sigma = 30\%$. What are the parameters u, d, and p of the binomial lattice? Since $\Delta = \frac{1}{52}$ is small, we can use the second set of formulas:

$$u \approx e^{0.30/\sqrt{52}} = 1.0425$$

$$d \approx e^{-0.30/\sqrt{52}} = 0.9592$$

$$p \approx \frac{1}{2} \left(1 + \frac{0.10}{0.30\sqrt{52}} \right) = 0.523.$$

15.5   Exercises

Exercise 15.1 Apply Theorem 15.5 to show that the price of a European call option and an American call option with the same strike price and expiration date are the same in the binomial lattice model. Why does Theorem 15.5 not apply for put options?

Exercise 15.2 Compute the value of an American call option on a stock with current price equal to $100, strike price equal to $102, and expiration date four weeks from today. The yearly volatility of the logarithm of the stock return is $\sigma = 0.30$. The risk-free interest rate is 4%. Use a binomial lattice with $N = 4$.

Exercise 15.3 Compute the value of an American put option on a stock with current price equal to $100, strike price equal to $98, and expiration date five weeks from today. The yearly volatility of the logarithm of the stock return is $\sigma = 0.30$. The risk-free interest rate is 4%. Use a binomial lattice with $N = 4$.

Exercise 15.4 This is a computational exercise. Repeat Exercises 15.2 and 15.3 using a binomial lattice with $N = 10$, $N = 100$, and $N = 1000$. Compare the results obtained for the call option with those given by the Black–Scholes–Merton formula (15.7).

16 Multi-Stage Stochastic Programming

Stochastic programming is a computational approach to stochastic optimization. Stochastic programs have been studied for several decades (see, e.g., Birge and Louveaux, 1997; Shapiro et al., 2009). Typically, the approach hinges on reformulation of the stochastic optimization problem as a deterministic one via a *scenario tree*. Computational and algorithmic advances have made stochastic programming techniques applicable to various classes of real-world problems.

16.1 Multi-Stage Stochastic Programming

Multi-stage stochastic optimization can be seen as a generalization of the general class of stochastic optimization model discussed in Chapter 10. Let $0, 1, \ldots,$ index a set of stages where decisions are to be made. Assume that between two consecutive stages $t - 1$ and t some random outcome ω_t is revealed. At each stage $t = 0, 1, \ldots, T$ we make a set of *non-anticipatory* decisions $\mathbf{x}_t$ that can only depend on the random information revealed up until that stage. Schematically, the process can be seen as follows:

$$\underset{\mathbf{x}_0}{\text{decision}} \rightsquigarrow \underset{\text{draw } \omega_1}{\text{random}} \rightsquigarrow \underset{\mathbf{x}_1}{\text{decision}} \rightsquigarrow \underset{\text{draw } \omega_2}{\text{random}} \rightsquigarrow \cdots \rightsquigarrow \underset{\text{draw } \omega_T}{\text{random}} \rightsquigarrow \underset{\mathbf{x}_T}{\text{decision}}$$

A multi-stage stochastic minimization problem is the following kind of *multi-fold* version of the two-stage stochastic model (10.2) discussed in Chapter 10:

$$\min_{\mathbf{x}_0} \quad g_0(\mathbf{x}_0) + \mathbb{E}[Q_1(\mathbf{x}_0, \omega_1)] \tag{16.1}$$
$$\mathbf{x}_0 \in \mathcal{X}_0,$$

where the recourse term $Q_1(\mathbf{x}_0, \omega_1)$ similarly depends on the decisions to be made at later stages:

$$Q_1(\mathbf{x}_0, \omega_1) := \min_{\mathbf{x}_1} \quad g_1(\mathbf{x}_1, \omega_1) + \mathbb{E}[Q_2(\mathbf{x}_1, \omega_2)]$$
$$\mathbf{x}_1 \in \mathcal{X}_1(\mathbf{x}_0, \omega_1),$$

with

$$Q_t(\mathbf{x}_{t-1}, \omega_t) := \min_{\mathbf{x}_t} \quad g_t(\mathbf{x}_t, \omega_t) + \mathbb{E}[Q_{t+1}(\mathbf{x}_t, \omega_{t+1})]$$
$$\mathbf{x}_t \in \mathcal{X}_t(\mathbf{x}_{t-1}, \omega_t),$$

for $t = 2, \ldots, T-1$, and the last-stage recourse term $Q_T(\mathbf{x}_{T-1}, \omega_T)$ is of the form

$$Q_T(\mathbf{x}_{T-1}, \omega_T) := \min_{\mathbf{x}_T} \quad g_T(\mathbf{x}_T, \omega_T)$$
$$\mathbf{x}_T \in \mathcal{X}_T(\mathbf{x}_{T-1}, \omega_T).$$

The multi-stage optimization problem (16.1) can also be written as

$$\min_{\mathbf{x}_0 \in \mathcal{X}_0} g_0(\mathbf{x}_0) + \mathbb{E}\left[\min_{\mathbf{x}_1 \in \mathcal{X}_1(\mathbf{x}_0, \omega_1)} g_1(\mathbf{x}_1, \omega_1) + \cdots + \mathbb{E}\left[\min_{\mathbf{x}_T \in \mathcal{X}_T(\mathbf{x}_{T-1}, \omega_T)} g_T(\mathbf{x}_T, \omega_T)\right]\right].$$

Consider the special case of linear multi-stage stochastic optimization, where the components are linear. More precisely, each $g_t(\mathbf{x}_t, \omega_t) = \mathbf{c}_t^\mathsf{T}\mathbf{x}_t$ for some vector $\mathbf{c}_t$ and each inter-temporal constraint $\mathbf{x}_t \in \mathcal{X}_t(\mathbf{x}_{t-1}, \omega_t)$ is of the form

$$\mathbf{B}_t\mathbf{x}_{t-1} + \mathbf{A}_t\mathbf{x}_t = \mathbf{b}_t, \quad \mathbf{x}_t \geq 0,$$

where $\omega_t = (\mathbf{c}_t, \mathbf{A}_t, \mathbf{B}_t, \mathbf{b}_t)$ is only revealed at stage t. In this case we often write

$$
\begin{aligned}
\min_{\mathbf{x}_0, \mathbf{x}_1, \ldots, \mathbf{x}_T} \quad & \mathbb{E}\left[\mathbf{c}_0^\mathsf{T}\mathbf{x}_0 + \mathbf{c}_1^\mathsf{T}\mathbf{x}_1 + \cdots + \mathbf{c}_T^\mathsf{T}\mathbf{x}_T\right] \\
\text{s.t.} \quad & \mathbf{A}_0\mathbf{x}_0 && = \mathbf{b}_0 \\
& \mathbf{B}_1\mathbf{x}_0 + \mathbf{A}_1\mathbf{x}_1 && = \mathbf{b}_1 \\
& \quad\quad\quad \mathbf{B}_2\mathbf{x}_1 + \mathbf{A}_2\mathbf{x}_2 && = \mathbf{b}_2 \quad (16.2) \\
& \quad\quad\quad\quad\quad\quad\quad \vdots && \vdots \\
& \quad\quad\quad\quad\quad \mathbf{B}_T\mathbf{x}_{T-1} + \mathbf{A}_T\mathbf{x}_T && = \mathbf{b}_T \\
& \mathbf{x}_0, \quad\quad \mathbf{x}_1, \quad\quad \cdots \quad\quad \mathbf{x}_T && \geq \quad 0.
\end{aligned}
$$

Example 16.1 (Financial planning example) Assume an investor has initial wealth W_0 at $t = 0$. At stage t she can invest in two asset classes: bonds and stocks. The (random) gross return on bonds from time $t-1$ to t is $R_{b,t}$ and the (random) gross return on stocks from $t-1$ to t is $R_{s,t}$. Assume that the investor needs to meet liabilities L_t at times $t = 1, \ldots, T$. She wants to maximize her expected wealth at time T (after covering the liabilities). Assume no shorting is allowed.

Formulation of the financial planning example
Variables:

x_t: amount of money invested in bonds at stage t, for $t = 0, \ldots, T-1$;
y_t: amount of money invested in stocks at stage t, for $t = 0, \ldots, T-1$;
W_T: wealth at time T

$$
\begin{aligned}
\max \quad & \mathbb{E}(W_T) \\
\text{s.t.} \quad & x_0 + y_0 = W_0 \\
& R_{b,t}x_{t-1} + R_{s,t}y_{t-1} = L_t + x_t + y_t, \quad t = 1, \ldots, T-1 \\
& R_{b,T}x_{T-1} + R_{s,T}y_{T-1} = L_T + W_T \\[4pt]
& x_t, y_t \geq 0, \quad t = 0, 1, \ldots, T-1 \\
& W_T \geq 0.
\end{aligned}
$$

Notice that in this model the parameters $R_{b,t}$ and $R_{s,t}$ are unknown prior to time t.

16.2   Scenario Optimization

As discussed in Chapter 10 for the two-stage case, a multi-stage optimization model can be recast as a *deterministic equivalent* if each of the random outcomes has a discrete distribution. In this case for each stage $t = 1, 2, \ldots, T$ there is a finite set of possible values or realizations $\{\omega_t^1, \ldots, \omega_t^S\}$ for the random outcome ω_t. These sets of realizations can be described by an *event tree* as depicted in Figure 16.1 for a problem with three stages. In this particular tree, the random variables ω_1 and ω_2 have two- and five-valued discrete distributions respectively. The tree structure is associated with the discrete filtration generated by the discrete-time random process $\omega_{[t]} := (\omega_1, \omega_2, \ldots, \omega_t)$, $t = 1, \ldots, T$. In particular, each possible value of ω_2 has a unique *predecessor* value of ω_1. We further elaborate on this tree structure below.

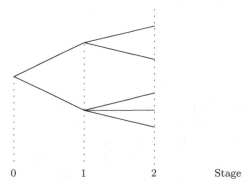

$$0 \qquad\qquad 1 \qquad\qquad 2 \qquad\qquad \text{Stage}$$

Figure 16.1 Event tree for a three-stage model

The set of scenarios described by the event tree in turn yield a deterministic equivalent of a multi-stage stochastic optimization model. We next illustrate this equivalence for the stochastic optimization model described in Example 16. in a particularly simple event tree. We subsequently describe the deterministic equivalent for linear multi-stage stochastic programs in more general event trees.

Example 16.2 (Financial planning revisited)   Consider the model described in Example 16.1. Suppose $T = 2$ and there are two equally likely outcomes ("H" and "T") for the joint returns $(R_{b,t}, R_{s,t})$ over each period, say

$$(R_{b,t}(H), R_{s,t}(H)) = (1.14, 1.25) \quad \text{and} \quad (R_{b,t}(T), R_{s,t}(T)) = (1.1, 1.06).$$

Figure 16.2 illustrates the corresponding scenario tree. In this event tree th labels "H" and "T" on the edges indicate the specific outcome between tw consecutive stages. Observe that each of the four scenarios HH, HT, TH, TT in the event tree occurs with probability $1/4$.

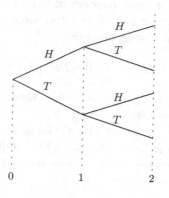

Figure 16.2 Event tree for financial planning model

The scenario tree in turn yields a deterministic equivalent formulation for the financial planning stochastic optimization model. In the deterministic equivalent the stage 0 decisions are made at the root of the tree and thus may not depend on any of the random outcomes. The stage 1 decisions may depend on the initial path H or T realized up to stage 1 in the event tree. Finally, the stage 2 decisions may depend on the path HH, HT, TH, or TT realized up to stage 2 in the event tree. The corresponding adaptiveness of the variables and constraints is explicitly reflected in the following deterministic equivalent formulation.

Scenario optimization model
Variables:

x_0, y_0: money in bonds and stocks at $t = 0$;
$x_1(H), x_1(T), y_1(H), y_1(T)$: money in bonds and stocks at $t = 1$;
$W_2(HH), W_2(HT), W_2(TH), W_2(TT)$: wealth at $t = 2$

$$\max \quad \frac{1}{4} \cdot (W_2(HH) + W_2(HT) + W_2(TH) + W_2(TT))$$

s.t. $\quad x_0 + y_0 = W_0$ \hfill (stage 0)

$1.14x_0 + 1.25y_0 = L_1 + x_1(H) + y_1(H)$ \hfill (stage 1, path H)
$1.1x_0 + 1.06y_0 = L_1 + x_1(T) + y_1(T)$ \hfill (stage 1, path T)

$1.14x_1(H) + 1.25y_1(H) = L_2 + W_2(HH)$ \hfill (stage 2, HH)
$1.1x_1(H) + 1.06y_1(H) = L_2 + W_2(HT)$ \hfill (stage 2, HT)
$1.14x_1(T) + 1.25y_1(T) = L_2 + W_2(TH)$ \hfill (stage 2, TH)
$1.1x_1(T) + 1.06y_1(T) = L_2 + W_2(TT)$ \hfill (stage 2, TT)

$x_0, y_0, x_1(H), x_1(T), y_1(H), y_1(T) \geq 0$
$W_2(HH), W_2(HT), W_2(TH), W_2(TT) \geq 0.$

The above scenario optimization approach is quite flexible. In particular, consider a variation of the above financial planning model where the objective is $\max \mathbb{E}\left(U(W_T)\right)$ for some concave utility function $U(W)$. The corresponding deterministic equivalent has exactly the same variables and constraints as the one above and the following objective:

$$\max \frac{1}{4}\left(U(W_2(HH)) + U(W_2(TH)) + U(W_2(HT)) + U(W_2(TT))\right).$$

Furthermore, if $U(\cdot)$ is piecewise linear, then the problem can be recast as a linear program. (See Exercise 16.1.)

Consider now the general multi-stage linear stochastic program (16.2). Suppose each random vector $\omega_t = (\mathbf{c}_t, \mathbf{A}_t, \mathbf{B}_t, \mathbf{b}_t)$ has a discrete distribution and consider their event tree representation. The description of the deterministic equivalent relies on the following notation. Let

$$\Omega_t := \{\omega_t^k = (\mathbf{c}_t^k, \mathbf{A}_t^k, \mathbf{B}_t^k, \mathbf{b}_t^k) : k = 1, \ldots, S_t\}$$

be the set of possible realizations of the random variable ω_t for some integer $S_t \geq 1$ and for each stage $t = 1, \ldots, T$. Let $p_t^k = \mathbb{P}(\omega_t = \omega_t^k)$, with $k = 1, \ldots, S_t$, $t = 1, \ldots, T$. The set $\Omega_t = \{\omega_t^1, \ldots, \omega_t^{S_t}\}$ corresponds to the nodes in layer t of the event tree, which can be conveniently denoted $(t, 1), \ldots, (t, S_t)$ as illustrated in Figure 16.3.

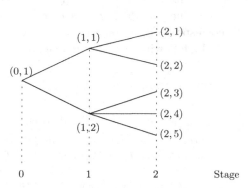

Figure 16.3 Event tree for a three-stage model with node labels

Observe that in the event tree there is always a single root node $(0, 1)$ in layer 0. Furthermore, for each $t = 1, \ldots, T - 1$ each node (t, k) has a *unique* predecessor $(t-1, \hat{k})$ in the immediately preceding layer of the tree. For instance, in the event tree depicted in Figure 16.3 the predecessors of each of the nodes in layer 2 is a unique node in layer 1 as follows

$$\hat{1} = \hat{2} = 1, \ \hat{3} = \hat{4} = \hat{5} = 2.$$

Observe that the probability of a non-terminal node equals the combined probability of its direct descendants; that is, for $t = 1, \ldots, T$ and for every node

$(t-1, \ell)$ we have

$$p_{t-1}^{\ell} = \sum_{(t,k):\hat{k}=\ell} p_t^k.$$

Consider the multi-stage linear stochastic program (16.2) and assume the random outcomes are described via a suitable event tree. We next detail the variables, objective, and constraints of the corresponding deterministic linear program equivalent.

Variables: Stage 0 variables: $\mathbf{x}_0$.

Stage t variables can be adapted to the S_t possible paths up to stage t; that is,

$$\mathbf{x}_t^k, \quad k = 1, \ldots, S_t.$$

Objective: The deterministic equivalent of the objective function

$$\min \mathbb{E}\left[\mathbf{c}_0^{\mathsf{T}}\mathbf{x}_0 + \mathbf{c}_1^{\mathsf{T}}\mathbf{x}_1 + \cdots + \mathbf{c}_T^{\mathsf{T}}\mathbf{x}_T\right] = \min \mathbb{E}\left[\mathbf{c}_0^{\mathsf{T}}\mathbf{x}_0 + \sum_{t=1}^{T} \mathbf{c}_t^{\mathsf{T}}\mathbf{x}_t\right]$$

is

$$\min \mathbf{c}_0^{\mathsf{T}}\mathbf{x}_0 + \sum_{t=1}^{T} \sum_{k=1}^{S_t} p_t^k (\mathbf{c}_t^k)^{\mathsf{T}} \mathbf{x}_t^k.$$

Constraints: The deterministic equivalent of each inter-temporal constraint

$$\mathbf{B}_t \mathbf{x}_{t-1} + \mathbf{A}_t \mathbf{x}_t = \mathbf{b}_t, \quad \mathbf{x}_t \geq 0,$$

is the set of constraints

$$\mathbf{B}_t^k \mathbf{x}_{t-1}^{\hat{k}} + \mathbf{A}_t^k \mathbf{x}_t^k = \mathbf{b}_t, \quad \mathbf{x}_t^k \geq 0, \quad \text{for } k = 1, \ldots, S_t.$$

Observe that these constraints link the stage t variables $\mathbf{x}_t^k$ associated with the layer t nodes (t, k) with the variable associated with their predecessor $(t-1, \hat{k})$.

Thus the complete deterministic equivalent of (16.2) is as follows:

$$\min \; c_0^{\mathsf{T}}\mathbf{x}_0 + \sum_{k=1}^{S_1} p_1^k(\mathbf{c}_1^k)^{\mathsf{T}}\mathbf{x}_1^k + \quad \cdots \quad + \sum_{k=1}^{S_T} p_T^k(\mathbf{c}_T^k)^{\mathsf{T}}\mathbf{x}_T^k \tag{16.3}$$

$$
\begin{aligned}
\text{s.t. } \mathbf{A}_0\mathbf{x}_0 & & & = \mathbf{b}_0 \\
\mathbf{B}_1^k\mathbf{x}_0 + & \mathbf{A}_1^k\mathbf{x}_1^k & & = \mathbf{b}_1^k, \; k = 1, \ldots, S_1 \\
& \mathbf{B}_2^k\mathbf{x}_1^{\hat{k}} + \; \mathbf{A}_2^k\mathbf{x}_2^k & & = \mathbf{b}_2, \; k = 1, \ldots, S_2 \\
& \quad\;\; \vdots & & \quad \vdots \\
& \mathbf{B}_T^k\mathbf{x}_{T-1}^{\hat{k}} + & \mathbf{A}_T^k\mathbf{x}_T^k = \mathbf{b}_T^k, & \; k = 1, \ldots, S_T \\
\mathbf{x}_0, & \quad \mathbf{x}_1^k, \quad \cdots & \mathbf{x}_T^k \geq 0. &
\end{aligned}
$$

For example, if $T = 2$ and the event tree is as depicted in Figure 16.3, the the deterministic equivalent is

$$\min \ \mathbf{c}_0^\mathsf{T}\mathbf{x}_0 \ + \ p_1^1(\mathbf{c}_1^1)^\mathsf{T}\mathbf{x}_1^1 \ + \ p_1^2(\mathbf{c}_1^2)^\mathsf{T}\mathbf{x}_1^2 \ + \ p_2^1(\mathbf{c}_2^1)^\mathsf{T}\mathbf{x}_2^1 \ + \ p_2^2(\mathbf{c}_2^2)^\mathsf{T}\mathbf{x}_2^2 \ + \ p_2^3(\mathbf{c}_2^3)^\mathsf{T}$$
$$+ \ p_2^4(\mathbf{c}_2^4)^\mathsf{T}\mathbf{x}_2^4 \ + \ p_2^5(\mathbf{c}_2^5)^\mathsf{T}$$

s.t.
$$
\begin{aligned}
\mathbf{A}_0\mathbf{x}_0 &= \mathbf{b} \\
\mathbf{B}_1^1\mathbf{x}_0 + \mathbf{A}_1^1\mathbf{x}_1^1 &= \mathbf{b} \\
\mathbf{B}_1^2\mathbf{x}_0 + \mathbf{A}_1^2\mathbf{x}_1^2 &= \mathbf{b} \\
\mathbf{B}_2^1\mathbf{x}_1^1 + \mathbf{A}_2^1\mathbf{x}_2^1 &= \mathbf{b} \\
\mathbf{B}_2^2\mathbf{x}_1^1 + \mathbf{A}_2^2\mathbf{x}_2^2 &= \mathbf{b} \\
\mathbf{B}_2^3\mathbf{x}_1^2 + \mathbf{A}_2^3\mathbf{x}_2^3 &= \mathbf{b} \\
\mathbf{B}_2^4\mathbf{x}_1^2 + \mathbf{A}_2^4\mathbf{x}_2^4 &= \mathbf{b} \\
\mathbf{B}_2^5\mathbf{x}_1^2 + \mathbf{A}_2^5\mathbf{x}_2^5 &= \mathbf{b}
\end{aligned}
$$
$$\mathbf{x}_0, \quad \mathbf{x}_1^1, \quad \mathbf{x}_1^2, \quad \mathbf{x}_2^1, \quad \mathbf{x}_2^2, \quad \mathbf{x}_2^3, \quad \mathbf{x}_2^4, \quad \mathbf{x}_2^5 \geq 0.$$

Observe that the constraint matrix in the above model has the followir structure:

$$
\begin{bmatrix}
\mathbf{A}_0 & & & & & & & \\
\mathbf{B}_1^1 & \mathbf{A}_1^1 & & & & & & \\
\mathbf{B}_1^2 & & \mathbf{A}_1^2 & & & & & \\
& \mathbf{B}_2^1 & & \mathbf{A}_2^1 & & & & \\
& \mathbf{B}_2^2 & & & \mathbf{A}_2^2 & & & \\
& & \mathbf{B}_2^3 & & & \mathbf{A}_2^3 & & \\
& & \mathbf{B}_2^4 & & & & \mathbf{A}_2^4 & \\
& & \mathbf{B}_2^5 & & & & & \mathbf{A}_2^5
\end{bmatrix}.
$$

The constraint matrix for the general deterministic equivalent (16.3) has a simila type of structure.

There are alternative ways of labeling the nodes and branches in the even tree. In particular, the simple binary branching in Example 16.2 readily sugges the natural edge labeling depicted in Figure 16.2. In that case the nodes in laye t can alternatively be labeled via the t-long sequence of labels along the pat from the root node. This kind of edge labeling may appear more intuitive br it could become cumbersome in other cases when there are different numbers of branches at each non-terminal node.

Note that the size of the deterministic equivalent of a multi-stage stochast program increases rapidly with the number of stages. For example, for a problem with 11 stages and a binary event tree, there are $2^{10} = 1024$ scenarios an therefore the linear program (16.3) may have several thousand constraints an variables, depending on the number of variables and constraints at each node Modern commercial codes can handle such large linear programs, but a moderat increase in the number of stages or in the number of branches at each stage coul make (16.3) too large to solve by standard linear programming solvers. When this happens, it is critical to exploit the special structure of (16.3) to solve the model efficiently.

The Benders decomposition (or L-shaped method) introduced in Section 10.5 can also be used for multi-stage problems (16.3) in a straightforward way: The stages are partitioned into a first set that gives rise to the "master problem" and a second set that gives rise to the "recourse problems". For example in a six-stage problem, the variables of the first two stages could define the master problem. When these variables are fixed, (16.3) decomposes into separate linear programs each involving variables of the last four stages. The solutions of these recourse linear programs provide optimality or feasibility cuts that can be added to the master problem. As discussed in Section 10.5, upper and lower bounds are computed at each iteration and the algorithm stops when the difference drops below a given tolerance. Using this approach, Gondzio and Kouwenberg (2001) were able to solve an asset liability management problem with over four million scenarios, whose linear programming formulation (16.3) had 12 million constraints and 24 million variables. This linear program was so large that storage space on the computer became an issue. The scenario tree had six levels and 13 branches at each node. In order to apply Benders' decomposition, Gondzio and Kouwenberg divided the six-period problem into a first-stage problem containing the first three periods and a second-stage problem containing periods four to six. This resulted in 2197 recourse linear programs, each involving 2197 scenarios. These recourse linear programs were solved by an interior-point algorithm. Note that Benders' decomposition is ideally suited for parallel computations since the recourse linear programs can be solved simultaneously. When the solution of all the recourse linear programs is completed (which takes the bulk of the time), the master problem is then solved on one processor while the other processors remain idle temporarily. Gondzio and Kouwenberg tested a parallel implementation on a computer with 16 processors and they obtained an almost perfect speedup, that is a speedup factor of almost k when using k processors.

16.3 Scenario Generation

A key aspect of multi-stage stochastic programming is the generation of scenarios so that the deterministic equivalent formulation (16.3) accurately represents the underlying stochastic optimization problem.

There are two separate issues. First, one needs to model the correlation over time among the random parameters. For a pension fund, such a model might relate wage inflation (a random parameter that influences the liability side) to interest rates and stock prices (random parameters that influence the asset side). Below we discuss a simple autoregressive model that can be used for this purpose. A second issue is the construction of a scenario tree from these inter-temporal statistical models: A finite number of scenarios must reflect as accurately as

possible the random processes modeled in the previous step, suggesting the nee
for a large number of scenarios. On the other hand, the linear program (16.3) ca
only be solved if the size of the scenario tree is reasonable, suggesting a limite
number of scenarios. To reconcile these two conflicting objectives, it might l
crucial to use variance reduction techniques. We address these issues in th
section.

Autoregressive Model

In order to generate the random parameters underlying the stochastic progra
one needs to construct an economic model reflecting the correlation between t
parameters. Historical data may be available. The goal is to generate meaningf
time series for constructing the scenarios. One approach is to use an autoregre
sive model.

Specifically, if $\mathbf{r}_t$ denotes the random vector of parameters in period t, a
autoregressive model is defined by

$$\mathbf{r}_t = \mathbf{D}_0 + \mathbf{D}_1\mathbf{r}_{t-1} + \cdots + \mathbf{D}_p\mathbf{r}_{t-p} + \boldsymbol{\epsilon}_t,$$

where p is the number of lags used in the regression, $\mathbf{D}_0, \mathbf{D}_1, \ldots, \mathbf{D}_p$ are tim
independent constant matrices, which are estimated through statistical metho
such as maximum likelihood, and $\boldsymbol{\epsilon}_t$ is a vector of i.i.d. random disturbanc
with mean zero.

To illustrate this, consider a problem where the vector $\mathbf{r}_t$ consists of thr
random parameters: s_t, b_t, and m_t are the rates of return of stocks, bonds, ar
the money market, respectively, in year t. An autoregressive model with $p =$
has the form:

$$\begin{bmatrix} s_t \\ b_t \\ m_t \end{bmatrix} = \begin{bmatrix} d_1 \\ d_2 \\ d_3 \end{bmatrix} + \begin{bmatrix} d_{11} & d_{12} & d_{13} \\ d_{21} & d_{22} & d_{23} \\ d_{31} & d_{32} & d_{33} \end{bmatrix} \begin{bmatrix} s_{t-1} \\ b_{t-1} \\ m_{t-1} \end{bmatrix} + \begin{bmatrix} \epsilon_t^s \\ \epsilon_t^b \\ \epsilon_t^m \end{bmatrix}, \quad t = 2, \ldots, T.$$

Assuming independent error terms ϵ_t^s, ϵ_t^b, and ϵ_t^m, and using historical dat
one can find the parameters $d_1, d_{11}, d_{12}, d_{13}$ in the first equation,

$$s_t = d_1 + d_{11}s_{t-1} + d_{12}b_{t-1} + d_{13}m_{t-1} + \epsilon_t^s,$$

using standard linear regression tools that minimize the sum of squared erro
ϵ_t^s. Useful statistics, such as the standard error σ_s of the estimates s_t, can al
be obtained. Similarly for b_t and m_t.

Constructing Scenario Trees

The random distributions relating the various parameters of a stochastic pr
gram must be discretized to generate a set of scenarios that is adequate for i
deterministic equivalent. Too few scenarios may lead to approximation error
On the other hand, too many scenarios will lead to an explosion in the size

the scenario tree, leading to an excessive computational burden. In this section, we discuss a simple random sampling approach and two variance reduction techniques: adjusted random sampling and tree fitting. Unfortunately, scenario trees constructed by these methods could contain spurious arbitrage opportunities. We end this section with a procedure to test that this does not occur.

Random Sampling

One can generate scenarios directly from the autoregressive model introduced in the previous section:

$$\mathbf{r}_t = \mathbf{D}_0 + \mathbf{D}_1\mathbf{r}_{t-1} + \cdots + \mathbf{D}_p\mathbf{r}_{t-p} + \boldsymbol{\epsilon}_t,$$

where $\epsilon_t \sim N(\mathbf{0}, \Sigma)$ are independently distributed multivariate normal distributions with mean 0 and covariance matrix Σ.

In our example with three random parameters s_t, b_t, and m_t, and independent error terms ϵ_t^s, ϵ_t^b, ϵ_t^m, the matrix Σ is a 3×3 diagonal matrix, with diagonal entries σ_s, σ_b, σ_m. Thirty branches or so may be needed to get a reasonable approximation of the distribution of the rates of return in stage 1. For a problem with three stages, 30 branches at each stage represent 27,000 scenarios. With more stages, the size of the linear program (16.3) explodes. Kouwenberg (2001) performed tests on scenario trees with fewer branches at each node (such as a five-stage problem with branching structure 10–6–6–4–4, meaning ten branches at the root, then six branches at each node in the next stage and so on) and he concluded that random sampling on such trees leads to unstable investment strategies. This occurs because the approximation error made by representing parameter distributions by random samples can be significant in a small scenario tree. As a result the optimal solution of (16.3) is not optimal for the actual parameter distributions. How can one construct a scenario tree that more accurately represents these distributions, without blowing up the size of the linear program (16.3)?

Adjusted Random Sampling

An easy way of improving upon random sampling is as follows. Assume that each node of the scenario tree has an even number $K = 2k$ of branches. Instead of generating $2k$ random samples from the autoregressive model, generate k random samples only and use the negative of their error terms to compute the values on the remaining k branches. This will fit all the odd moments of the distributions correctly. In order to fit the variance of the distributions as well, one can scale the sampled values. The sampled values are all scaled by a multiplicative factor until their variance fits that of the corresponding parameter.

Tree Fitting

How can one best approximate a continuous distribution by a discrete distribution with K values? In other words, how should one choose values v_k and their probabilities p_k, for $k = 1, \ldots, K$, in order to approximate the given distribution as accurately as possible? A natural answer is to match as many of the moments as possible. In the context of a scenario tree, the problem is somewhat more complicated since there are several correlated parameters at each node and there is interdependence between periods as well. Hoyland and Wallace (2001) propose to formulate this fitting problem as a nonlinear program. The fitting problem can be solved either at each node separately or on the overall tree. We explain the fitting problem at a node. Let S_l be the values of the statistical properties of the distributions that one desires to fit, for $l = 1, \ldots, s$. These might be the expected values of the distributions, the correlation matrix, the skewness and kurtosis. Let $\mathbf{v}_k$ and p_k denote the vector of values on branch k and its probability, respectively, for $k = 1, \ldots, K$. Let $f_l(\mathbf{v}, \mathbf{p})$ be the mathematical expression of property l for the discrete distribution (for example, the mean of the vectors $\mathbf{v}_k$, their correlation, skewness, and kurtosis). Each property has a positive weight w_l indicating its importance in the desired fit. Hoyland and Wallace formulate the fitting problem as

$$\min_{\mathbf{v}, \mathbf{p}} \quad \sum_l w_l (f_l(\mathbf{v}, \mathbf{p}) - S_l)^2$$

$$\text{s.t.} \quad \sum_k p_k = 1 \tag{16.}$$

$$\mathbf{p} \geq 0.$$

One might want some statistical properties to match exactly. As an example, consider again the autoregressive model:

$$\mathbf{r}_t = \mathbf{D}_0 + \mathbf{D}_1 \mathbf{r}_{t-1} + \cdots + \mathbf{D}_p \mathbf{r}_{t-p} + \epsilon_t,$$

where $\epsilon_t \sim N(\mathbf{0}, \Sigma)$ are independently distributed multivariate normal distributions with mean 0 and covariance matrix Σ. To simplify notation, let us write instead of ϵ_t. The random vector ϵ has distribution $N(\mathbf{0}, \Sigma)$ and we would like to approximate this continuous distribution by a finite number of disturbance vectors ϵ^k occuring with probability p_k, for $k = 1, \ldots, K$. Let ϵ_q^k denote the qth component of vector ϵ^k. One might want to fit the mean of ϵ exactly and its covariance matrix as well as possible. In this case, the fitting problem is

$$\min_{\epsilon^1, \ldots, \epsilon^K, \mathbf{p}} \quad \sum_{q=1}^l \sum_{r=1}^l \left(\sum_{k=1}^K p_k \epsilon_q^k \epsilon_r^k - \Sigma_{qr} \right)^2$$

$$\text{s.t.} \quad \sum_{k=1}^K p_k \epsilon^k = \mathbf{0}$$

$$\sum_k p_k = 1$$

$$\mathbf{p} \geq 0.$$

Arbitrage-Free Scenario Trees
Approximating the continuous distributions of the uncertain parameters by a finite number of scenarios in the linear program (16.3) typically creates modeling

errors. In fact, if the scenarios are not chosen properly or if their number is too small, the supposed "linear programming equivalent" could be far from being equivalent to the original stochastic optimization problem. One of the most disturbing aspects of this phenomenon is the possibility of creating arbitrage opportunities when constructing the scenario tree. When this occurs, model (16.3) is flawed as it would be distorted by the arbitrage opportunities. Klaassen (2002) was the first to address this issue. In particular, he shows how arbitrage opportunities can be detected *ex post* in a scenario tree. See Exercise 16.3 for details. When arbitrage opportunities exist, a simple solution is to discard the scenario tree and to construct a new one with more branches. Klaassen also discusses what constraints to add to the nonlinear program (16.4) in order to preclude arbitrage opportunities *ex ante*. The additional constraints are nonlinear, thus increasing the difficulty of solving (16.4).

16.4 Exercises

Exercise 16.1 Consider the following variation of the "financial planning" problem discussed in Section 16.1:

- Assume there are four stages, 0 through 3 (i.e., $T = 3$).
- Over each period we have two equally likely outcomes for joint returns of bonds and stocks: (14%, 25%) and (10%, 6%).
- Initial wealth $W_0 = 55$. $L_1 = L_2 = 0$. Final liability $L_3 = 70$.

(a) Use a multi-stage scenario optimization approach to determine the sequence of investment decisions so that the liability is met at $T = 3$, and the expected value of the remaining wealth is maximized. Your investment decisions at each stage must be non-anticipatory. That is, they can only depend on the scenario path up to that stage.

(b) Modify your model to solve the following variation: Instead of the single liability at stage 3, the following sequence of liabilities must be met at stages 1, 2, 3:

$$L_1 = 20, \ L_2 = 20, \ L_3 = 25.$$

(c) Assume this time that $L_1 = L_2 = 0$ and that the final liability is $L_3 = 90$. Since this is too high, it is clear that, regardless of the investment decisions, the final wealth W_3 will be negative in some scenarios. Modify your model to maximize instead $\mathbb{E}(U(W_3))$, where the utility function $U(W)$ is as follows:

$$U(W) = \begin{cases} W & \text{if } W \geq 0 \\ 3W & \text{if } W < 0. \end{cases}$$

It is preferable for your model to be a linear program for computational purposes. For that purpose, you need to recast the objective

$$\max \ \mathbb{E}(U(W_T))$$

so that the resulting model is a linear program. To do so, observe tha $U(W) = \min\{W, 3W\}$ and use a suitable set of new variables.

Exercise 16.2   Consider the following dynamic portfolio problem.

- At time $t = 0$ you have an initial endowment W_0.
- At time $t = 0, 1, \ldots, T - 1$ you invest a fraction x_t of your wealth in a risk asset and the remaining fraction $1 - x_t$ in a risk-free asset. The risk-free an risky asset returns between t and $t + 1$ are $r_{f,t+1}$ and $r_{s,t+1}$ respectively.
- At time $t = 1, \ldots, T$ you receive an exogenous and deterministic income of I

Let W_t denote your wealth at time $t = 0, 1, \ldots, T$. Your goal is to maximiz utility of final wealth $\mathbb{E}(U(W_T))$.

(a) Write the law of motion for W_t.
(b) Assume that between t and $t + 1$ there are two equally likely outcomes and T. The risky return $r_{s,t+1}$ in each of these scenarios is as follows:
  - In outcome H: $r_{s,t+1} = 0.5$.
  - In outcome T: $r_{s,t+1} = -0.4$.
  Assume a zero risk-free return, $r_{f,t+1} = 0$, in both scenarios.
  Suppose $T = 2$. Write down the "law of motion" for W_1 and W_2 in eac relevant scenario using the above numerical values for $r_{f,t+1}, r_{s,t+1}$.
(c) Assume $W_0 = 1$ and $I_1 = I_2 = 0.1$. Use scenario optimization and Exce Solver or MATLAB to solve the two-period portfolio optimization proble

$$\max_{x_0, x_1} \mathbb{E}\left[\frac{W_2^{1-\gamma}}{1-\gamma}\right]$$

for $\gamma = 0.4, 0.7, 0.9$.
(d) Repeat part (c) but this time assume $W_0 = 1$ and $I_1 = I_2 = 0$.
(e) Repeat part (c) but this time assume $W_0 = 1$ and $I_1 = I_2 = 0.2$.
(f) Can you infer anything from the numerical results in (c), (d), and (e) abou long-term investment when you know you will receive income along th investment horizon?

Exercise 16.3   Recall from Section 4.2 in Chapter 4 that an arbitrage oppo tunity is an opportunity to make money without any cost and without any ris Consider a particular node at some stage $t - 1 \geq 0$ in a scenario tree whose se of immediate descendants in stage t is K. For each $k \in K$ let $\mathbf{r}^k \in \mathbb{R}^n$ denote th vector of asset returns of a set of n assets realized in branch k between stage $t - 1$ and t.

(a) Show that an arbitrage opportunity exists if there is an asset allocatio $\mathbf{x} = \begin{bmatrix} x_1 & \cdots & x_n \end{bmatrix}$ such that

$$\sum_{j=1}^{n} x_j \leq 0, \quad \sum_{j=1}^{n} (\mathbf{r}^k)^{\mathsf{T}} \mathbf{x} \geq 0,$$

where at least one inequality is strict.

(b) Show that the condition in part (a) holds if and only if the following condition does not hold: there exist $y_k > 0$, for $k \in K$, such that

$$\sum_{k \in K} y_k \mathbf{r}^k = 1.$$

Hint: Apply the same kind of reasoning used in Section 4.2.

(c) Use part (a) and part (b) above to modify the nonlinear program (16.4) in order to formulate a fitting problem at a node that does not contain any arbitrage opportunities.

17 Stochastic Programming Models: Asset–Liability Management

17.1   Asset–Liability Management

The financial health of any company, and in particular of financial institution is reflected in the balance sheet of the company. Proper management of t company requires attention to both sides of the balance sheet – assets a liabilities. *Asset–liability management* offers sophisticated mathematical too for an integrated management of assets and liabilities.

Asset–liability management recognizes that static, one-period investment pla ning models (such as mean–variance optimization) fail to incorporate the mul period nature of the liabilities faced by the company. A multi-period model th emphasizes the need to meet liabilities in each period for a finite (or possib infinite) horizon is often required. Since liabilities and asset returns usually ha random components, their optimal management requires techniques to optimi under uncertainty. In particular, stochastic programming approaches have bee effective for these kinds of problems.

The main components of the asset–liability management problem are t stream of (random) liabilities faced by the firm, spread out over time, and t (random) returns of the assets that the firm may use for investments. Positio can be adjusted at each intermediate stage, adapting to the information reveal up to that stage. This is closely related to the financial planning example pr sented in Example 16.1.

The model assumes a planning horizon of T periods. Let $R_{i,t}$ denote the gro return of asset i between time $t-1$ and t, for $i = 1, \ldots, n$ and $t = 1, \ldots, T$. L L_t denote the liability at time $t = 1, \ldots, T$. Suppose we want to maximize t expected wealth of the firm at time T.

Multi-stage stochastic programming formulation
Variables:
$x_{i,t}$: amount invested in asset i at time t, for $i = 1, \ldots, n$ and $t = 0, 1, \ldots, T-$

Objective:

$$\max \quad \mathbb{E}\left[\sum_{i=1}^{n} R_{i,T} x_{i,T-1} - L_T\right]$$

$$\text{s.t.} \quad \sum_{i=1}^{n} R_{i,t} x_{i,t-1} = L_t + \sum_{i=1}^{n} x_{i,t}, \quad \text{for } t = 1, \ldots, T-1$$

$$x_{i,t} \geq 0, \quad \text{for } i = 1, \ldots, n, \ t = 1, \ldots, T-1.$$

The equality constraint in this formulation states that the surplus left after liability L_t is covered will be invested in the amounts $x_{i,t}$ in asset i for $i = 1, \ldots, n$.

The objective selected in the model above is to maximize the expected wealth at the end of the planning horizon. In practice, one might have a different objective. For example, in some cases, minimizing value at risk (VaR) or conditional value at risk (CVaR) might be more appropriate. Other priorities may dictate other objective functions.

To address the issue of the most appropriate objective function, one must understand the role of liabilities. Pension funds and insurance companies are among the most typical arenas for the integrated management of assets and liabilities.

17.2 The Case of an Insurance Company

We consider the case of a Japanese insurance company, the Yasuda Fire and Marine Insurance Co. Ltd., following the work of Cariño et al. (1994). In this case, the liabilities are mainly savings-oriented policies issued by the company. Each new policy sold represents a deposit, or inflow of funds. Interest is periodically credited to the policy until maturity, typically three to five years, at which time the principal amount plus credited interest is refunded to the policyholder. The crediting rate is typically adjusted each year in relation to a market index like the prime rate. Therefore, we cannot say with certainty what the future liabilities will be. Insurance business regulations stipulate that interest credited to some policies be earned from investment income, not capital gains. So, in addition to ensuring that the maturity cash flows are met, the firm must seek to avoid interim shortfalls in income earned versus interest credited. In fact, it is the risk of not earning adequate income quarter by quarter that the decision makers view as the primary component of risk at Yasuda.

The problem is to determine the optimal allocation of the deposited funds into several asset categories: cash, fixed-rate and floating-rate loans, bonds, equities, real estate, and other assets. Since we can revise the portfolio allocations over time, the decision we make is not just among allocations today but among

allocation strategies over time. A realistic dynamic asset–liability model mu
also account for the payment of taxes. This is made possible by distinguishir
between interest income and price return.

A stochastic linear program is used to model the problem. The linear progra
has uncertainty in many coefficients. This uncertainty is modeled through a fini
number of scenarios. In this fashion, the problem is transformed into a ver
large-scale linear program of the form (16.3). The random elements include pri
return and interest income for each asset class, as well as policy crediting rate
We next describe the main components of the multi-stage stochastic prograr
ming model.

Stages: The stages of the model are indexed by $t = 0, 1, \ldots, T$.

Variables:

$x_{i,t}$ = market value in asset i at stage t for $i = 1, \ldots, n$ and $t = 0, 1, \ldots, T$.

w_t = interest income shortfall at stage for $t = 1, \ldots, T$.

v_t = interest income surplus at stage for $t = 1, \ldots, T$.

Random parameters in the stochastic linear program:

$RP_{i,t}$ = price return of asset i between stage $t - 1$ and stage t, for $i = 1, \ldots,$
and $t = 1, \ldots, T$.

$RI_{i,t}$ = interest income of asset i between stage $t - 1$ and stage t, for i
$1, \ldots, n$ and $t = 1, \ldots, T$.

F_t = deposit inflow between stage $t - 1$ and stage t, for $t = 1, \ldots, T$.

P_t = principal payout between stage $t - 1$ and stage t, for $t = 1, \ldots, T$.

I_t = interest payout between stage $t - 1$ and stage t, for $t = 1, \ldots, T$.

g_t = rate at which interest is credited to policies between stage $t - 1$ and stag
t, for $t = 1, \ldots, T$.

L_t = liability valuation at stage t.

Parameterized objective function components:

$c_t(\cdot)$ = piecewise linear convex penalty for shortfall at time t.

The goal of the model is to allocate funds among available assets to maximis
expected wealth at the end of the planning horizon T minus the expected pena
ized shortfall accumulated through the planning horizon. The problem can b

formulated as the following multi-stage stochastic program:

$$\max \quad \mathbb{E}\left[\sum_{i=1}^{n} x_{i,T} - \sum_{t=1}^{T} c_t(w_t)\right]$$

$$\text{s.t.} \quad \sum_{i=1}^{n} x_{i,t} - \sum_{i=1}^{n}(1 + RP_{i,t} + RI_{i,t})x_{i,t-1} = F_t - P_t - I_t \quad \text{for } t = 1, \ldots, T$$
$$\text{asset accumulation}$$

$$\sum_{i=1}^{n} RI_{i,t}x_{i,t-1} + w_t - v_t = g_t L_{t-1} \quad \text{for } t = 1, \ldots, T$$
$$\text{interest income shortfall}$$

$$L_t = (1 + g_t)L_{t-1} + F_t - P_t - I_t \quad \text{for } t = 1, \ldots, T$$
$$\text{liability accumulation}$$

$$x_{i,t} \geq 0, \quad w_t \geq 0, \quad v_t \geq 0. \tag{17.1}$$

In the model discussed in Cariño et al. (1994), the stochastic linear program (17.1) is converted into a large linear program using a finite number of scenarios to deal with the random elements in the data. Creation of scenario inputs is made in stages using a tree. The tree structure can be described by the number of branches at each stage. For example, a 1–8–4–4–2–1 tree has 256 scenarios. Stage $t = 0$ is the initial stage. Stage $t = 1$ may be chosen to be the end of Quarter 1 and has eight different branches in this example. Stage $t = 2$ may be chosen to be the end of Year 1, with each of the previous eight branches giving rise to four new branches, and so on. For the Yasuda Fire and Marine Insurance Co. Ltd., a problem with seven asset classes and six stages gives rise to a stochastic linear program (17.1) with 12 constraints (other than non-negativity) and 54 variables. Using 256 scenarios, this stochastic program is converted into a linear program with several thousand constraints and over 10,000 variables. Solving this model yielded extra income estimated to be about US $80 million per year for the company.

17.3 Option Pricing via Stochastic Programming

The option pricing problem discussed in Chapter 15 and modeled via the binomial lattice can alternatively be formulated as a stochastic programming problem. As should be expected, the two approaches are equivalent under the assumptions made for the binomial lattice model. However, there is additional flexibility in the stochastic programming approach that makes it applicable under less restrictive assumptions. In particular, we will discuss how the stochastic programming approach can easily model transaction costs. This is an important practical issue that cannot be incorporated in the binomial lattice model.

We will work with the following similar setting to that in Chapter 15. Let S_t, for $t = 0, 1, \ldots, N$, denote the share price of a risky asset at times $t = 0, 1, \ldots, N$. Assume the economy also has a risk-free asset whose interest rate is r in each period $[t - 1, t]$ for $t = 1, \ldots, N$.

European Options

Consider a European option contract that matures at time N with payoff $g(S_N)$ for some function $g(\cdot)$ of the underlying asset price S_N. The following stochastic program provides a model for the lowest-cost portfolio of the underlying asset and the risk-free asset that can be constructed at time 0 and be subsequently rebalanced to super-replicate the payoff $g(S_N)$ of the European option contract.

Variables:

    $x_t =$ amount of shares of the risky asset at time t for $t = 0, \ldots, N-1$.
    $y_t =$ amount of money in the risk-free asset at time t for $t = 0, \ldots, N-1$.

Objective:

$$
\begin{aligned}
\min \quad & S_0 x_0 + y_0 \\
\text{s.t.} \quad & S_N x_{N-1} + (1+r)y_{N-1} \geq g(S_N) \\
& S_t x_{t-1} + (1+r)y_{t-1} \geq S_t x_t + y_t, \quad t = 1, \ldots, N-1.
\end{aligned}
\tag{17.}
$$

Consider the following *binomial* tree model for the risky prices. Assume that there are exactly two possible random outcomes ("H" and "T") between time $t-1$ and t for $t = 1, \ldots, N$. For the simplest case $N = 1$, the binomial tree model yields the following deterministic equivalent of (17.2):

$$
\begin{aligned}
\min \quad & S_0 x_0 + y_0 \\
\text{s.t.} \quad & S_1(H)x_0 + (1+r)y_0 \geq g(S_1(H)) \\
& S_1(T)x_0 + (1+r)y_0 \geq g(S_1(T)).
\end{aligned}
\tag{17.}
$$

Observe that the linear programming dual of (17.3) is

$$
\begin{aligned}
\max \quad & g(S_1(H))v(H) + g(S_1(T))v(T) \\
\text{s.t.} \quad & S_1(H)v(H) + S_1(T)v(T) = S_0 \\
& (1+r)v(H) + (1+r)v(T) = 1 \\
& v(H), v(T) \geq 0,
\end{aligned}
\tag{17.}
$$

which in turn can be rewritten via the change of variables $\tilde{p} := (1+r)v(H)$ and $\tilde{q} := (1+r)v(T)$ as

$$
\begin{aligned}
\max \quad & \frac{1}{1+r}(g(S_1(H))\tilde{p} + g(S_1(T))\tilde{q}) \\
\text{s.t.} \quad & \frac{1}{1+r}(S_1(H)\tilde{p} + S_1(T)\tilde{q}) = S_0 \\
& \tilde{p} + \tilde{q} = 1 \\
& \tilde{p}, \tilde{q} \geq 0.
\end{aligned}
\tag{17.}
$$

Without loss of generality assume $S_1(H) \geq S_1(T)$. Furthermore, assume $S_1(H) > S_1(T)$ as otherwise the pricing problem of the option contract is trivial. It follows that (17.5) is feasible if and only if $S_1(T) \leq (1+r)S_0 \leq S_1(H)$. In this case the only feasible solution to (17.5) is

$$
\tilde{p} = \frac{(1+r)S_0 - S_1(T)}{S_1(H) - S_1(T)}, \quad \tilde{q} = \frac{S_1(H) - (1+r)S_0}{S_1(H) - S_1(T)},
$$

and thus the optimal value of (17.3) and (17.4) is

$$\frac{1}{1+r}(\tilde{p}g(S_1(H)) + \tilde{q}g(S_1(T))).$$

Observe that this price is exactly the same (as it should be) as the one obtained via the one-period binomial model discussed in Section 4.3 when the single-period economy has no arbitrage. A similar duality argument shows that in the absence of arbitrage the stochastic programming model (17.2) is equivalent to the binomial lattice approach for the general multi-period case; that is, when $N > 1$. The following example illustrates this equivalence.

Example 17.1 Suppose $n = 2$, $r = \frac{1}{4}$, and the prices S_0, S_1, S_2 of the risky asset are as indicated at the nodes of the binomial tree depicted in Figure 17.1. Assume that the two branches emerging from each node are equally likely. Determine the price of a European put option maturing at time $N = 2$ with strike price 50; that is, with payoff $g(S_2) = (50 - S_2)^+$.

In this case the deterministic equivalent of (17.2) is

$$
\begin{aligned}
\min \quad & 40x_0 + y_0 \\
\text{s.t.} \quad & 80x_0 + 1.25y_0 \geq 80x_1(H) + y_1(H) \\
& 20x_0 + 1.25y_0 \geq 20x_1(T) + y_1(T) \\
& 160x_1(H) + 1.25y_1(H) \geq (50 - 160)^+ = 0 \\
& 40x_1(H) + 1.25y_1(H) \geq (50 - 40)^+ = 10 \\
& 40x_1(T) + 1.25y_1(T) \geq (50 - 40)^+ = 10 \\
& 10x_1(T) + 1.25y_1(T) \geq (50 - 10)^+ = 40.
\end{aligned}
$$

The optimal solution to this linear program is

$$x_1(H) = -0.0833, \; y_1(H) = 10.6666, \; x_1(T) = -1, \; y_1(T) = 40, \qquad (17.6)$$
$$x_0 = -0.2666, \; y_0 = 20.2666$$

and thus its optimal value is 9.6.

On the other hand, the binomial lattice approach would yield the risk-neutral probabilities $\tilde{p} = \tilde{q} = \frac{1}{2}$. Consequently, the value $V_1(S_1)$ of the option at time 1 is

$$V_1(80) = \frac{1}{1.25} \cdot \frac{0 + 10}{2} = 4, \; V_1(20) = \frac{1}{1.25} \cdot \frac{10 + 40}{2} = 20;$$

and the value $V_0(S_0)$ of the option at time 0 is

$$V_0(40) = \frac{1}{1.25} \cdot \frac{4 + 20}{2} = 9.6.$$

The (super-)replicating portfolio (17.6) can also be recovered via delta-hedging.

American Options

Consider now an American option contract that can be exercised at any time $t = 0, 1, \ldots, N$ with payoff $g(S_t)$ for some function $g(\cdot)$ of the underlying asset price

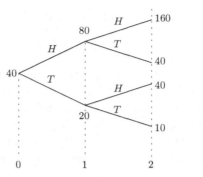

Figure 17.1 Binomial tree for option pricing example

S_t. The stochastic program (17.2) has the following straightforward modificati‹ for finding a lowest-cost portfolio of the underlying asset and the risk-free ass‹ that can be constructed at time 0 and be subsequently rebalanced to supe‹ replicate the payoff of the American option contract:

$$
\begin{aligned}
\min \quad & S_0 x_0 + y_0 \\
\text{s.t.} \quad & S_N x_{N-1} + (1+r) y_{N-1} \geq g(S_N) \\
& S_t x_{t-1} + (1+r) y_{t-1} \geq \max\{S_t x_t + y_t, g(S_t)\}, \quad t = 1, \ldots, N-1.
\end{aligned}
$$

The latter problem in turn can be equivalently stated as follows:

$$
\begin{aligned}
\min \quad & S_0 x_0 + y_0 \\
\text{s.t.} \quad & S_N x_{N-1} + (1+r) y_{N-1} \geq g(S_N) \\
& S_t x_{t-1} + (1+r) y_{t-1} \geq S_t x_t + y_t, \quad t = 1, \ldots, N-1 \\
& S_t x_{t-1} + (1+r) y_{t-1} \geq g(S_t), \quad t = 1, \ldots, N-1.
\end{aligned}
\tag{17.}
$$

Again for a binomial event tree model the above stochastic programmi‹ approach is equivalent to the binomial lattice approach discussed in Chapter in the absence of arbitrage.

Transaction Costs

The stochastic programming models (17.2) and (17.7) can be readily extend‹ to incorporate proportional transaction costs. Observe that in the absence transaction costs a transaction to sell w shares of the risky asset when its price S will generate a revenue equal to wS. By contrast, if a proportional transacti‹ cost θ applies to the sell transaction then the revenue would instead be $(1-\theta)w$‹ Similarly, if a proportional transaction cost θ applies to a buy transaction of shares, then the cost of the transaction would be $(1+\theta)wS$.

The stochastic programming model (17.2) can be modified as follows account for a proportional transaction cost θ applicable to each buy or s‹

transaction of the risky asset:

$$
\begin{aligned}
\min \quad & S_0 x_0 + \theta |S_0 x_0| + y_0 \\
\text{s.t.} \quad & S_N x_{N-1} + (1+r) y_{N-1} - \theta |S_N x_{N-1}| \geq g(S_N) \\
& S_t x_{t-1} + (1+r) y_{t-1} - \theta |S_t (x_t - x_{t-1})| \geq S_t x_t + y_t, \quad t = 1, \ldots, N-1.
\end{aligned}
\tag{17.8}
$$

Similarly, the stochastic programming model (17.7) can also be modified to account for the same kind of transaction costs as follows:

$$
\begin{aligned}
\min \quad & S_0 x_0 + \theta |S_0 x_0| + y_0 \\
\text{s.t.} \quad & S_N x_{N-1} + (1+r) y_{N-1} - \theta |S_N x_{N-1}| \geq g(S_N) \\
& S_t x_{t-1} + (1+r) y_{t-1} - \theta |S_t (x_t - x_{t-1})| \geq S_t x_t + y_t, \quad t = 1, \ldots, N-1 \\
& S_t x_{t-1} + (1+r) y_{t-1} - \theta |S_t x_t| \geq g(S_t), \quad t = 1, \ldots, N-1.
\end{aligned}
\tag{17.9}
$$

Observe that the stochastic programs (17.9) and (17.8) include some term with absolute values in the objective and constraints. The models can be recast as linear stochastic programs by introducing some extra variables and constraints, as the following example illustrates.

Example 17.2 Suppose $n = 2$, $r = \frac{1}{4}$, and the prices S_0, S_1, S_2 of the risky asset are as indicated at the nodes of the binomial tree depicted in Figure 17.1. Assume that the two branches emerging from each node are equally likely. Determine the price of a European put option maturing at time $N = 2$ with strike price 50; that is, with payoff $g(S_2) = (50 - S_2)^+$. Assume a proportional transaction cost θ applies to every buy or sell transaction.

In this case the deterministic equivalent of (17.8) is

$$
\begin{aligned}
\min \quad & 40 x_0 + y_0 + 40\theta u_0 \\
\text{s.t.} \quad & 80 x_0 + 1.25 y_0 - 80\theta v_1(H) \geq 80 x_1(H) + y_1(H) \\
& 20 x_0 + 1.25 y_0 - 20\theta v_1(T) \geq 20 x_1(T) + y_1(T) \\
& 160 x_1(H) + 1.25 y_1(H) - 160\theta w_1(H) \geq (50 - 160)^+ = 0 \\
& 40 x_1(H) + 1.25 y_1(H) - 40\theta w_1(H) \geq (50 - 40)^+ = 10 \\
& 40 x_1(T) + 1.25 y_1(T) - 40\theta w_1(T) \geq (50 - 40)^+ = 10 \\
& 10 x_1(T) + 1.25 y_1(T) - 10\theta w_1(T) \geq (50 - 10)^+ = 40 \\
& u_0 \geq x_0, \quad u_0 \geq -x_0 \\
& v_1(H) \geq x_1(H) - x_0, \quad v_1(T) \geq -x_1(T) + x_0 \\
& w_1(H) \geq x_1(H), \quad w_1(H) \geq -x_1(H) \\
& w_1(T) \geq x_1(T), \quad w_1(T) \geq -x_1(T).
\end{aligned}
$$

Table 17.1 shows the optimal value and holdings $x_0, y_0, x_1(H), y_1(H), x_1(T), y_1(T)$ of the optimal super-replicating portfolio for various levels of transaction cost θ.

Table 17.1

θ	Optimal value	x_0	y_0	$x_1(H)$	$y_1(H)$	$x_1(T)$	$y_1(T$
0	9.6	−0.26666	20.26666	−0.08333	10.66666	−1	40
0.01	9.93044	−0.26879	20.57443	−0.08251	10.66666	−0.9901	40
0.05	11.24337	−0.27546	21.71077	−0.07937	10.66666	−0.95238	40
0.1	12.84987	−0.28052	22.94857	−0.07576	10.66666	−0.90909	40

17.4 Synthetic Options

An important issue in portfolio selection is the potential decline of the portfol
value below some critical limit. How can we control the risk of downside losses?
possible answer is to create a payoff structure similar to a European call optio

While a corporate investor may be able to construct a diversified portfoli
there may be no option market available on this portfolio. One solution may l
to use index options. However, exchange-traded options with sufficient liquidi
are limited to maturities of about three months. This makes the cost of long-ter
protection expensive, requiring the purchase of a series of highly priced shor
term options. For large institutional or corporate investors, a cheaper soluti
is to artificially produce the desired payoff structure using available resourc
This is called a *synthetic option strategy*. A model of this kind was proposed l
Zhao and Ziemba (2001) and can be described as follows.

Problem parameters:

W_0 = investor's initial wealth

T = investor's planning horizon

R = gross return of a riskless asset for one period

$R_{i,t}$ = gross return for asset i at time t

$\theta_{i,t}$ = transaction cost for purchases and sales of asset i at time t.

The gross returns $R_{i,t}$ above are random, but their distributions are known.

Variables:

$x_{i,t}$ = amount allocated to asset i at time t

$A_{i,t}$ = amount of asset i bought at time t

$D_{i,t}$ = amount of asset i sold at time t

y_t = amount allocated to riskless asset at time t.

We formulate a stochastic program that produces the desired payoff at th
end of the planning horizon T, much in the flavor of the stochastic progran
developed in the previous section. Let us first discuss the constraints.

The initial portfolio must satisfy

$$y_0 + x_{1,0} + \ldots + x_{n,0} = W_0.$$

Similarly, the portfolio at time t must satisfy

$$x_{i,t} = R_{i,t}x_{i,t-1} + A_{i,t} - D_{i,t} \qquad \text{for } t = 1, \dots, T$$

$$y_t = Ry_{t-1} - \sum_{i=1}^{n}(1 + \theta_{i,t})A_{i,t} + \sum_{i=1}^{n}(1 - \theta_{i,t})D_{i,t} \qquad \text{for } t = 1, \dots, T.$$

One can also impose upper bounds on the proportion of any risky asset in the portfolio:

$$0 \le x_{i,t} \le m_t \left(y_t + \sum_{j=1}^{n} x_{j,t} \right),$$

where m_t is chosen by the investor.

The value of the portfolio at the end of the planning horizon is

$$v = Ry_{T-1} + \sum_{i=1}^{n}(1 - \theta_{i,T})R_{i,T}x_{i,T-1},$$

where the summation term is the value of the risky assets at time T.

To construct the desired synthetic option, we split v into the riskless value of the portfolio Z and a surplus $z \ge 0$ which depends on random events. Using a scenario approach to the stochastic program, Z is the worst-case payoff over all the scenarios. The surplus z is a random variable that depends on the scenario. Thus

$$v = Z + z, \qquad z \ge 0.$$

We consider Z and z as variables of the problem, and we optimize them together with the asset allocations x and other variables described earlier. The objective function of the stochastic program is

$$\max \mathbb{E}(z) + \mu Z,$$

where $\mu \ge 1$ is the risk aversion of the investor. The risk aversion μ is given data. When $\mu = 1$, the objective is to maximize expected return. When μ is very large, the objective is to maximize "riskless profit".

Example 17.3 Consider an investor with initial wealth $W_0 = 1$ who wants to construct a portfolio comprising one risky asset and one riskless asset using the "synthetic option" model described above. We next describe the deterministic equivalent of this model for a two-period planning horizon, i.e. $T = 2$, and an event tree with four scenarios. The construction is similar to that in Example 16.2. Suppose the return on the riskless asset is a non-random value R per period and there are two equally likely possible random outcomes ("H" and "T") over each time period. Let $R_t(H)$ and $R_t(T)$ denote the return of the risky asset in the period $[t-1, t]$ when the outcome is H and T respectively. Suppose the transaction cost for purchases and sales of the risky asset is a non-random value θ.

The scenario tree in this case is identical to that depicted in Figure 16.1 f
Example 16.2. The deterministic equivalent of the multi-stage stochastic line
program in this case is as follows:

$$
\max \quad \tfrac{1}{4}(z(HH) + z(HT) + z(TH) + z(TT)) + \mu Z
$$

$$
\text{s.t.} \quad y_0 + x_0 = 1
$$

$$
x_1(H) = R_1(H)x_0 + A_1(H) - D_1(H)
$$

$$
x_1(T) = R_1(T)x_0 + A_1(T) - D_1(T)
$$

$$
y_1(H) = Ry_0 - (1 + \theta)A_1(H) + (1 - \theta)D_1(H)
$$

$$
y_1(T) = Ry_0 - (1 + \theta)A_1(T) + (1 - \theta)D_1(T)
$$

$$
z(HH) + Z = Ry_1(H) + (1 - \theta)R_2(H)x_1(H)
$$

$$
z(HT) + Z = Ry_1(H) + (1 - \theta)R_2(T)x_1(H)
$$

$$
z(TH) + Z = Ry_1(T) + (1 - \theta)R_2(H)x_1(T)
$$

$$
z(TT) + Z = Ry_1(T) + (1 - \theta)R_2(T)x_1(T)
$$

$$
x_0, y_0, x_1(H), x_1(T), y_1(H), y_1(T), A_1(H), D_1(H), A_1(T), A_2(T) \geq 0
$$

$$
z(HH), z(HT), z(TH), z(TT) \geq 0
$$

$$
Z \quad \text{free.}
$$

Zhao and Ziemba (2001) introduce and apply the above generic synthet
option model to an example with three assets (cash, bonds, and stocks) ar
four periods (a one-year horizon with quarterly portfolio reviews). The quarter
return on cash is constant at $\rho = 0.0095$. For stocks and bonds, the expecte
logarithmic rates of returns are $s = 0.04$ and $b = 0.019$ respectively. Transactic
costs are assumed to be 0.5% for stocks and 0.1% for bonds. The scenarios neede
in the stochastic program are generated using an autoregression model which
constructed based on historical data (quarterly returns from 1985 to 1998; tl
Salomon Brothers bond index and S&P 500 index respectively). Specifically, tl
autoregression model is

$$
s_t = 0.037 - 0.193s_{t-1} + 0.418b_{t-1} - 0.172s_{t-2} + 0.517b_{t-2} + \epsilon_t
$$

$$
b_t = 0.007 - 0.140s_{t-1} + 0.175b_{t-1} - 0.023s_{t-2} + 0.122b_{t-2} + \eta_t,
$$

where the pair (ϵ_t, η_t) characterizes uncertainty. Zhao and Ziemba used a rando
sampling approach to estimate the joint distribution of (ϵ_t, η_t). From this joi
distribution of (ϵ_t, η_t) a set of 20 pairs can be selected to estimate the empiric
distribution of (ϵ_t, η_t). In this way, a scenario tree with 160,000 ($= 20 \times 20$
20×20) paths describing possible outcomes of asset returns is generated for tl
four periods.

The authors solved the resulting large deterministic linear program. We discu
some of the results obtained when this linear program is solved for a risk aversic
of $\mu = 2.5$. The value of the terminal portfolio is always at least 4.6% more tha
the initial portfolio wealth and the distribution of terminal portfolio values
skewed to larger values because of dynamic downside risk control. The expecte
return is 16.33% and the volatility is 7.2%. It is interesting to compare the
values with those obtained from a static Markowitz model. The expected retur
is 15.4% for the same volatility but no minimum return is guaranteed. In fac

in some scenarios, the value of the Markowitz portfolio is 5% *less* at the end of the one-year horizon than it was at the beginning.

It is also interesting to look at a typical portfolio (one of the 160,000 paths) generated by the synthetic option model (the linear program was set up with an upper bound of 70% placed on the fraction of stocks or bonds in the portfolio).

Quarter t	Cash	Stocks	Bonds	Portfolio value at the end of Quarter t
1	12%	18%	70%	103
2		41%	59%	107
3		70%	30%	112
4	30%		70%	114

17.5   Exercises

Exercise 17.1   For a non-dividend paying stock, collect data on four or five call options for the nearest maturity (but at least one month). Calculate the *implied volatility* for each option; that is, the value of σ that makes equation (15.7) hold for the market prices of the call options. Solve the option pricing problem (17.7) when the number of stages is seven using the implied volatility of the at-the-money option to construct the tree.

Exercise 17.2   Repeat Exercise 17.1 allowing for transaction costs, with different values of θ, to see if the volatility smile can be explained by transaction costs. Specifically, given a value for σ and for θ, calculate option prices and see how they match up to observed prices. Try $\theta = 0.001, 0.005, 0.01, 0.02, 0.05$.

Exercise 17.3   Develop a synthetic option model in the spirit of that used by Zhao and Ziemba (2001), adapted to the size limitation of your linear programming solver. Compare with a static model.

Part IV

Other Optimization Techniques

18 Conic Programming: Theory and Algorithms

Conic programming refers to a class of convex optimization problems that generalizes linear and quadratic programming. The gist of conic programming is to replace the non-negativity constraint with a *conic* constraint.

18.1 Conic Programming

A *conic program* in standard form is an optimization problem of the form

$$
\begin{aligned}
\min_{\mathbf{x}} \quad & \mathbf{c}^\mathsf{T}\mathbf{x} \\
\text{s.t.} \quad & \mathbf{A}\mathbf{x} = \mathbf{b} \\
& \mathbf{D}\mathbf{x} - \mathbf{d} \in \mathcal{K}
\end{aligned}
\tag{18.1}
$$

for some vectors and matrices $\mathbf{c} \in \mathbb{R}^n$, $\mathbf{b} \in \mathbb{R}^m$, $\mathbf{d} \in \mathbb{R}^p$, $\mathbf{A} \in \mathbb{R}^{m \times n}$, $\mathbf{D} \in \mathbb{R}^{p \times n}$ and some closed convex cone $\mathcal{K} \subseteq \mathbb{R}^p$.

When $\mathcal{K} = \mathbb{R}^p_+$ the problem (18.1) is a linear program. However, conic programming is far more general. We next discuss two particularly important classes of conic programs, namely *second-order* and *semidefinite* programming.

18.1.1 Second-Order Programming

The *second-order* cone, also known as the *Lorenz* cone or the *ice-cream* cone, is defined as follows:

$$
\mathbb{L}_n = \left\{ \mathbf{x} = \begin{bmatrix} x_0 \\ \bar{\mathbf{x}} \end{bmatrix} \in \mathbb{R}^n : \|\bar{\mathbf{x}}\|_2 \le x_0 \right\}.
$$

See Figure 18.1.

A *second-order program* is a problem of the form (18.1) where $\mathcal{K}$ is a direct product of second-order cones; that is,

$$
\mathcal{K} = \mathbb{L}_{n_1} \times \cdots \times \mathbb{L}_{n_r}
$$

for some positive integers $n_1, \ldots, n_r$.

We next illustrate the modeling power of second-order programming by showing that a convex quadratically constrained quadratic program can be recast as a second-order program. In particular, second-order programming generalizes both linear programming and convex quadratic programming.

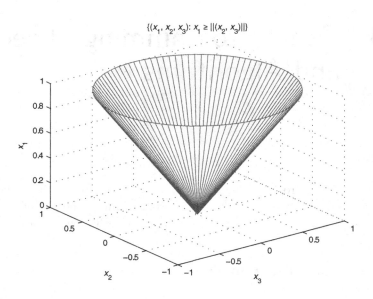

Figure 18.1 Second-order cone

Consider a convex quadratically constrained quadratic program of the form

$$\min_{\mathbf{x}} \quad \mathbf{c}_0^\mathsf{T}\mathbf{x} + \tfrac{1}{2}\mathbf{x}^\mathsf{T}\mathbf{Q}_0\mathbf{x}$$
$$\text{s.t.} \quad \mathbf{c}_i^\mathsf{T}\mathbf{x} + \tfrac{1}{2}\mathbf{x}^\mathsf{T}\mathbf{Q}_i\mathbf{x} \le b_i, \ i = 1, \ldots, m,$$

(18.2)

where $\mathbf{c}_i \in \mathbb{R}^n$, $\mathbf{Q}_i \in \mathbb{S}_+^n$ for $i = 0, 1, \ldots, m$ and $b_i \in \mathbb{R}$ for $i = 1, \ldots, m$. Here $\mathbb{S}$ denotes the family of $n \times n$ positive semidefinite matrices.

Observe that (18.2) can be rewritten as

$$\min_{\mathbf{x},t} \quad t$$
$$\text{s.t.} \quad \mathbf{c}_0^\mathsf{T}\mathbf{x} + \tfrac{1}{2}\mathbf{x}^\mathsf{T}\mathbf{Q}_0\mathbf{x} \le t$$

(18.3)

$$\mathbf{c}_i^\mathsf{T}\mathbf{x} + \tfrac{1}{2}\mathbf{x}^\mathsf{T}\mathbf{Q}_i\mathbf{x} \le b_i, \ i = 1, \ldots, m.$$

The following step is key in the formulation of (18.2) as a second-order program given $\mathbf{Q} \in \mathbb{S}_+^n$, $\mathbf{c} \in \mathbb{R}^n$, $b \in \mathbb{R}$ the quadratic inequality

$$\mathbf{c}^\mathsf{T}\mathbf{x} + \tfrac{1}{2}\mathbf{x}^\mathsf{T}\mathbf{Q}\mathbf{x} \le b$$

can be formulated as a second-order cone constraint. To see that, observe that because $\mathbf{Q} \in \mathbb{S}_+^n$ there exists $\mathbf{L} \in \mathbb{R}^{n \times p}$ such that $\mathbf{Q} = \mathbf{L}\mathbf{L}^\mathsf{T}$ (in particular the Cholesky factorization satisfies this requirement). Therefore

$$\mathbf{c}^\mathsf{T}\mathbf{x} + \tfrac{1}{2}\mathbf{x}^\mathsf{T}\mathbf{Q}\mathbf{x} \le b \quad \Leftrightarrow \quad \mathbf{c}^\mathsf{T}\mathbf{x} + \tfrac{1}{2}\|\mathbf{L}^\mathsf{T}\mathbf{x}\|^2 \le b \quad \Leftrightarrow \quad \begin{bmatrix} b - \mathbf{c}^\mathsf{T}\mathbf{x} + 1 \\ b - \mathbf{c}^\mathsf{T}\mathbf{x} - 1 \\ \sqrt{2}\,\mathbf{L}^\mathsf{T}\mathbf{x} \end{bmatrix} \in \mathbb{L}_{p+2}.$$

It thus follows that (18.3) and in turn (18.2) can be rewritten as the following second-order program:

$$\min_{\mathbf{x},t} \quad t$$

$$\text{s.t.} \quad \begin{bmatrix} t - \mathbf{c}_0^\mathsf{T}\mathbf{x} + 1 \\ t - \mathbf{c}_0^\mathsf{T}\mathbf{x} - 1 \\ \sqrt{2}\,\mathbf{L}_0^\mathsf{T}\mathbf{x} \end{bmatrix} \in \mathbb{L}_{p_0+2}$$

$$\begin{bmatrix} b_i - \mathbf{c}_i^\mathsf{T}\mathbf{x} + 1 \\ b_i - \mathbf{c}_i^\mathsf{T}\mathbf{x} - 1 \\ \sqrt{2}\,\mathbf{L}_i^\mathsf{T}\mathbf{x} \end{bmatrix} \in \mathbb{L}_{p_i+2}, \ i = 1, \ldots, m,$$

where $\mathbf{L}_i \in \mathbb{R}^{n \times p_i}$ such that $\mathbf{Q}_i = \mathbf{L}_i\mathbf{L}_i^\mathsf{T}$ for $i = 0, 1, \ldots, m$.

Tracking Error and Volatility Constraints

In the context of quantitative asset management, portfolios are typically chosen relative to some predetermined benchmark, as we discussed in Section 6.5. As a consequence, it is common to use a constraint on the active risk (also known as tracking error) instead of, or in addition to, the total risk.

More precisely, suppose $\mathbf{x}$ denotes the vector of percentage holdings in a portfolio. Let $\mathbf{r}$ and r_B denote respectively the vector of asset returns and the benchmark return. Recall that the active return is the difference $\mathbf{r}^\mathsf{T}\mathbf{x} - r_B$ between the portfolio return and the benchmark return. The active risk is the variance of the active return. If $\mathbf{x}^B$ denotes the vector of percentage holdings in the benchmark, then the active risk can be written as

$$\text{var}(\mathbf{r}^\mathsf{T}(\mathbf{x} - \mathbf{x}^B)) = (\mathbf{x} - \mathbf{x}^B)^\mathsf{T}\mathbf{V}(\mathbf{x} - \mathbf{x}^B),$$

where $\mathbf{V}$ is the covariance matrix of asset returns.

A typical mean–variance model for benchmark-relative portfolio management has the following form:

$$\begin{aligned} \max_{\mathbf{x}} \quad & \boldsymbol{\alpha}^\mathsf{T}\mathbf{x} \\ \text{s.t.} \quad & (\mathbf{x} - \mathbf{x}^B)^\mathsf{T}\mathbf{V}(\mathbf{x} - \mathbf{x}^B) \leq \bar{\psi}^2 \\ & \mathbf{A}\mathbf{x} = \mathbf{b} \\ & \mathbf{C}\mathbf{x} \leq \mathbf{d}. \end{aligned} \tag{18.4}$$

Note that this is not a quadratic program because it has a nonlinear constraint. However, the problem (18.4) is a convex quadratically constrained quadratic program of the form (18.2) discussed in Section 18.1.1. Therefore, it has a straightforward formulation as a second-order conic program.

The above model can be readily extended to include multiple measures of risk. For instance, the following model, which is an extension of the model discussed by Jorion (2003) that enforces upper bound constraints on both total risk and tracking error, also has a straightforward second-order conic programming formulation:

$$\begin{aligned}
\max_{\mathbf{x}} \quad & \boldsymbol{\alpha}^\mathsf{T}\mathbf{x} \\
\text{s.t.} \quad & (\mathbf{x} - \mathbf{x}^B)^\mathsf{T}\mathbf{V}(\mathbf{x} - \mathbf{x}^B) \leq \bar{\psi}^2 \\
& \mathbf{x}^\mathsf{T}\mathbf{V}\mathbf{x} \leq \bar{\sigma}^2 \\
& \mathbf{A}\mathbf{x} = \mathbf{b} \\
& \mathbf{C}\mathbf{x} \leq \mathbf{d}.
\end{aligned}$$

(18.)

18.1.2   Semidefinite Programming

Some applications, such as the approximation of covariance matrices discusse
below, lead to conic optimization models involving the space of symmetric matr
ces and the cone of positive semidefinite matrices described next. Let $\mathbb{S}^n$ deno
the space of $n \times n$ symmetric matrices. Although this space is equivalent
$\mathbb{R}^{n(n+1)/2}$, it is more convenient and customary to treat it as a space of matrice
A matrix $\mathbf{X} \in \mathbb{S}^n$ is *positive semidefinite* if

$$\mathbf{u}^\mathsf{T}\mathbf{X}\mathbf{u} \geq 0 \text{ for all } \mathbf{u} \in \mathbb{R}^n.$$

It is a common convention to write $\mathbf{X} \succeq \mathbf{0}$ to indicate that $\mathbf{X} \in \mathbb{S}^n$ is positiv
semidefinite. The cone of positive semidefinite matrices $\mathbb{S}^n_+$ is defined as

$$\mathbb{S}^n_+ := \{\mathbf{X} \in \mathbb{S}^n : \mathbf{X} \succeq \mathbf{0}\}.$$

See Figure 18.2.

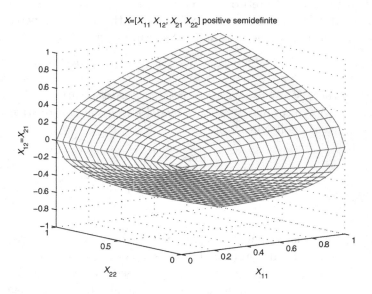

$X=[X_{11}\ X_{12};\ X_{21}\ X_{22}]$ positive semidefinite

Figure 18.2 Cone of positive semidefinite matrices

A *semidefinite program* is a problem of the form (18.1) where $\mathcal{K}$ is the cone
positive semidefinite matrices.

Endow the space $\mathbb{S}^n$ of symmetric $n \times n$ matrices with the following *Frobenius inner product*. For $\mathbf{X}, \mathbf{S} \in \mathbb{S}^n$ let

$$\mathbf{X} \bullet \mathbf{S} := \text{trace}(\mathbf{XS}) = \sum_{i,j} X_{ij} S_{ij}.$$

A semidefinite program is in *standard form* if it is written as

$$\begin{aligned}
\min_{\mathbf{X}} \quad & \mathbf{C} \bullet \mathbf{X} \\
\text{s.t.} \quad & \mathbf{AX} = \mathbf{b} \\
& \mathbf{X} \succeq \mathbf{0},
\end{aligned}$$

where $\mathbf{C} \in \mathbb{S}^n$, $\mathbf{b} \in \mathbb{R}^m$, and $\mathbf{A} : \mathbb{S}^n \to \mathbb{R}^m$ is a linear mapping.

Approximating Covariance Matrices

We next illustrate the modeling power of semidefinite programming by showing that a covariance estimation problem can be recast as a semidefinite program. To that end, recall that any proper covariance matrix must be symmetric and positive semidefinite. Suppose $\hat{\mathbf{V}} \in \mathbb{S}^n$ is an estimate of a covariance matrix that is not necessarily positive semidefinite and consider the problem of finding the positive semidefinite matrix that is closest to $\hat{\mathbf{V}}$; that is,

$$\begin{aligned}
\min_{\mathbf{X}} \quad & \|\mathbf{X} - \hat{\mathbf{V}}\| \\
\text{s.t.} \quad & \mathbf{X} \succeq \mathbf{0}.
\end{aligned} \tag{18.6}$$

This problem can be formulated as

$$\begin{aligned}
\min_{\mathbf{X},t} \quad & t \\
\text{s.t.} \quad & \|\mathbf{X} - \hat{\mathbf{V}}\| \le t \\
& \mathbf{X} \succeq \mathbf{0}.
\end{aligned}$$

If the norm $\|\mathbf{X} - \hat{\mathbf{V}}\|$ is the Frobenius norm

$$\|\mathbf{X} - \hat{\mathbf{V}}\|_F := \sqrt{\sum_{i,j} (X_{ij} - \hat{V}_{ij})^2},$$

then the constraint $\|\mathbf{X} - \hat{\mathbf{V}}\|_F \le t$ can be written as a second-order cone constraint and it follows that the above covariance estimation problem (18.6) can be written as a conic program over the Cartesian product of a second-order cone and a semidefinite cone. As we detail in the exercises at the end of this chapter, problem (18.6) can also be formulated as a conic program over a suitable semidefinite cone for other choices of norms such as the operator norm or the infinity norm.

18.2   Numerical Conic Programming Solvers

During the last three decades there have been major advancements in the theor
algorithms, and software for conic programming. In particular, the softwa
packages SeDuMi and SDPT3 are MATLAB-based, freely available solvers for con
programs. The commercial software vendors Gurobi and MOSEK also offer solve
for conic programs. These solvers constitute the engine behind the MATLA
modeling language CVX that we have discussed in previous chapters.

Both SeDuMi and SDPT3 as well as most other software packages are designe
to solve a conic program in the following *standard* form:

$$\min_{\mathbf{x}} \quad \mathbf{c}^\mathsf{T}\mathbf{x}$$
$$\text{s.t.} \quad \mathbf{A}\mathbf{x} = \mathbf{b}$$
$$\mathbf{x} \in \mathcal{K},$$

where the cone $\mathcal{K}$ is a Cartesian product of the form $\mathbb{R}^f \times \mathbb{R}^\ell_+ \times \mathbb{L}_{n_1} \times \cdots$
$\mathbb{L}_{n_r} \times \mathbb{S}^{d_1}_+ \times \cdots \times \mathbb{S}^{d_k}_+$. In other words, the conic constraint $\mathbf{x} \in \mathcal{K}$ models a vect
$\mathbf{x}$ that has a block of f free components, followed by a block of ℓ non-negativ
components, followed by r blocks of second-order cone-constrained component
and finally followed by k blocks of semidefinite-constrained components. Th
package SeDuMi uses the following syntax:

```
>> [x,y,info] = sedumi(A,b,c,K) ;
```

Here K is a MATLAB structure with fields K.f, K.l, K.q, K.s detailing th
dimensions of the above blocks. The matrix and vectors A, b, c should be of th
appropriate dimensions.

The package SDPT3 uses a similar syntax. It should be noted that althoug
the process of formatting a particular problem in the appropriate SeDuMi
SDPT3 format is relatively routine, the particular details and steps could in som
cases introduce errors. The modeling environment provided by CVX performs tha
formatting in an automated fashion.

18.3   Duality and Optimality Conditions

As in linear and quadratic programming, there is a *dual* conic program associate
with every *primal* conic program. The construction of the dual conic progra
relies on the following more fundamental construction. Let $\mathcal{K} \subseteq \mathbb{R}^p$ be a close
convex cone. The dual cone $\mathcal{K}^* \subseteq \mathbb{R}^p$ of $\mathcal{K}$ is defined as

$$\mathcal{K}^* := \{\mathbf{s} \in \mathbb{R}^p : \mathbf{s}^\mathsf{T}\mathbf{x} \geq 0 \text{ for all } \mathbf{x} \in \mathcal{K}\}.$$

It is easy to see that $\mathcal{K}^* \subseteq \mathbb{R}^p$ is also a closed convex cone.

Just as we did for linear and quadratic programming, the dual problem ca
be derived via the following *Lagrangian function* associated with (18.1):

$$L(\mathbf{x}, \mathbf{y}, \mathbf{s}) := \mathbf{c}^\mathsf{T}\mathbf{x} + \mathbf{y}^\mathsf{T}(\mathbf{b} - \mathbf{A}\mathbf{x}) + \mathbf{s}^\mathsf{T}(\mathbf{d} - \mathbf{D}\mathbf{x}).$$

The constraints of (18.1) can be encoded via the Lagrangian function. For a given vector $\mathbf{x}$

$$\max_{\substack{\mathbf{y},\mathbf{s}\\ \mathbf{s}\in\mathcal{K}^*}} L(\mathbf{x},\mathbf{y},\mathbf{s}) = \begin{cases} \mathbf{c}^\mathsf{T}\mathbf{x} & \text{if } \mathbf{A}\mathbf{x}=\mathbf{b} \text{ and } \mathbf{D}\mathbf{x}-\mathbf{d}\in\mathcal{K} \\ +\infty & \text{otherwise.} \end{cases}$$

Therefore the primal problem (18.1) can be written as

$$\min_{\mathbf{x}} \max_{\substack{\mathbf{y},\mathbf{s}\\ \mathbf{s}\in\mathcal{K}^*}} L(\mathbf{x},\mathbf{y},\mathbf{s}).$$

The dual problem is obtained by flipping the order of the min and max operations:

$$\max_{\substack{\mathbf{y},\mathbf{s}\\ \mathbf{s}\in\mathcal{K}^*}} \min_{\mathbf{x}} L(\mathbf{x},\mathbf{y},\mathbf{s}).$$

It is easy to see that the dual problem can be written as follows:

$$\begin{aligned} \max_{\mathbf{y},\mathbf{s}} \quad & \mathbf{b}^\mathsf{T}\mathbf{y}+\mathbf{d}^\mathsf{T}\mathbf{s} \\ \text{s.t.} \quad & \mathbf{A}^\mathsf{T}\mathbf{y}+\mathbf{D}^\mathsf{T}\mathbf{s}=\mathbf{c} \\ & \mathbf{s}\in\mathcal{K}^*. \end{aligned} \tag{18.7}$$

In particular, when the primal problem is in the following *standard form,*

$$\begin{aligned} \min_{\mathbf{x}} \quad & \mathbf{c}^\mathsf{T}\mathbf{x} \\ \text{s.t.} \quad & \mathbf{A}\mathbf{x}=\mathbf{b} \\ & \mathbf{x}\in\mathcal{K}, \end{aligned}$$

the dual problem is

$$\begin{aligned} \max_{\mathbf{y},\mathbf{s}} \quad & \mathbf{b}^\mathsf{T}\mathbf{y} \\ \text{s.t.} \quad & \mathbf{A}^\mathsf{T}\mathbf{y}+\mathbf{s}=\mathbf{c} \\ & \mathbf{s}\in\mathcal{K}^*. \end{aligned}$$

Observe that the dual problem of a conic program is again a conic program. As in linear and quadratic programming, there is a deep connection between the primal problem (18.1) and its dual (18.7). The following result follows by construction.

Theorem 18.1 (Weak duality) *Assume $\mathbf{x}$ is a feasible point for (18.1) and $(\mathbf{y},\mathbf{s})$ is a feasible point for (18.7). Then*

$$\mathbf{b}^\mathsf{T}\mathbf{y}+\mathbf{d}^\mathsf{T}\mathbf{s}\le\mathbf{c}^\mathsf{T}\mathbf{x}.$$

Proof If $\mathbf{x}$ and $(\mathbf{y},\mathbf{s})$ satisfy the above assumptions then

$$\begin{aligned} \mathbf{b}^\mathsf{T}\mathbf{y}+\mathbf{d}^\mathsf{T}\mathbf{s} &\le (\mathbf{A}\mathbf{x})^\mathsf{T}\mathbf{y}+(\mathbf{D}\mathbf{x})^\mathsf{T}\mathbf{s} \\ &= (\mathbf{A}^\mathsf{T}\mathbf{y}+\mathbf{D}^\mathsf{T}\mathbf{s})^\mathsf{T}\mathbf{x} \\ &= \mathbf{c}^\mathsf{T}\mathbf{x}. \end{aligned}$$

$\square$

In contrast to linear and quadratic programming, strong duality does n
always hold for conic programming. Fortunately, strong duality holds under
mild regularity assumption. The conic program (18.1) satisfies the *Slater* cond
tion if there exists $\mathbf{x} \in \mathbb{R}^n$ such that

$$\mathbf{Ax} = \mathbf{b}, \ \mathbf{Dx} - \mathbf{d} \in \operatorname{relint}(\mathcal{K}),$$

where $\operatorname{relint}(\mathcal{K})$ denotes the relative interior of the set $\mathcal{K}$.

Similarly, the conic program (18.7) satisfies the *Slater* condition if there exis
$\mathbf{y} \in \mathbb{R}^m$ and $\mathbf{s} \in \operatorname{relint}(\mathcal{K}^*)$ such that

$$\mathbf{A}^\mathsf{T}\mathbf{y} + \mathbf{D}^\mathsf{T}\mathbf{s} = \mathbf{c}.$$

Theorem 18.2 (Strong duality)   *Suppose the problems* (18.1) *and* (18.7) *satis*
the Slater condition. Then both problems have optimal solutions and their optim
values are the same.

We refer the reader to Güler (2010) or Renegar (2001) for a proof of Theore
18.2. The following characterization of the solutions to both (18.1) and (18.
readily follows from Theorem 18.1 and Theorem 18.2.

Theorem 18.3 (Optimality conditions)   *The vectors* $\mathbf{x} \in \mathbb{R}^n$ *and* $(\mathbf{y}, \mathbf{s}) \in \mathbb{R}^m$
$\mathbb{R}^n$ *are optimal solutions to* (18.1) *and* (18.7) *respectively if*

$$\begin{aligned}
\mathbf{c} - \mathbf{A}^\mathsf{T}\mathbf{y} - \mathbf{D}^\mathsf{T}\mathbf{s} &= \mathbf{0} \\
\mathbf{Ax} - \mathbf{b} &= \mathbf{0} \\
\mathbf{Dx} - \mathbf{d} &\in \mathcal{K} \\
\mathbf{s} &\in \mathcal{K}^* \\
\mathbf{s}^\mathsf{T}(\mathbf{Dx} - \mathbf{d}) &= 0.
\end{aligned}$$

(18.

The following partial converse also holds: if (18.1) *and* (18.7) *satisfy the Slat*
condition then they both have optimal solutions $\mathbf{x}$ *and* $(\mathbf{y}, \mathbf{s})$ *that satisfy* (18.8

For a conic program in standard form, the optimality conditions (18.8) can l
written as follows:

$$\begin{aligned}
\mathbf{A}^\mathsf{T}\mathbf{y} + \mathbf{s} &= \mathbf{c} \\
\mathbf{Ax} &= \mathbf{b} \\
\mathbf{x} &\in \mathcal{K} \\
\mathbf{s} &\in \mathcal{K}^* \\
\mathbf{s}^\mathsf{T}\mathbf{x} &= 0.
\end{aligned}$$

(18.

18.4   Algorithms

By relying on the structure of the cone $\mathcal{K}$, the main algorithmic template
interior-point methods, such as the one described in Chapter 2 and in Chapt
5, can be extended to a larger class of conic programs. The central idea is
generate a sequence of points that converges to a solution to (18.9). We ne:

sketch the gist of interior-point methods for semidefinite programming. For a full and in-depth treatment of this interesting material we refer the reader to the seminal articles by Nesterov and Todd (1997, 1998) and Schmieta and Alizadeh (2001, 2003) and the textbooks by Nesterov (2004), Nesterov and Nemirovskii (1994), and Renegar (2001).

For convenience of exposition, we consider a semidefinite program in standard form:

$$\begin{aligned} \min_{\mathbf{X}} \quad & \mathbf{C} \bullet \mathbf{X} \\ \text{s.t.} \quad & \mathbf{AX} = \mathbf{b} \\ & \mathbf{X} \succeq \mathbf{0}, \end{aligned} \tag{18.10}$$

where $\mathbf{C} \in \mathbb{S}^n$, $\mathbf{b} \in \mathbb{R}^m$, and $\mathbf{A} : \mathbb{S}^n \to \mathbb{R}^m$ is a linear mapping. As the exercises at the end of this chapter detail, the dual of (18.10) is the semidefinite program

$$\begin{aligned} \min_{\mathbf{y}, \mathbf{S}} \quad & \mathbf{b}^\mathsf{T}\mathbf{y} \\ \text{s.t.} \quad & \mathbf{A}^*\mathbf{y} + \mathbf{S} = \mathbf{C} \\ & \mathbf{S} \succeq \mathbf{0}, \end{aligned} \tag{18.11}$$

where $\mathbf{A}^* : \mathbb{R}^m \to \mathbb{S}^n$ denotes the *adjoint* of $\mathbf{A}$; that is, the unique linear mapping satisfying

$$(\mathbf{AX})^\mathsf{T}\mathbf{y} = \mathbf{X} \bullet \mathbf{A}^*\mathbf{y}$$

for all $\mathbf{X} \in \mathbb{S}^n$ and all $\mathbf{y} \in \mathbb{R}^m$.

We will rely on the following key property of positive semidefinite matrices:

$$\mathbf{X}, \mathbf{S} \in \mathbb{S}^n_+ \text{ and } \mathbf{X} \bullet \mathbf{S} = \mathbf{0} \quad \Rightarrow \quad \mathbf{XS} = \mathbf{0}. \tag{18.12}$$

From Theorem 18.3 and (18.12) it follows that $\mathbf{X}$ and $(\mathbf{y}, \mathbf{S})$ are optimal solutions to (18.10) and (18.11) respectively if

$$\begin{aligned} \mathbf{A}^*\mathbf{y} + \mathbf{S} &= \mathbf{C} \\ \mathbf{AX} &= \mathbf{b} \\ \mathbf{XS} &= \mathbf{0} \\ \mathbf{X}, \mathbf{S} &\succeq \mathbf{0}. \end{aligned} \tag{18.13}$$

As in the linear programming case, interior-point methods for semidefinite programming generate a sequence of iterates that satisfy $\mathbf{X}, \mathbf{S} \succ \mathbf{0}$. Each iteration of the algorithm aims to make progress towards satisfying $\mathbf{A}^*\mathbf{y} + \mathbf{S} = \mathbf{C}$, $\mathbf{AX} = \mathbf{b}$, and $\mathbf{XS} = \mathbf{0}$.

Given $\mu > 0$, let $(\mathbf{X}(\mu), \mathbf{y}(\mu), \mathbf{S}(\mu))$ be the solution to the following perturbed version of the above optimality conditions:

$$\begin{bmatrix} \mathbf{A}^*\mathbf{y} + \mathbf{S} - \mathbf{C} \\ \mathbf{AX} - \mathbf{b} \\ \mathbf{XS} \end{bmatrix} = \begin{bmatrix} \mathbf{0} \\ \mathbf{0} \\ \mu\mathbf{I} \end{bmatrix}, \ \mathbf{X}, \mathbf{S} \succ \mathbf{0}.$$

The first condition above can be written as $\mathbf{r}_\mu(\mathbf{X}, \mathbf{y}, \mathbf{S}) = \mathbf{0}$ for the *residual vector*:

$$r_\mu(\mathbf{X}, \mathbf{y}, \mathbf{S}) := \begin{bmatrix} \mathbf{A}^*\mathbf{y} + \mathbf{S} - \mathbf{C} \\ \mathbf{A}\mathbf{X} - \mathbf{b} \\ \mathbf{X}\mathbf{S} - \mu\mathbf{I} \end{bmatrix}.$$

The *central path* is the set $\{(\mathbf{X}(\mu), \mathbf{y}(\mu), \mathbf{S}(\mu)) : \mu > 0\}$. It is intuitively cle. that $(\mathbf{X}(\mu), \mathbf{y}(\mu), \mathbf{S}(\mu))$ converges to an optimal solution to both (18.10) and (18.11). This suggests the following algorithmic strategy: suppose $(\mathbf{X}, \mathbf{y}, \mathbf{S})$ "near" $(\mathbf{X}(\mu), \mathbf{y}(\mu), \mathbf{S}(\mu))$ for some $\mu > 0$. Use $(\mathbf{X}, \mathbf{y}, \mathbf{S})$ to move to a bett point $(\mathbf{X}^+, \mathbf{y}^+, \mathbf{S}^+)$ "near" $(\mathbf{X}(\mu^+), \mathbf{y}(\mu^+), \mathbf{S}(\mu^+))$ for some $\mu^+ < \mu$.

It can be shown that if a point $(\mathbf{X}, \mathbf{y}, \mathbf{S})$ is on the central path, then th corresponding value of μ satisfies $\mathbf{X} \bullet \mathbf{S} = n\mu$. Likewise, given $\mathbf{X}, \mathbf{S} \succ \mathbf{0}$, define

$$\mu(\mathbf{X}, \mathbf{S}) := \frac{\mathbf{X} \bullet \mathbf{S}}{n}.$$

To move from a current point $(\mathbf{X}, \mathbf{y}, \mathbf{S})$ to a new point, we use a suitable *Newto step*; that is, the solution to the following system of equations obtained as linearization of the system of nonlinear equations $r_\mu(\mathbf{X}, \mathbf{y}, \mathbf{S}) = \mathbf{0}$:

$$\begin{bmatrix} \mathbf{0} & \mathbf{A}^* & \mathbf{I} \\ \mathbf{A} & \mathbf{0} & \mathbf{0} \\ \mathbf{F} & \mathbf{0} & \mathbf{G} \end{bmatrix} \begin{bmatrix} \Delta\mathbf{X} \\ \Delta\mathbf{y} \\ \Delta\mathbf{S} \end{bmatrix} = \begin{bmatrix} \mathbf{C} - \mathbf{A}^\mathsf{T}\mathbf{y} - \mathbf{S} \\ \mathbf{b} - \mathbf{A}\mathbf{X} \\ \mu\mathbf{X}^{-1} - \mathbf{S} \end{bmatrix} \tag{18.1}$$

for some suitably chosen mappings $\mathbf{F}, \mathbf{G}$ that depend on the current $\mathbf{X}, \mathbf{S}$. Th details of these mappings are somewhat technical and related to nuances cor cerning the space of symmetric $n \times n$ matrices. Further details can be foun in Renegar (2001) and in the exercises at the end of the chapter.

Algorithm 18.1 presents a template for an interior-point method for semide: nite programming.

Algorithm 18.1 Interior-point method for semidefinite programming

---
1: choose $\mathbf{X}^0, \mathbf{S}^0 \succ \mathbf{0}$
2: **for** $k = 0, 1, \ldots$ **do**
3:     solve the Newton system (18.14) for $(\mathbf{X}, \mathbf{y}, \mathbf{S}) = (\mathbf{X}^k, \mathbf{y}^k, \mathbf{S}^k)$ and $\mu :=$ $0.1\mu(\mathbf{X}^k, \mathbf{S}^k)$
4:     choose a step length $\alpha \in (0, 1]$ and set $(\mathbf{X}^{k+1}, \mathbf{y}^{k+1}, \mathbf{S}^{k+1}) =$ $(\mathbf{X}^k, \mathbf{y}^k, \mathbf{S}^k) + \alpha(\Delta\mathbf{X}, \Delta\mathbf{y}, \Delta\mathbf{S})$
5: **end for**

The step length α in step 4 should be chosen so that $\mathbf{X}^{k+1}, \mathbf{S}^{k+1} \succ \mathbf{0}$ an the size of $r_\mu(\mathbf{X}^{k+1}, \mathbf{y}^{k+1}, \mathbf{S}^{k+1})$ is sufficiently smaller than $r_\mu(\mathbf{X}^k, \mathbf{y}^k, \mathbf{S}^k)$. line-search procedure such as the one described in Algorithm 2.4 in Chapter can be used for choosing the step length α.

18.5   Notes

The seminal works of Alizadeh (1991) and Nesterov and Nemirovskii (1994) triggered a massive burst of research activity in optimization. This eventually led to a mature theory and computational technology for solving important classes of conic programs, notably second-order and semidefinite programming. A particularly important development was the extension of primal–dual interior-point methods to conic programs over *symmetric cones* by Nesterov and Todd (1997, 1998). Such cones include the non-negative orthant, the second-order cone, the semidefinite cone, and Cartesian products of them. The textbook by Renegar (2001) gives an excellent exposition of the main advances in conic programming.

The software packages SeDuMi and SDPT3 were developed respectively by the late Sturm (1999) and by Toh et al. (1999). These two packages are some of the default engines used by the MATLAB-based modeling language CVX.

18.6   Exercises

Exercise 18.1   Recall that the trace of a square matrix $\mathbf{M} \in \mathbb{R}^{n \times n}$ is

$$\text{trace}(\mathbf{M}) = \sum_{i=1}^{n} M_{ii}.$$

Suppose $\mathbf{A} \in \mathbb{R}^{m \times n}, \mathbf{B} \in \mathbb{R}^{n \times p}$, and $\mathbf{C} \in \mathbb{R}^{p \times m}$. Show that the trace satisfies the following property:

$$\text{trace}(\mathbf{ABC}) = \text{trace}(\mathbf{CAB}).$$

Exercise 18.2

(a) Suppose $\mathbf{X} \in \mathbb{S}_+^n$. Show that

$$\text{trace}(\mathbf{X}) = 0 \quad \Rightarrow \quad \mathbf{X} = \mathbf{0}.$$

(b) Suppose $\mathbf{L} \in \mathbb{R}^{n \times m}$. Show that $\mathbf{LL}^\mathsf{T} \in \mathbb{S}_+^n$. The converse is true also but it is harder to show: if $\mathbf{X} \in \mathbb{S}_+^n$ then there exists $\mathbf{L} \in \mathbb{R}^{n \times m}$ for some m such that $\mathbf{X} = \mathbf{LL}^\mathsf{T}$.

(c) Show an example of two matrices $\mathbf{X}, \mathbf{S} \in \mathbb{S}^n$ such that $\mathbf{XS} \notin \mathbb{S}^n$.

(d) Prove (18.12). That is, show that

$$\mathbf{X}, \mathbf{S} \in \mathbb{S}_+^n, \text{ trace}(\mathbf{XS}) = 0 \quad \Rightarrow \quad \mathbf{XS} = \mathbf{0}.$$

Hint: Observe that this is not immediate from part (a) because by part (c) *a priori* we do not even know if $\mathbf{XS} \in \mathbb{S}^n$. To get around this difficulty, use part (b), Exercise 18.1, and part (a).

Exercise 18.3 Show that the Lorenz cone and the semidefinite cone are "se[l]
dual." In other words, show that

$$(\mathbb{L}_n)^* = \mathbb{L}_n$$

and

$$(\mathbb{S}_+^n)^* = \mathbb{S}_+^n.$$

Exercise 18.4 This exercise shows that semidefinite programming includes [a]
special cases both linear and second-order programming.

(a) Suppose $\mathbf{x} \in \mathbb{R}^n$ and $\mathbf{X} = \mathrm{diag}(\mathbf{x}) \in \mathbb{S}^n$. Show that

$$\mathbf{x} \geq \mathbf{0} \quad \Leftrightarrow \quad \mathbf{X} \succeq \mathbf{0}.$$

(b) Suppose $\mathbf{x} = \begin{bmatrix} x_0 \\ \bar{\mathbf{x}} \end{bmatrix} \in \mathbb{R}^n$ and $\mathbf{X} = \begin{bmatrix} x_0 & \bar{\mathbf{x}}^\mathsf{T} \\ \bar{\mathbf{x}} & x_0 \mathbf{I}_{n-1} \end{bmatrix} \in \mathbb{S}^n$. Show that

$$\mathbf{x} \in \mathbb{L}_n \quad \Leftrightarrow \quad \mathbf{X} \succeq \mathbf{0}.$$

(c) Use (a) and (b) to conclude that any linear program or second-order progra[m]
can be recast as a semidefinite program.

19 Robust Optimization

In many optimization models the inputs to the problem are either not known at the time the problem must be solved, are computed inaccurately, or are otherwise uncertain. Since the solutions obtained can be quite sensitive to these inputs, one serious concern is that we are solving the wrong problem, and that the solution we find is far from optimal for the correct problem. Robust optimization is an approach to optimization problems with data uncertainty to obtain solutions that are *good* for *all* or *most* possible realizations of the uncertain parameters.

19.1 Uncertainty Sets

In robust optimization, the description of the parameter uncertainty is formalized via *uncertainty sets*. Uncertainty sets can represent or may be formed by difference of opinions on the possible values of problem parameters, alternative estimates of parameters generated via statistical techniques from historical data, Bayesian, or other estimation techniques. The size of the uncertainty set is typically determined by the level of desired robustness.

Some of the most common types of uncertainty sets encountered in robust optimization models include the following:

- Uncertainty sets representing a finite number of scenarios generated for the possible values of the parameters:

$$\mathcal{U} = \{\mathbf{p}_1, \mathbf{p}_2, \ldots, \mathbf{p}_k\}.$$

- Uncertainty sets representing the convex hull of a finite number of scenarios generated for the possible values of the parameters (these are sometimes called polytopic uncertainty sets):

$$\mathcal{U} = \text{conv}\{\mathbf{p}_1, \mathbf{p}_2, \ldots, \mathbf{p}_k\}.$$

- Uncertainty sets representing an interval description for each uncertain parameter:

$$\mathcal{U} = \{\mathbf{p} : \mathbf{l} \leq \mathbf{p} \leq \mathbf{u}\}.$$

Confidence intervals encountered frequently in statistics can be the source of such uncertainty sets.

- Ellipsoidal uncertainty sets:

$$\mathcal{U} = \{\mathbf{p} : \mathbf{p} = \mathbf{p}_0 + \mathbf{M}\mathbf{u}, \|\mathbf{u}\| \leq 1\}.$$

These uncertainty sets can also arise from statistical estimation in the for
of confidence regions; see Goldfarb and Iyengar (2003). In addition to the
mathematically compact description, ellipsoidal uncertainty sets have tl
nice property that they smooth the optimal value function (Werner, 201C

It is a non-trivial task to determine the uncertainty set that is appropriate f
a particular model as well as the type of uncertainty sets that lead to tractab
problems. As a general guideline, the shape of the uncertainty set will ofte
depend on the sources of uncertainty as well as the sensitivity of the solutions
these uncertainties. The size of the uncertainty set, on the other hand, will ofte
be chosen based on the desired level of robustness. When uncertain paramete
reflect the "true" values of moments of random variables, as is the case
mean–variance portfolio optimization, we simply have no way of knowing the
unobservable true values exactly. In such cases, after making some assumptio:
about the stationarity of these random processes, we can generate estimates
these true parameters using statistical procedures. Goldfarb and Iyengar (200:
for example, show that if we use a linear factor model for the multivariate retur:
of several assets and estimate the factor loading matrices via linear regression, tl
confidence regions generated for these parameters are ellipsoidal sets and the
advocate their use in robust portfolio selection as uncertainty sets. To genera
interval-type uncertainty sets, Tütüncü and Koenig (2004) use bootstrappi:
strategies as well as moving averages of returns from historical data. The sha]
and the size of the uncertainty set can significantly affect the robust solutio:
generated. However, with few guidelines backed by theoretical and empiric
studies, their choice remains a mix of art and science.

19.2   Different Flavors of Robustness

As we next describe, different types of robustness arise depending on wh
parameters of a problem are uncertain, and depending also on what exact
constitutes a "good" robust solution.

Constraint Robustness

Constraint robustness refers to situations where the uncertainty is in the co
straints and we seek solutions that remain feasible for all possible values
the uncertain inputs. This type of solution is required in many engineeri:
applications. Typical instances include multi-stage problems where the uncerta
outcomes of earlier stages have an effect on the decisions of the later stages ar
the decision variables must be chosen to satisfy constraints no matter wh.
happens with the uncertain parameters of the problem.

Here is a precise mathematical model for finding constraint-robust solutions. Consider an optimization problem of the form

$$\min_{\mathbf{x}} \quad f(\mathbf{x})$$
$$\text{s.t.} \quad G(\mathbf{x}, \mathbf{p}) \in K. \tag{19.1}$$

In this problem $\mathbf{x}$ is the vector of decision variables, $f(\mathbf{x})$ is the (certain) objective function, G and K are the structural elements of the constraints assumed to be certain, and $\mathbf{p}$ is the vector of possibly uncertain parameters of the problem. Consider an uncertainty set $\mathcal{U}$ that contains all possible values of the uncertain parameters $\mathbf{p}$. Then, a constraint-robust optimal solution can be found by solving the following problem:

$$\min_{\mathbf{x}} \quad f(\mathbf{x})$$
$$\text{s.t.} \quad G(\mathbf{x}, \mathbf{p}) \in K, \quad \text{for all} \ \mathbf{p} \in \mathcal{U}. \tag{19.2}$$

The feasible set in the robust optimization model (19.2) is the intersection of the feasible sets:

$$S(\mathbf{p}) = \{\mathbf{x} : G(\mathbf{x}, \mathbf{p}) \in K\}, \quad \mathbf{p} \in \mathcal{U}.$$

We note that there are no uncertain parameters in the objective function of the problem (19.1). However, this is not a restrictive assumption. An optimization problem with uncertain parameters in both the objective function and constraints can be easily reformulated to fit the form in (19.1). Indeed, the problem

$$\min_{\mathbf{x}} \quad f(\mathbf{x}, \mathbf{p})$$
$$\text{s.t.} \quad G(\mathbf{x}, \mathbf{p}) \in K$$

is equivalent to the problem

$$\min_{\mathbf{x}, t} \quad t$$
$$\text{s.t.} \quad f(\mathbf{x}, \mathbf{p}) \le t$$
$$\qquad G(\mathbf{x}, \mathbf{p}) \in K.$$

This last problem has all its uncertainties in its constraints only.

Objective Robustness

Another important robustness concept is objective robustness. This refers to solutions that will remain close to optimal for all possible realizations of the uncertain problem parameters. Since such solutions may be difficult to obtain, especially when uncertainty sets are relatively large, an alternative goal for objective robustness is to find solutions whose worst-case behavior is optimized. The worst-case behavior of a solution corresponds to the value of the objective function for the worst possible realization of the uncertain data for that particular solution. We now develop a mathematical model that addresses objective robustness. Consider an optimization problem of the form:

$$\min_{\mathbf{x} \in S} f(\mathbf{x}, \mathbf{p}).$$

Here, S is a (certain) feasible set and $f(\mathbf{x}, \mathbf{p})$ is the objective function th depends on uncertain parameters $\mathbf{p}$. As before, let $\mathcal{U}$ denote the uncertainty that contains all possible values of the uncertain parameters $\mathbf{p}$. Then an objecti robust solution can be obtained by solving the *saddle-point problem*

$$\min_{\mathbf{x} \in S} \max_{\mathbf{p} \in \mathcal{U}} f(\mathbf{x}, \mathbf{p}).$$

As indicated at the end of the previous subsection, objective robustness can seen as a special case of constraint robustness via a suitable reformulation. Ho ever, it is important to distinguish between these two problem variants as th "natural" robust formulations lead to two different classes of optimization fo mulations. Robust-constraint problems naturally lead to optimization proble with infinitely many constraints whereas robust-objective problems natura lead to saddle-point problems. There are different methodologies available f each of these two problem classes.

Relative Robustness

The focus of constraint and objective robustness models on an *absolute* measu of worst-case performance is not consistent with risk tolerances of many decisi makers. Instead, we may prefer to measure the worst case in a *relative* mann relative to the best possible solution under each scenario. This leads us to t notion of relative robustness. Consider the optimization problem

$$\min_{\mathbf{x} \in S} f(\mathbf{x}, \mathbf{p}), \tag{19.}$$

where $\mathbf{p}$ is uncertain with uncertainty set $\mathcal{U}$. To simplify the description, restrict our attention to the case with objective uncertainty and assume that t constraints are certain. Given a fixed $\mathbf{p} \in \mathcal{U}$, let $z^*(\mathbf{p})$ denote the optimal val function

$$z^*(\mathbf{p}) := \min_{\mathbf{x} \in S} f(\mathbf{x}, \mathbf{p}).$$

Furthermore, define the optimal solution map

$$\mathbf{x}^*(\mathbf{p}) = \arg\min_{\mathbf{x} \in S} f(\mathbf{x}, \mathbf{p}).$$

Note that $z^*(\mathbf{p})$ can be extended-valued and $\mathbf{x}^*(\mathbf{p})$ can be set-valued. To motiva the notion of relative robustness we first define a measure of regret associat with a decision after the uncertainty is resolved. If we choose $\mathbf{x}$ as our vect and $\mathbf{p}$ is the realized value of the uncertain parameter, the *regret* associated wi choosing $\mathbf{x}$ instead of an element of $\mathbf{x}^*(\mathbf{p})$ is defined as

$$r(\mathbf{x}, \mathbf{p}) = f(\mathbf{x}, \mathbf{p}) - z^*(\mathbf{p}) = f(\mathbf{x}, \mathbf{p}) - f(\mathbf{x}^*(\mathbf{p}), \mathbf{p}).$$

Note that the regret function is always non-negative and can also be regarded as a measure of the "benefit of hindsight". Now, for a given $\mathbf{x} \in S$ consider the maximum regret function:

$$R(\mathbf{x}) := \max_{\mathbf{p} \in \mathcal{U}} r(\mathbf{x}, \mathbf{p}) = \max_{\mathbf{p} \in \mathcal{U}} f(\mathbf{x}, \mathbf{p}) - z^*(\mathbf{p}).$$

A relative robust solution to problem (19.3) is a vector $\mathbf{x}$ that minimizes the maximum regret:

$$\min_{\mathbf{x} \in S} \max_{\mathbf{p} \in \mathcal{U}} f(\mathbf{x}, \mathbf{p}) - z^*(\mathbf{p}). \tag{19.4}$$

While they are intuitively attractive, relative robust formulations can also be significantly more difficult than the standard absolute robust formulations. Indeed, since $z^*(\mathbf{p})$ is the optimal value function and involves an optimization problem itself, the problem (19.4) is a three-level optimization problem as opposed to the two-level problems in absolute robust formulations. Furthermore, the optimal value function $z^*(\mathbf{p})$ is rarely available in analytic form, is typically non-smooth, and is often hard to analyze. Another difficulty is that if f is linear in $\mathbf{p}$, as is often the case, then $z^*(\mathbf{p})$ is a concave function. Therefore, the inner maximization problem in (19.4) is a convex *maximization* problem and is difficult for most $\mathcal{U}$.

A simpler variant of (19.4) can be constructed by deciding on the maximum level of regret to be tolerated beforehand and by solving a feasibility problem instead with this level imposed as a constraint. For example, if we decide to limit the maximum regret to R, then the problem to solve becomes the following: find an $\mathbf{x}$ satisfying $\mathbf{x} \in S$ such that

$$f(\mathbf{x}, \mathbf{p}) - z^*(\mathbf{p}) \le R, \quad \text{for all } \mathbf{p} \in \mathcal{U}.$$

If desired, one can then perform a bisection search on R to find its optimal value. Another variant of relative robustness models arises when we measure the regret in terms of the proximity of our chosen solution to the optimal solution set rather than in terms of the optimal objective values. For this model, consider the following distance function for a given $\mathbf{x}$ and $\mathbf{p}$:

$$d(\mathbf{x}, \mathbf{p}) = \inf_{\mathbf{x}^* \in \mathbf{x}^*(\mathbf{p})} \|\mathbf{x} - \mathbf{x}^*\|.$$

When the solution set is a singleton, there is no optimization involved in the definition. As above, we then consider the maximum distance function

$$D(\mathbf{x}) = \max_{\mathbf{p} \in \mathcal{U}} d(\mathbf{x}, \mathbf{p}) = \max_{\mathbf{p} \in \mathcal{U}} \inf_{\mathbf{x}^* \in \mathbf{x}^*(\mathbf{p})} \|\mathbf{x} - \mathbf{x}^*\|.$$

For relative robustness in this new sense, we seek $\mathbf{x}$ that solves

$$\min_{\mathbf{x} \in S} \max_{\mathbf{p} \in \mathcal{U}} d(\mathbf{x}, \mathbf{p}). \tag{19.5}$$

This variant is an attractive model for cases where we have time to revise our decision variables $\mathbf{x}$, perhaps only slightly, once $\mathbf{p}$ is revealed. In such cases, we

will want to choose an $\mathbf{x}$ that will not need much perturbation under any scenar
i.e., we seek the solution to (19.5). This model can also be useful for multi-peri
problems where revisions of decisions between periods can be costly. Portfo
rebalancing problems with transaction costs are examples of such settings.

19.3   Techniques for Solving Robust Optimization Models

In this section we review a few of the commonly used techniques for the soluti
of robust optimization problems. The tools we discuss are essentially reformu
tion strategies for robust optimization problems so that they can be rewritten a
deterministic optimization problem with no uncertainty. In these reformulatio
we look for economy, so that the new formulation is not much bigger than t
original, "uncertain" problem, and tractability, so that the new problem can
solved efficiently using standard optimization methods.

The variety of the robustness models and the types of uncertainty sets rule o
a unified approach. However, there are some common threads and the mater
in this section can be seen as a guide to the available tools which can
combined or appended with other techniques to solve a given problem in t
robust optimization setting.

Sampling

One of the simplest strategies for achieving robustness under uncertainty is
sample several scenarios for the uncertain parameters from a set that contai
possible values of these parameters. This sampling can be done with or wit
out using distributional assumptions on the parameters and produces a robu
optimization formulation with a finite uncertainty set. If uncertain paramete
appear in the constraints, we create a copy of each such constraint correspondi
to each scenario. Uncertainty in the objective function can be handled in a simil
manner. Consider the generic uncertain optimization problem

$$\min_{\mathbf{x}} \quad f(\mathbf{x}, \mathbf{p})$$
$$\text{s.t.} \quad G(\mathbf{x}, \mathbf{p}) \in K, \text{ for all } \mathbf{p} \in \mathcal{U}.$$

If the uncertainty set $\mathcal{U}$ is a finite set, i.e., $\mathcal{U} = \{\mathbf{p}_1, \mathbf{p}_2, \dots, \mathbf{p}_k\}$, the robu
formulation is obtained as follows:

$$\min_{\mathbf{x}, t} \quad t$$
$$\text{s.t.} \quad f(\mathbf{x}, \mathbf{p}_i) \leq t, \ i = 1, \dots, k$$
$$\quad G(\mathbf{x}, \mathbf{p}_i) \in K, \ i = 1, \dots, k.$$

Note that no reformulation is necessary in this case and the duplicat
constraints preserve the structural properties (linearity, convexity, etc.) of t
original constraints. Consequently, when the uncertainty set is a finite set t
resulting robust optimization problem is larger but theoretically no more difficu

than the non-robust version of the problem. The situation is somewhat similar to stochastic programming formulations. Examples of robust optimization formulations with finite uncertainty sets can be found, for example in Rustem and Howe (2002).

Conic Optimization

Moving from finite uncertainty sets to continuous sets such as intervals or ellipsoids presents a theoretical challenge. The robust version of an uncertain constraint that has to be satisfied for all values of the uncertain parameters in a continuous set results in a semi-infinite optimization formulation. These problems are called semi-infinite since there are infinitely many constraints but only finitely many variables.

Fortunately, for some types of uncertainty sets, it is possible to reformulate their robust semi-infinite programming versions using a *finite* set of conic constraints. To illustrate this, consider the following simple linear program:

$$
\begin{aligned}
\max_{\mathbf{x}} \quad & \mathbf{r}^\mathsf{T}\mathbf{x} \\
\text{s.t.} \quad & \mathbf{1}^\mathsf{T}\mathbf{x} = 1 \\
& \mathbf{x} \geq \mathbf{0}.
\end{aligned}
\tag{19.6}
$$

What is the optimal solution to this linear program?

Now suppose the objective coefficients are uncertain with ellipsoidal uncertainty, e.g., suppose the objective coefficient vector $\mathbf{r}$ can be any element in the uncertainty set

$$
\mathcal{U} = \left\{ \mathbf{r} : \|\mathbf{r} - \boldsymbol{\mu}\|_2 \leq \delta \right\},
$$

where $\boldsymbol{\mu}$ is the "nominal" value of $\mathbf{r}$. The robust version of (19.6) is

$$
\begin{aligned}
\max_{\mathbf{x}} \min_{\mathbf{r} \in \mathcal{U}} \quad & \mathbf{r}^\mathsf{T}\mathbf{x} \\
\text{s.t.} \quad & \mathbf{1}^\mathsf{T}\mathbf{x} = 1 \\
& \mathbf{x} \geq \mathbf{0}.
\end{aligned}
$$

Some simple calculations show that for a given $\mathbf{x}$

$$
\min_{\mathbf{r} \in \mathcal{U}} \mathbf{r}^\mathsf{T}\mathbf{x} = \boldsymbol{\mu}^\mathsf{T}\mathbf{x} - \delta \cdot \|\mathbf{x}\|_2.
$$

Thus the robust version of (19.6) is

$$
\begin{aligned}
\max_{\mathbf{x}} \quad & \boldsymbol{\mu}^\mathsf{T}\mathbf{x} - \delta \cdot \|\mathbf{x}\|_2 \\
\text{s.t.} \quad & \mathbf{1}^\mathsf{T}\mathbf{x} = 1 \\
& \mathbf{x} \geq \mathbf{0}.
\end{aligned}
$$

The latter problem can be rewritten as the following conic program:

$$\max_{\mathbf{x},t} \quad \boldsymbol{\mu}^{\mathsf{T}}\mathbf{x} - \delta \cdot t$$
$$\text{s.t.} \quad \mathbf{1}^{\mathsf{T}}\mathbf{x} = 1$$
$$\begin{bmatrix} t \\ \mathbf{x} \end{bmatrix} \in \mathbb{L}_{n+1}$$
$$\mathbf{x} \geq \mathbf{0}.$$

More generally, suppose $\mathbf{r}$ has the following *ellipsoidal* uncertainty set:

$$\mathcal{U} = \{\mathbf{r} : (\mathbf{r} - \boldsymbol{\mu})^{\mathsf{T}}\Sigma^{-1}(\mathbf{r} - \boldsymbol{\mu}) \leq \delta^2\}$$

for some symmetric and positive definite matrix Σ. Then the robust version (19.6) is

$$\max_{\mathbf{x}} \quad \boldsymbol{\mu}^{\mathsf{T}}\mathbf{x} - \delta \cdot \sqrt{\mathbf{x}^{\mathsf{T}}\Sigma\mathbf{x}}$$
$$\text{s.t.} \quad \mathbf{1}^{\mathsf{T}}\mathbf{x} = 1$$
$$\mathbf{x} \geq \mathbf{0},$$

which again can be formulated as a conic program. Observe the resemblance between the latter model and Markowitz's mean–variance model.

The machinery of robust optimization can be applied to mean–variance portfolio optimization to mitigate the effects of estimation errors in the expected returns and/or in the covariance matrix (Ceria and Stubbs, 2006; Goldfarb and Iyengar, 2003). The basic idea is to consider the mean–variance optimization problem in one of its forms, e.g.,

$$\max_{\mathbf{x}} \quad \boldsymbol{\mu}^{\mathsf{T}}\mathbf{x} - \tfrac{1}{2}\gamma \cdot \mathbf{x}^{\mathsf{T}}\mathbf{V}\mathbf{x}$$
$$\text{s.t.} \quad \mathbf{A}\mathbf{x} = \mathbf{b} \qquad \qquad (19.$$
$$\mathbf{C}\mathbf{x} \leq \mathbf{d},$$

and assume $\boldsymbol{\mu}$ belongs to some uncertainty set,

$$\mathcal{U} = \{\boldsymbol{\mu} : (\boldsymbol{\mu} - \hat{\boldsymbol{\mu}})^{\mathsf{T}}\Sigma^{-1}(\boldsymbol{\mu} - \hat{\boldsymbol{\mu}}) \leq \delta^2\}.$$

Then the robust version of (19.7) is

$$\max_{\mathbf{x}} \quad \hat{\boldsymbol{\mu}}^{\mathsf{T}}\mathbf{x} - \delta \cdot \sqrt{\mathbf{x}^{\mathsf{T}}\Sigma\mathbf{x}} - \tfrac{1}{2}\gamma \cdot \mathbf{x}^{\mathsf{T}}\mathbf{V}\mathbf{x}$$
$$\text{s.t.} \quad \mathbf{A}\mathbf{x} = \mathbf{b} \qquad \qquad (19.$$
$$\mathbf{C}\mathbf{x} \leq \mathbf{d}.$$

We next show that (19.8) is a conic program. To that end, it suffices to find a conic representation of the objective function. Let $\mathbf{R} \in \mathbb{R}^{n \times p}, \mathbf{L} \in \mathbb{R}^{n \times q}$ such that $\Sigma = \mathbf{R}\mathbf{R}^{\mathsf{T}}$ and $\mathbf{V} = \mathbf{L}\mathbf{L}^{\mathsf{T}}$. Both $\mathbf{R}$ and $\mathbf{L}$ exist because Σ and $\mathbf{V}$ are positive semidefinite. By introducing new variables s, t, the problem (19.8) can be rewritten as the following conic program:

$$\max_{\mathbf{x},s,t} \quad \hat{\mu}^{\mathsf{T}}\mathbf{x} - \delta \cdot s - \tfrac{1}{2}\gamma \cdot t$$

$$\text{s.t.} \quad \begin{bmatrix} s \\ \mathbf{Rx} \end{bmatrix} \in \mathbb{L}_{p+1}$$

$$\begin{bmatrix} t+1 \\ t-1 \\ 2\mathbf{Lx} \end{bmatrix} \in \mathbb{L}_{q+2}$$

$$\mathbf{Ax} = \mathbf{b}$$
$$\mathbf{Cx} \le \mathbf{d}.$$

Saddle-Point Characterizations

For the solution of problems arising from objective uncertainty, the robust solution can be characterized using saddle-point conditions when the original problem satisfies certain convexity assumptions. The benefit of this characterization is that we can then use algorithms such as interior-point methods already developed and available for saddle-point problems. As an example of this strategy, consider the objective-robust formulation discussed in Section 19.2:

$$\min_{\mathbf{x} \in S} \max_{\mathbf{p} \in \mathcal{U}} f(\mathbf{x}, \mathbf{p}). \tag{19.9}$$

We note that the dual of this robust optimization problem is obtained by changing the order of the minimization and maximization problems:

$$\max_{\mathbf{p} \in \mathcal{U}} \min_{\mathbf{x} \in S} f(\mathbf{x}, \mathbf{p}). \tag{19.10}$$

Under mild assumptions on $f, S, \mathcal{U}$, there exists a *saddle-point* solution $(\mathbf{x}^*, \mathbf{p}^*) \in S \times \mathcal{U}$ such that

$$f(\mathbf{x}^*, \mathbf{p}) \le f(\mathbf{x}^*, \mathbf{p}^*) \le f(\mathbf{x}, \mathbf{p}^*) \quad \text{for all} \quad \mathbf{x} \in S, \ \mathbf{p} \in \mathcal{U}.$$

This characterization is the basis of the robust optimization algorithms given in Tütüncü and Koenig (2004).

19.4   Some Robust Optimization Models in Finance

Since many financial optimization problems involve future values of security prices, interest rates, exchange rates, etc., which are not known in advance but can only be forecasted or estimated, such problems fit perfectly into the framework of robust optimization. We next describe some examples of robust optimization formulations for a variety of financial optimization problems.

Robust Profit Opportunities in Risky Portfolios

Consider an investment environment with n financial securities whose futu price vector $\mathbf{r} \in \mathbb{R}^n$ is a random variable. Let $\mathbf{p} \in \mathbb{R}^n$ represent the curre prices of these securities. If the investor chooses a portfolio $\mathbf{x} = \begin{bmatrix} x_1 & \cdots & x \end{bmatrix}$ that satisfies

$$\mathbf{p}^\mathsf{T}\mathbf{x} < 0$$

and the realization of the random variable $\mathbf{r}$ satisfies

$$\mathbf{r}^\mathsf{T}\mathbf{x} \geq 0 \tag{19.1}$$

then there is an arbitrage opportunity: an investor could make money b constructing the portfolio $\mathbf{x}$ with negative cash flow (pocketing money) an subsequently collecting the non-negative cash flow $\mathbf{r}^\mathsf{T}\mathbf{x}$ of the portfolio $\mathbf{x}$.

Since arbitrage opportunities generally do not persist in financial markets, on might be interested in the alternative and weaker profitability notion where th non-negativity of the portfolio is only guaranteed to occur with high probabilit More precisely, consider the following relaxation of (19.11):

$$\mathbf{P}(\mathbf{r}^\mathsf{T}\mathbf{x} \geq 0) \geq 0.99. \tag{19.1}$$

Let $\boldsymbol{\mu}$ and $\mathbf{Q}$ represent the expected future price vector and covariance matr of the random vector $\mathbf{r}$. Then $\mathbb{E}(\mathbf{r}) = \boldsymbol{\mu}^\mathsf{T}\mathbf{x}$ and $\mathrm{stdev}(\mathbf{r}^\mathsf{T}\mathbf{x}) = \sqrt{\mathbf{x}^\mathsf{T}\mathbf{Q}\mathbf{x}}$. If th random vector $\mathbf{r}$ is Gaussian, then (19.12) is equivalent to

$$\boldsymbol{\mu}^\mathsf{T}\mathbf{x} - \theta \cdot \sqrt{\mathbf{x}^\mathsf{T}\mathbf{Q}\mathbf{x}} \geq 0,$$

where $\theta = \Phi^{-1}(0.99)$ and Φ is the standard normal cumulative distributio Therefore, if we find an $\mathbf{x}$ satisfying

$$\boldsymbol{\mu}^\mathsf{T}\mathbf{x} - \theta \cdot \sqrt{\mathbf{x}^\mathsf{T}\mathbf{Q}\mathbf{x}} \geq 0, \quad \mathbf{p}^\mathsf{T}\mathbf{x} < 0,$$

for a large enough positive value of θ, we have an approximation of an arbitra opportunity. Note that, by relaxing the constraint $\mathbf{p}^\mathsf{T}\mathbf{x} < 0$ as $\mathbf{p}^\mathsf{T}\mathbf{x} \leq 0$ or $\mathbf{p}^\mathsf{T}\mathbf{x} \leq -\varepsilon$, we obtain a conic feasibility system. Therefore, the resulting syste can be solved using the conic optimization approaches.

We next explore some portfolio selection models that incorporate the unce tainty of problem inputs.

Robust Portfolio Selection

This section is adapted from Tütüncü and Koenig (2004). Recall that Markowitz mean–variance optimization problem can be stated in the following form, whi combines the reward and risk in the objective function:

$$\max_{\mathbf{x} \in \mathcal{X}} \boldsymbol{\mu}^\mathsf{T}\mathbf{x} - \frac{\gamma}{2} \cdot \mathbf{x}^\mathsf{T}\mathbf{Q}\mathbf{x}. \tag{19.1}$$

Here $\boldsymbol{\mu}$ and $\mathbf{Q}$ are respectively estimates of the vector of expected values an covariance of returns of a universe of securities, and γ is a risk-aversion consta used to trade off the reward (expected return) and risk (portfolio variance). The set $\mathcal{X}$ is the set of feasible portfolios which may carry information on short-sale restrictions, sector distribution requirements, etc. Since such restrictions are predetermined, we can assume that the set $\mathcal{X}$ is known without any uncertainty at the time the problem is solved.

Recall also that solving the problem above for different values of γ we obtain the *efficient frontier* of the set of feasible portfolios. The optimal portfolio will be different for individuals with different risk-taking tendencies, but it will always be on the efficient frontier.

One of the limitations of this model is its need to accurately estimate the expected returns and covariances. In Bawa et al. (1979), the authors argue that using estimates of the unknown expected returns and covariances leads to an *estimation risk* in portfolio choice, and that methods for optimal selection of portfolios must take this risk into account. Furthermore, the optimal solution is sensitive to perturbations in these input parameters – a small change in the estimate of the return or the variance may lead to a large change in the corresponding solution; see, for example, Michaud and Michaud (2008). This attribute is unfavorable since the modeler may want to periodically rebalance the portfolio based on new data and may incur significant transaction costs to do so. Furthermore, using point estimates of the expected return and covariance parameters does not fulfill the needs of a conservative investor. Such an investor would not necessarily trust these estimates and would be more comfortable choosing a portfolio that will perform well under a number of different scenarios. Of course, such an investor cannot expect to get better performance on some of the more likely scenarios, but will have insurance for more extreme cases. All these arguments point to the need of a portfolio optimization formulation that incorporates robustness and tries to find a solution that is relatively insensitive to inaccuracies in the input data. Since all the uncertainty is in the objective function coefficients, we seek an objective robust portfolio, as outlined in Section 19.2.

For *robust portfolio optimization* we consider a model that allows return and covariance matrix information to be given in the form of intervals. For example, this information may take the form "the expected return on security j is between 8% and 10%" rather than claiming that it is 9%. Mathematically, we will represent this information as membership in the following set:

$$\mathcal{U} = \{(\boldsymbol{\mu}, \mathbf{Q}) : \boldsymbol{\mu}^L \leq \boldsymbol{\mu} \leq \boldsymbol{\mu}^U, \quad \mathbf{Q}^L \leq \mathbf{Q} \leq \mathbf{Q}^U, \quad \mathbf{Q} \succeq \mathbf{0}\}, \qquad (19.14)$$

where $\boldsymbol{\mu}^L, \boldsymbol{\mu}^U, \mathbf{Q}^L, \mathbf{Q}^U$ are the extreme values of the intervals we just mentioned. The restriction $\mathbf{Q} \succeq \mathbf{0}$ is necessary since $\mathbf{Q}$ is a covariance matrix and, therefore, must be positive semidefinite. These intervals may be generated in different ways. An extremely cautious modeler may want to use historical lows and highs of certain input parameters as the range of their values. One may generate different estimates using different scenarios on the general economy and then combine the resulting estimates. Different analysts may produce different estimates for these parameters and one may choose the extreme estimates as the endpoints of the intervals. One may choose a confidence level and then generate estimates covariance and return parameters in the form of prediction intervals.

We want to find a portfolio that maximizes the objective function in (19.13) the worst-case realization of the input parameters μ and $\mathbf{Q}$ from their uncertain set $\mathcal{U}$ in (19.14). Given these considerations the robust optimization proble takes the following form

$$\max_{\mathbf{x} \in \mathcal{X}} \min_{(\mu, \mathbf{Q}) \in \mathcal{U}} \mu^\mathsf{T} \mathbf{x} - \frac{\gamma}{2} \cdot \mathbf{x}^\mathsf{T} \mathbf{Q} \mathbf{x}. \tag{19.1}$$

This problem can be expressed as a *saddle-point problem* and be solved usin the technique outlined in Halldórsson and Tütüncü (2003).

Relative Robustness in Portfolio Selection

We consider the following simple three-asset portfolio model from Ceria ar Stubbs (2006):

$$\begin{aligned}
\max \quad & \mu^\mathsf{T} \mathbf{x} \\
\text{s.t.} \quad & TE(\mathbf{x}) \leq 0.1 \\
& \mathbf{1}^\mathsf{T} \mathbf{x} = 1 \\
& \mathbf{x} \geq \mathbf{0},
\end{aligned} \tag{19.1}$$

where $\mathbf{x} = \begin{bmatrix} x_1 & x_2 & x_3 \end{bmatrix}$ and

$$TE(\mathbf{x}) = \sqrt{ \begin{bmatrix} x_1 - 0.5 \\ x_2 - 0.5 \\ x_3 \end{bmatrix}^\mathsf{T} \begin{bmatrix} 0.1764 & 0.09702 & 0 \\ 0.9702 & 0.1089 & 0 \\ 0 & 0 & 0 \end{bmatrix} \begin{bmatrix} x_1 - 0.5 \\ x_2 - 0.5 \\ x_3 \end{bmatrix} }.$$

This is essentially a two-asset portfolio optimization problem where the thi asset represents the proportion of the funds that are not invested. The first tw assets have standard deviations of 42% and 33% respectively and a correlatic coefficient of 0.7. The "benchmark" is the portfolio that invests funds half-an half in the two assets. The function $TE(\mathbf{x})$ represents the tracking error of t portfolio with respect to the half-and-half benchmark and the first constrain indicates that this tracking error should not exceed 10%. The second constrain is the budget constraint; the third enforces no shorting. We depict the projectic of the feasible set of this problem onto the space spanned by variables x_1 and a in Figure 19.1.

We now build a relative robustness model for this portfolio problem. W assume that the covariance matrix estimate is certain. We consider a very simp uncertainty set for the expected return estimates consisting of three scenaric represented with the three arrows in Figure 19.2. These three scenarios co respond to the following values for μ: $(6, 4, 0)$, $(5, 5, 0)$, and $(4, 6, 0)$. Whe $\mu = (6, 4, 0)$ the optimal solution is $(0.831, 0.169, 0)$ with objective value 5.66 Similarly, when $\mu = (4, 6, 0)$ the optimal solution is $(0.169, 0.831, 0)$ with obje tive value 5.662. When $\mu = (5, 5, 0)$ all points between the previous two optim

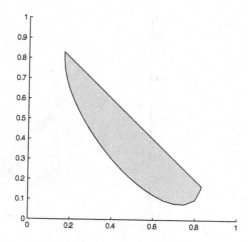

Figure 19.1 The feasible set of the mean–variance model (19.16)

solutions are optimal with a shared objective value of 5.0. Therefore, the relative robust formulation for this problem can be written as follows:

$$
\begin{aligned}
\min_{\mathbf{x},t} \quad & t \\
\text{s.t.} \quad & 5.662 - (6x_1 + 4x_2) \leq t \\
& 5.662 - (4x_1 + 6x_2) \leq t \\
& 5 - (5x_1 + 5x_2) \leq t \\
& TE(\mathbf{x}) \leq 0.1 \\
& \mathbf{1}^\mathsf{T}\mathbf{x} = 1 \\
& \mathbf{x} \geq \mathbf{0}.
\end{aligned} \tag{19.17}
$$

Instead of solving the problem where the optimal regret level is a variable (t in the formulation), an easier strategy is to choose a level of regret that can be tolerated and find portfolios that do not exceed this level of regret in any scenario. For example, choosing a maximum tolerable regret level of 0.75 we get the following feasibility problem:

$$
\begin{aligned}
& 5.662 - (6x_1 + 4x_2) \leq 0.75 \\
& 5.662 - (4x_1 + 6x_2) \leq 0.75 \\
& 5 - (5x_1 + 5x_2) \leq 0.75 \\
& TE(\mathbf{x}) \leq 0.1 \\
& \mathbf{1}^\mathsf{T}\mathbf{x} = 1 \\
& \mathbf{x} \geq \mathbf{0}.
\end{aligned}
$$

This problem and its feasible set of solutions is illustrated in Figure 19.2.

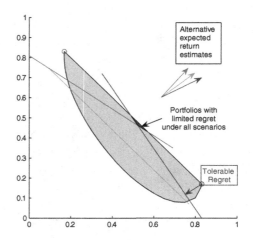

Figure 19.2 Set of solutions with regret less than 0.75 for the mean–variance model (19.16)

19.5   Notes

Robust optimization was introduced by Ben-Tal and Nemirovski (1998, 2000) and independently by El Ghaoui and Lebret (1997) and El Ghaoui et al. (1998). The textbook by Ben-Tal et al. (2009) gives a thorough discussion on the subject including an extensive list of references.

Although robust optimization is widely popular in a variety of disciplines, is not as widespread in financial optimization yet. There are strong supporters of its potential in finance (Ceria and Stubbs, 2006; Goldfarb and Iyengar, 2003; Tütüncü and Koenig, 2004). There are also some skeptics (Scherer, 2007).

19.6   Exercises

Exercise 19.1   Consider the optimization problem

$$\max \quad \boldsymbol{\mu}^\mathsf{T}\mathbf{x} - \tfrac{1}{2}\gamma \cdot \mathbf{x}^\mathsf{T}\mathbf{V}\mathbf{x} \\ \mathbf{1}^\mathsf{T}\mathbf{x} = 1, \tag{19.18}$$

where $\mathbf{V}$ is a positive definite covariance matrix of asset returns, $\boldsymbol{\mu}$ is a vector of expected returns, and $\gamma > 0$ is a risk-aversion constant. Assume $\mathbf{V}$ is certain but $\boldsymbol{\mu}$ is uncertain.

(a) Let $z(\boldsymbol{\mu})$ denote the optimal value of (19.18). Show that $\boldsymbol{\mu} \mapsto z(\boldsymbol{\mu})$ is quadratic convex function.

(b) Let $\mathcal{U}$ denote the uncertainty set for $\boldsymbol{\mu}$. Formulate both the absolute and relative robust optimization versions of (19.18).

(c) Show that the absolute and relative robust optimization versions for the uncertainty sets $\mathcal{U} = \{\mu^1, \ldots, \mu^k\}$ and $\mathcal{U} = \text{conv}\{\mu^1, \ldots, \mu^k\}$ are equivalent.

Exercise 19.2 (Robust least squares) Let $\mathbf{P} \in \mathbb{R}^{m \times n}$, with $\mathbf{q} \in \mathbb{R}^m$, and consider the least-squares problem

$$\min_{\mathbf{v}} \|\mathbf{P}\mathbf{v} - \mathbf{q}\|_2. \tag{19.19}$$

The purpose of this exercise is to compare the above usual least-squares problem with a robust version.

(a) Suppose the matrix $\mathbf{P}$ can take any value in the ellipsoidal uncertainty set

$$\mathcal{U} = \{\mathbf{P} : \|\mathbf{P} - \bar{\mathbf{P}}\| = \rho\},$$

where $\|\mathbf{P} - \bar{\mathbf{P}}\|$ is either the operator norm or the Frobenius norm of the matrix $\mathbf{P} - \bar{\mathbf{P}}$.

Show that the robust version $\min_{\mathbf{P} \in \mathcal{U}} \min_{\mathbf{v}} \|\mathbf{P}\mathbf{v} - \mathbf{q}\|_2$ of (19.19) is equivalent to the problem

$$\min_{\mathbf{v}} \|\bar{\mathbf{P}}\mathbf{v} - \mathbf{q}\|_2 + \rho\|\mathbf{v}\|_2. \tag{19.20}$$

(b) Set $\mathbf{P}, \mathbf{q}$ as follows:
```
> P = [1, 0 ; 1, 0.001 ; 10, -0.01] ;
> q = [1 2 0]' ;
```
Use the MATLAB command

```
> v =  P \ q
```

to find the solution $\mathbf{v}^*$ to (19.19) for the above values of $\mathbf{P}, \mathbf{q}$.

What is the value of $\mathbf{v}^*$ and the value of $\|\mathbf{P}\mathbf{v}^* - \mathbf{q}\|_2$ that you found?

(c) Now consider the matrix $\mathbf{Q}$ obtained by adding a small random perturbation to $\mathbf{P}$

```
> Q = P + 0.05*randn(3,2)
```

What is the value of $\|\mathbf{Q}\mathbf{v}^* - \mathbf{q}\|_2$ for the solution $\mathbf{v}$ found in (a)? Repeat this a few (two or three) times. What do you observe?

(d) Now use the robust formulation (19.20) to find a robust solution $\mathbf{v}^*$ to (19.19) for $\rho = 0.1$.

What is the value of $\mathbf{v}^*$ and the value of $\|\mathbf{P}\mathbf{v}^* - \mathbf{q}\|_2$ that you found? How different are they from the usual least-squares answers found in part (b)?

(e) Repeat part (c) for the robust solution $\mathbf{v}$ that you found in (d). How different is the behavior now?

Exercise 19.3 Consider the convex quadratic inequality

$$\mathbf{x}^\top(\mathbf{A}\mathbf{A}^\top)\mathbf{x} - 2\mathbf{b}^\top\mathbf{x} + \gamma \leq 0,$$

where the parameters $(\mathbf{A}, \mathbf{b}, \gamma)$ belong to the uncertainty set

$$\mathcal{U} = \left\{ (\mathbf{A}, \mathbf{b}, \gamma) = (\mathbf{A}^0, \mathbf{b}^0, \gamma^0) + \sum_{j=1}^{k} u_j(\mathbf{A}^j, \mathbf{b}^j, \gamma^j) : \|\mathbf{u}\|_2 \leq 1 \right\}$$

for some fixed $(\mathbf{A}^j, \mathbf{b}^j, \gamma^j)$, for $j = 0, 1, \ldots, k$.

Show that the (infinite) robust quadratic constraint

$$\mathbf{x}^{\mathsf{T}}(\mathbf{A}\mathbf{A}^{\mathsf{T}})\mathbf{x} - 2\mathbf{b}^{\mathsf{T}}\mathbf{x} + \gamma \leq 0 \quad \text{for all} \quad (\mathbf{A}, \mathbf{b}, \gamma) \in \mathcal{U}$$

holds if and only if there exist $\mathbf{z}^j, y^j$, for $j = 1, \ldots, k$, and λ such that

$$\mathbf{A}^j\mathbf{x} = \mathbf{z}^j, \ j = 0, 1, \ldots, k$$
$$(\mathbf{b}^j)^{\mathsf{T}}\mathbf{x} = y^j, \ j = 0, 1, \ldots, k$$
$$\lambda \geq 0$$
$$\begin{bmatrix} \gamma^0 + 2y^0 - \lambda & (\mathbf{y} + \frac{1}{2}\gamma)^{\mathsf{T}} & (\mathbf{z}^0)^{\mathsf{T}} \\ \mathbf{y} + \frac{1}{2}\gamma & \lambda\mathbf{I} & \mathbf{Z}^{\mathsf{T}} \\ \mathbf{z}^0 & \mathbf{Z} & \mathbf{I} \end{bmatrix} \succeq \mathbf{0},$$

where $\mathbf{y} = \begin{bmatrix} y^1 & \cdots & y^k \end{bmatrix}^{\mathsf{T}}$, $\gamma = \begin{bmatrix} \gamma^1 & \cdots & \gamma^k \end{bmatrix}^{\mathsf{T}}$, and $\mathbf{Z} = \begin{bmatrix} \mathbf{z}^1 & \cdots & \mathbf{z}^k \end{bmatrix}$.

Conclude that the robust version of the optimization problem

$$\min_{\mathbf{x}} \quad \mathbf{c}^{\mathsf{T}}\mathbf{x}$$
$$\text{s.t.} \quad \mathbf{x}^{\mathsf{T}}(\mathbf{A}\mathbf{A}^{\mathsf{T}})\mathbf{x} - 2\mathbf{b}^{\mathsf{T}}\mathbf{x} + \gamma \leq 0$$

for the above kind of uncertainty set $\mathcal{U}$ can be written as a semidefinite program.

20 Nonlinear Programming: Theory and Algorithms

It is sometimes necessary to consider more general nonlinear programs than the ones we have already studied: linear, quadratic, or conic programs. We give a brief introduction to this vast topic, and we discuss an application to estimating a volatility surface.

20.1 Nonlinear Programming

Consider a very general optimization problem of the form

$$\min_{\mathbf{x}} \quad f(\mathbf{x})$$
$$\text{s.t.} \quad g_j(\mathbf{x}) = b_j, \ \text{for} \ j = 1,\ldots,m$$
$$h_i(\mathbf{x}) \le d_i, \ \text{for} \ i = 1,\ldots,p,$$

or the equivalent more concise form

$$\min_{\mathbf{x}} \quad f(\mathbf{x})$$
$$\text{s.t.} \quad \mathbf{g}(\mathbf{x}) = \mathbf{b} \tag{20.1}$$
$$\mathbf{h}(\mathbf{x}) \le \mathbf{d},$$

where $f, g_j, h_i : \mathbb{R}^n \to \mathbb{R}$. In the special case when all functions f, g_j, h_i are linear, problem (20.1) is a linear program as discussed in Chapter 2. When some of the functions f, g_j, h_i are nonlinear, problem (20.1) is a *nonlinear program*.

Many practical problems are naturally formulated as nonlinear programs. We already saw quadratic programming and conic programming in earlier chapters. However, the family of problems that can be formulated as nonlinear programs is enormous, as many, if not most, imaginable kinds of constraints and objectives can be cast in terms of nonlinear functions. (For some noteworthy examples, see the exercises at the end of the chapter.) The immense modeling power of nonlinear programming comes at a cost: unlike linear, quadratic, and conic programming, which have a solid theory and are solvable via a few algorithmic templates, both the theory and methods to solve general nonlinear programs are far more complicated. Different types of nonlinear programs, determined by structural properties of the objective and constraint functions, are amenable to different types of algorithms. The subsequent sections sketch the main theory and most popular algorithmic ideas. A comprehensive treatment of this vast

topic is beyond the scope of this textbook. We refer the reader to the excelle
references Bertsekas (1999), Güler (2010), and Nocedal and Wright (2006) f
more details.

20.2   Numerical Nonlinear Programming Solvers

There are numerous software packages for solving nonlinear programs. The f
lowing are some popular ones. We list them according to the class of algorithm
(discussed in Section 20.4) that they are based on:

(1) CONOPT, GRG2, Excel SOLVER. These solvers are based on the *generaliz
    reduced-gradient method*.
(2) MATLAB optimization toolbox, SNOPT, NLPQL. These solvers are based
    sequential quadratic programming.
(3) MINOS, LANCELOT. These solvers are based on an *augmented Lagrangi
    approach*.
(4) MOSEK, LOQO, IPOPT. These solvers are based on *interior-point methods*.

20.3   Optimality Conditions

In this section we consider the class of nonlinear programs (20.1) where t
functions f, g_i, h_j are once or twice continuously differentiable. The optimali
conditions for linear and convex quadratic programs extend to this more gener
context, albeit some new technicalities arise. In particular, for a general nonline
program the theory described below applies to *local minima*.

Let $\mathcal{X} := \{\mathbf{x} \in \mathbb{R}^n : \mathbf{g}(\mathbf{x}) = \mathbf{b}, \ \mathbf{h}(\mathbf{x}) \leq \mathbf{d}\}$ denote the feasible set of (20.
A point $\mathbf{x}^* \in \mathcal{X}$ is a *local minimum* of (20.1) if there exists $r > 0$ such th
$f(\mathbf{x}^*) \leq f(\mathbf{x})$ for all $\mathbf{x} \in \mathbb{B}_r(\mathbf{x}^*) \cap \mathcal{X}$, where $\mathbb{B}_r(\mathbf{x}^*)$ denotes the ball of radius
around $\mathbf{x}^*$; that is,

$$\mathbb{B}_r(\mathbf{x}^*) := \{\mathbf{x} \in \mathbb{R}^n : \|\mathbf{x} - \mathbf{x}^*\| \leq r\}.$$

A point $\mathbf{x}^* \in \mathcal{X}$ is a *strict local minimum* of (20.1) if there exists $r > 0$ such th
$f(\mathbf{x}^*) < f(\mathbf{x})$ for all $\mathbf{x} \in \mathbb{B}_r(\mathbf{x}^*) \cap \mathcal{X}$.

Unconstrained Case

For ease of exposition we first describe the optimality conditions for the simpl
case without constraints. Consider the unconstrained optimization problem

$$\min_{\mathbf{x} \in \mathbb{R}^n} f(\mathbf{x}), \tag{20.}$$

where $f : \mathbb{R}^n \to \mathbb{R}$.

Theorem 20.1 (First-order necessary conditions) *Suppose f is continuously differentiable. If a point $\mathbf{x}^* \in \mathbb{R}^n$ is a local minimum of (20.2) then $\nabla f(\mathbf{x}^*) = \mathbf{0}$.*

The above necessary conditions can be sharpened when the objective function is twice differentiable.

Theorem 20.2 (Second-order necessary and sufficient conditions) *Suppose f is twice continuously differentiable.*

(a) *If a point $\mathbf{x}^* \in \mathbb{R}^n$ is a local minimum of (20.2) then $\nabla f(\mathbf{x}^*) = \mathbf{0}$ and $\nabla^2 f(\mathbf{x}^*) \succeq \mathbf{0}$.*

(b) *If $\mathbf{x}^* \in \mathbb{R}^n$ is such that $\nabla f(\mathbf{x}^*) = \mathbf{0}$ and $\nabla^2 f(\mathbf{x}^*) \succ \mathbf{0}$ then $\mathbf{x}^*$ is a strict local minimum of (20.2).*

Constrained Case

Consider now the general constrained problem (20.1). The optimality conditions for (20.1) rely on the following technical condition.

Definition 20.3 Let $\mathbf{x} \in \mathcal{X}$. Define $I(\mathbf{x}) := \{i : h_i(\mathbf{x}) = d_i\}$. The point $\mathbf{x}$ satisfies the *linear independence constraint qualification* if the set of gradient vectors

$$\{\nabla g_j(\mathbf{x}) : j = 1, \ldots, m\} \cup \{\nabla h_i(\mathbf{x}) : i \in I(\mathbf{x})\}$$

is linearly independent.

Theorem 20.4 (First-order necessary conditions) *Suppose f, g_i, h_j are continuously differentiable. If a point $\mathbf{x}^* \in \mathcal{X}$ is a local minimum of (20.1) and satisfies the linear independence constraint qualification, then there exist some Lagrange multipliers $\mathbf{y} \in \mathbb{R}^m$ and $\mathbf{s} \in \mathbb{R}^p$ such that*

$$\nabla f(\mathbf{x}^*) + \sum_{j=1}^{m} y_j \nabla g_j(\mathbf{x}^*) + \sum_{j=1}^{p} s_i \nabla h_i(\mathbf{x}^*) = \mathbf{0}$$
$$\mathbf{s} \geq \mathbf{0} \qquad\qquad (20.3)$$
$$s_i(h_i(\mathbf{x}^*) - d_i) = 0, \quad for\ i = 1, \ldots, p.$$

Observe that the first block of equations in (20.3) can be written as

$$\nabla_{\mathbf{x}} L(\mathbf{x}^*, \mathbf{y}, \mathbf{s}) = \nabla f(\mathbf{x}^*) + \nabla \mathbf{g}(\mathbf{x}^*)\mathbf{y} + \nabla \mathbf{h}(\mathbf{x}^*)\mathbf{s} = \mathbf{0},$$

where $L(\mathbf{x}, \mathbf{y}, \mathbf{s})$ is the following *Lagrangian function* for (20.1):

$$L(\mathbf{x}, \mathbf{y}, \mathbf{s}) := f(\mathbf{x}) + \mathbf{y}^{\mathsf{T}}(\mathbf{g}(\mathbf{x}) - \mathbf{b}) + \mathbf{s}^{\mathsf{T}}(\mathbf{h}(\mathbf{x}) - \mathbf{d}),$$

and where

$$\nabla \mathbf{g}(\mathbf{x}) = \begin{bmatrix} \nabla g_1(\mathbf{x}) & \cdots & \nabla g_m(\mathbf{x}) \end{bmatrix} \quad \text{and} \quad \nabla \mathbf{h}(\mathbf{x}) = \begin{bmatrix} \nabla h_1(\mathbf{x}) & \cdots & \nabla h_p(\mathbf{x}) \end{bmatrix}.$$

Observe the nice analogy to the first-order conditions for the unconstrained case. We next give second-order necessary and sufficient conditions. The precise

statements of the second-order conditions involve the following tangent subspace
Let $\mathbf{x} \in \mathcal{X}$. The *tangent subspace* $T(\mathbf{x})$ is defined as

$$T(\mathbf{x}) := \{\mathbf{d} \in \mathbb{R}^n : \nabla g_j(\mathbf{x})^\mathsf{T}\mathbf{d} = 0, \ j = 1, \ldots, m, \ \text{and} \ \nabla h_i(\mathbf{x})^\mathsf{T}\mathbf{d} = 0, \ i \in I(\mathbf{x}$$

Theorem 20.5 (Second-order necessary conditions) *Suppose f is twice continuously differentiable. If a point $\mathbf{x}^* \in \mathbb{R}^n$ is a local minimum of (20.2) and satisf the linear independence constraint qualification, then there exist $\mathbf{y} \in \mathbb{R}^m$ a $\mathbf{s} \in \mathbb{R}^p$ such that (20.3) holds and*

$$\mathbf{d}^\mathsf{T}(\nabla_{\mathbf{x}}^2 L(\mathbf{x}^*, \mathbf{y}, \mathbf{s}))\mathbf{d} \geq 0$$

for all $\mathbf{d} \in T(\mathbf{x}^*)$.

Theorem 20.6 (Second-order sufficient conditions) *Suppose f is twice contiuously differentiable and $\mathbf{x}^* \in \mathbb{R}^n$ satisfies the linear independence constra qualification. If there exist $\mathbf{y} \in \mathbb{R}^m$ and $\mathbf{s} \in \mathbb{R}^p$ such that (20.3) holds as well $h_i(\mathbf{x}^*) = d_i \Rightarrow s_i > 0$, and*

$$\mathbf{d}^\mathsf{T}(\nabla_{\mathbf{x}}^2 L(\mathbf{x}^*, \boldsymbol{\lambda}, \boldsymbol{\mu}))\mathbf{d} > 0$$

for all non-zero $\mathbf{d} \in T(\mathbf{x}^*)$, then $\mathbf{x}^*$ is a local minimum of (20.1).

Convex Case

As we detail next, the optimality conditions described above simplify a strengthen substantially when the underlying problem is convex.

Proposition 20.7 *Assume the objective function f in (20.2) is convex a differentiable. Then $\mathbf{x}^*$ is an optimal solution to (20.2) if and only if $\nabla f(\mathbf{x}^*) =$*

Proposition 20.8 *Assume the objective function f in (20.1) is convex, t equality constraint functions g_i are linear, and the inequality constraint functio h_j are convex. Furthermore assume all f, g_i, h_j are differentiable. Then $\mathbf{x}^*$ is optimal solution to (20.1) if and only if there exist $\mathbf{y} \in \mathbb{R}^m, \mathbf{s} \in \mathbb{R}^p$ such that*

$$\nabla_{\mathbf{x}} L(\mathbf{x}^*, \mathbf{y}, \mathbf{s}) = \mathbf{0}, \ \mathbf{g}(\mathbf{x}^*) = \mathbf{b}, \ \mathbf{h}(\mathbf{x}^*) \leq \mathbf{d}, \ \mathbf{s} \geq \mathbf{0}, \ \mathbf{s}^\mathsf{T}(\mathbf{h}(\mathbf{x}^*) - \mathbf{d}) = 0.$$

20.4   Algorithms

Unconstrained Case

We next describe three main algorithmic approaches to solving an unconstrain optimization problem of the form (20.2), namely the *gradient descent metho Newton's method,* and the *subgradient method*. These methods in turn provide t foundation for the more elaborate methods for solving constrained optimizati problems.

Gradient Descent

Suppose the objective function f in (20.2) is differentiable. In this case a simple method for solving (20.2) is based on going downhill on the graph of the function f. The gradient gives the direction of fastest initial increase and thus its negative is the direction of fastest initial decrease. This can also be motivated by the first-order Taylor approximation of f around a point $\mathbf{x}$: for $\mathbf{p}$ small we have

$$f(\mathbf{x} + \mathbf{p}) \approx f(\mathbf{x}) + \nabla f(\mathbf{x})^\mathsf{T} \mathbf{p}.$$

Among all $\mathbf{p}$ of fixed norm, the one pointing in the direction $-\nabla f(\mathbf{x})$ minimizes the right-hand side.

Algorithm 20.1 gives a formal description of the gradient descent method.

Algorithm 20.1 Gradient descent method

1: choose $\mathbf{x}^0 \in \mathbb{R}^n$
2: **for** $k = 0, 1, \ldots$ **do**
3:     choose a step length $\alpha_k > 0$ and set $\mathbf{x}^{k+1} = \mathbf{x}^k - \alpha_k \nabla f(\mathbf{x}^k)$
4: **end for**

The choice of step length α is a critical detail in the implementation of the gradient descent algorithm. If α is too large, the algorithm may fail to converge to a solution because the objective value could even increase after one iteration. On the other hand, if α is too small, the algorithm will be too slow. This issue applies not only to the gradient descent method but to any method that aims to move along a direction $\mathbf{p}$. Suppose $\mathbf{p} \in \mathbb{R}^n$ is a *descent direction* at the current point $\mathbf{x}^k$; that is, $\nabla f(\mathbf{x}^k)^\mathsf{T} \mathbf{p} < 0$. A popular approach is to choose the step length large enough and perform *backtracking*; that is, shrink α_k by a multiplicative constant smaller than one until the following sufficient decrease condition holds for some predetermined $\mu \in (0, 1)$:

$$f(\mathbf{x}^k + \alpha_k \mathbf{p}) \leq f(\mathbf{x}^k) + \alpha_k \cdot \mu \cdot \nabla f(\mathbf{x}^k)^\mathsf{T} \mathbf{p}. \tag{20.4}$$

The sufficient decrease requirement in (20.4) is called the *Armijo–Goldstein* condition. The first-order Taylor approximation for f around $\mathbf{x}$ ensures that (20.4) holds for sufficiently small α_k provided f is differentiable and $\mathbf{d}$ is a descent direction. Algorithm 20.2 describes this kind of backtracking. This type of backtracking is also often called *line search*.

Algorithm 20.2 Backtracking to select the step length α_k

1: choose $\alpha_k > 0$ and $\beta, \mu \in (0, 1)$
2: **while** (20.4) fails **do** $\alpha_k = \beta \cdot \alpha_k$
3: **end while**

Newton's Method

The gradient descent method uses only *first-order* information to choose t
descent direction at each main iteration. There are several approaches to inc
porate additional information and speed up convergence. Newton's method yiel
a substantially improved direction by incorporating *second-order* information.
its *pure* form each step of Netwon's method for (20.2) updates a trial point **x**
the new point

$$\mathbf{x}^+ = \mathbf{x} - \nabla^2 f(\mathbf{x})^{-1} \nabla f(\mathbf{x}).$$

The latter update can be motivated by considering the second-order Tayl
approximation to f around **x**:

$$f(\mathbf{x} + \mathbf{p}) \approx f(\mathbf{x}) + \nabla f(\mathbf{x})^{\mathsf{T}} \mathbf{p} + \frac{1}{2} \mathbf{p}^{\mathsf{T}} \nabla^2 f(\mathbf{x}) \mathbf{p}.$$

Observe that when $\nabla^2 f(\mathbf{x}) \succ \mathbf{0}$ the Newton step $\mathbf{p} := -\nabla^2 f(\mathbf{x})^{-1} \nabla f(\mathbf{x})$ mi
mizes the right-hand side.

Newton's method also applies to solving nonlinear equations. Consider t
system of nonlinear equations

$$F(\mathbf{x}) = \mathbf{0} \tag{20.}$$

where $F : \mathbb{R}^n \to \mathbb{R}^n$ is a differentiable function. Let $F'(\mathbf{x})$ denote the *Jacobi
matrix* of F, that is, the $n \times n$ matrix with (i, j) component

$$F'(\mathbf{x})_{ij} = \frac{\partial f_i(\mathbf{x})}{\partial x_j},$$

where $f_1(\mathbf{x}), \dots, f_n(\mathbf{x})$ are the components of $F(\mathbf{x})$.

In its pure form, Newton's method for (20.5) updates a trial point **x** to t
new point

$$\mathbf{x}^+ = \mathbf{x} - F'(\mathbf{x})^{-1} F(\mathbf{x}).$$

The latter update can be motivated by considering the first-order Taylor appro
imation to F around **x**:

$$F(\mathbf{x} + \mathbf{p}) \approx F(\mathbf{x}) + F'(\mathbf{x}) \mathbf{p}.$$

The Newton step $\mathbf{p} = -F'(\mathbf{x})^{-1} F(\mathbf{x})$ makes the above right-hand side equal
zero.

Observe that Newton's method for the unconstrained optimization proble
(20.2) is exactly the same as Newton's method for solving the system of nonline
equations $\nabla f(\mathbf{x}) = \mathbf{0}$.

Newton's method has a much faster rate of convergence than gradient desce
provided the initial iterate is sufficiently close to the solution. On the oth
hand, when the initial iterate is far from the solution, the above pure form
Newton's method may fail to converge. The latter drawback can be rectified l
performing some backtracking along the Newton step direction as described
Algorithm 20.3. The step length α_k can be chosen via the backtracking procedu

described in Algorithm 20.2 to ensure the Armijo–Goldstein sufficient decrease condition (20.4) holds. For the Newton direction $\mathbf{d} = -\nabla^2 f(\mathbf{x}_k)^{-1} \nabla f(\mathbf{x}^k)$, a natural and customary initial step length at each step is $\alpha_k = 1$.

Algorithm 20.3 Newton's method with backtracking

1: choose $\mathbf{x}^0 \in \mathbb{R}^n$
2: **for** $k = 0, 1, \ldots$ **do**
3:     choose a step length $\alpha_k \in (0, 1]$ via backtracking and set $\mathbf{x}^{k+1} = \mathbf{x}^k - \alpha_k \nabla^2 f(\mathbf{x}^k)^{-1} \nabla f(\mathbf{x}^k)$
4: **end for**

Subgradient Method

In the special case when the objective function f is convex, the gradient descent method can be extended to *non-smooth* functions; that is, functions that are not necessarily differentiable. Non-smooth functions arise often in optimization. In particular, the Lagrangian relaxation heuristic for (8.9) described in Section 8.3.3 yields the minimization of a non-smooth convex function.

Let $f : \mathbb{R}^n \to \mathbb{R}$ be a convex function. A point $\mathbf{g} \in \mathbb{R}^n$ is a *subgradient* of f at $\mathbf{x} \in \mathbb{R}^n$ if for all $\mathbf{y} \in \mathbb{R}^n$

$$f(\mathbf{y}) - f(\mathbf{x}) \geq \mathbf{g}^\top (\mathbf{y} - \mathbf{x}).$$

The *subdifferential* of f at $\mathbf{x}$, denoted $\partial f(\mathbf{x})$, is the set of subgradients of f at $\mathbf{x}$.

The subdifferential of a convex function is non-empty at every point. The following example illustrates the subdifferential of a simple non-smooth function. Consider the convex function $f : \mathbb{R} \to \mathbb{R}$ defined by $f(x) = |x|$. In this case we have

$$\partial f(x) = \begin{cases} 1 & \text{if } x > 0 \\ -1 & \text{if } x < 0 \\ [-1, 1] & \text{if } x = 0. \end{cases}$$

Algorithm 20.4 describes the subgradient method for (20.2) when f is a convex function. Observe that it is a natural extension of Algorithm 20.1.

Algorithm 20.4 Subgradient method

1: choose $\mathbf{x}^0 \in \mathbb{R}^n$
2: **for** $k = 0, 1, \ldots$ **do**
3:     choose $\mathbf{g}_k \in \partial f(\mathbf{x}^k)$ and a step length $\alpha_k > 0$, and set $\mathbf{x}^{k+1} = \mathbf{x}^k - \alpha_k \mathbf{g}_k$
4: **end for**

For non-smooth functions, the choice of step length α_k for the subgradient method cannot be chosen via a backtracking procedure as the Armijo–Goldstein condition (20.4) cannot be guaranteed in the absence of differentiability. Various

choices have been proposed in the literature. The following two generic types
step lengths are particularly simple and popular. The first one is to choose fix
sizes $\alpha_k = \alpha > 0$ for all k. The second one is to choose slowly diminishing siz
such that

$$\sum_{k=0}^{\infty} \alpha_k^2 < \infty, \quad \sum_{k=0}^{\infty} \alpha_k = \infty.$$

Constrained Case

Generalized Reduced Gradient
The main idea behind the *generalized reduced gradient* method is to reduce
constrained problem to a sequence of unconstrained problems in a space
lower dimension. To illustrate this procedure, consider the special case wh
the equality constraints are linear:

$$\min_{\mathbf{x}} \quad f(\mathbf{x})$$
$$\text{s.t.} \quad \mathbf{A}\mathbf{x} = \mathbf{b}$$

(20.

for some $\mathbf{A} \in \mathbb{R}^{m \times n}$. Without loss of generality we may assume that $\mathbf{A}$ h
full row rank as otherwise either some constraints are redundant or the proble
is infeasible. Since $\mathbf{A}$ has full rank, we can partition both $\mathbf{A}$ and $\mathbf{x}$ as follow
$\mathbf{A} = \begin{bmatrix} \mathbf{A}_B & \mathbf{A}_N \end{bmatrix}$ and $\mathbf{x} = \begin{bmatrix} \mathbf{x}_B \\ \mathbf{x}_N \end{bmatrix}$ for some subset $B \subseteq \{1, \ldots, n\}$ such that $\mathbf{A}$
is non-singular. Therefore

$$\mathbf{A}\mathbf{x} = \mathbf{b} \quad \Leftrightarrow \quad \mathbf{A}_B\mathbf{x}_B + \mathbf{A}_N\mathbf{x}_N = \mathbf{b} \quad \Leftrightarrow \quad \mathbf{x}_B = \mathbf{A}_B^{-1}(\mathbf{b} - \mathbf{A}_N\mathbf{x}_N).$$

Consequently, problem (20.6) is equivalent to the following *reduced space* unco
strained minimization problem:

$$\min_{\mathbf{x}_N} \hat{f}(\mathbf{x}_N)$$

where

$$\hat{f}(\mathbf{x}_N) = f(\mathbf{A}_B^{-1}(\mathbf{b} - \mathbf{A}_N\mathbf{x}_N), \mathbf{x}_N).$$

Consider a more general program with nonlinear equality constraints:

$$\min_{\mathbf{x}} \quad f(\mathbf{x})$$
$$\text{s.t.} \quad \mathbf{g}(\mathbf{x}) = \mathbf{b}.$$

(20.

We can extend the above approach by approximating the nonlinear equali
constraints with their first-order Taylor approximation. More precisely, suppo
the current point is $\mathbf{x}^k$. Consider the modification of (20.7) obtained by replaci
$\mathbf{g}(\mathbf{x}) = \mathbf{b}$ with its first-order Taylor approximation

$$\min_{\mathbf{x}} \quad f(\mathbf{x})$$
$$\text{s.t.} \quad \mathbf{g}(\mathbf{x}^k) + \nabla \mathbf{g}(\mathbf{x}^k)^\mathsf{T}(\mathbf{x} - \mathbf{x}^k) = \mathbf{b}.$$

(20.

Observe that the latter problem is of the form (20.6) and is thus amenable to the type of reduced space approach described above. Algorithm 20.5 describes a template for a generalized reduced gradient approach to problem (20.7). The step length α at each iteration is typically chosen to balance both goals of objective function reduction and constraint satisfaction.

Algorithm 20.5 Generalized reduced gradient

1: choose $\mathbf{x}^0$
2: **for** $k = 0, 1, \ldots$ **do**
3:　solve the linearized constraints problem (20.8) to find a search direction $\Delta \mathbf{x}^k$
4:　choose a step length $\alpha > 0$ and set $\mathbf{x}^{k+1} = \mathbf{x}^k + \alpha \Delta \mathbf{x}^k$
5: **end for**

The generalized reduced gradient approach can be extended to deal with inequality constraints as well via an *active-set approach* like that discussed in Chapter 5. The basic idea is that the active inequalities can be treated as equality constraints. The challenge of course is to determine the correct set of active inequalities at the optimal solution.

Sequential Quadratic Programming
The central idea of *sequential quadratic programming* is to capitalize on algorithms for quadratic programming to solve more general nonlinear programming problems of the form (20.1). Given a current iterate $\mathbf{x}^k$, problem (20.1) can be approximated with the following quadratic program:

$$\min_{\mathbf{x}} \quad f(\mathbf{x}^k) + \nabla f(\mathbf{x}^k)^\mathsf{T}(\mathbf{x} - \mathbf{x}^k) + \tfrac{1}{2}(\mathbf{x} - \mathbf{x}^k)^\mathsf{T} \mathbf{B}_k (\mathbf{x} - \mathbf{x}^k)$$
$$\text{s.t.} \quad g(\mathbf{x}^k) + \nabla g(\mathbf{x}^k)^\mathsf{T}(\mathbf{x} - \mathbf{x}^k) = \mathbf{b}$$
$$h(\mathbf{x}^k) + \nabla h(\mathbf{x}^k)^\mathsf{T}(\mathbf{x} - \mathbf{x}^k) \le \mathbf{d}, \tag{20.9}$$

where

$$\mathbf{B}_k = \nabla^2_{\mathbf{xx}} L(\mathbf{x}^k, \mathbf{y}^k, \mathbf{s}^k)$$

is the Hessian of the Lagrangian function with respect to the $\mathbf{x}$ variables and $(\mathbf{y}^k, \mathbf{s}^k)$ is the current estimate of the vector of Lagrange multipliers.

Algorithm 20.6 describes a template for a sequential quadratic programming approach to problem (20.7). Once again, the step length α at each iteration is typically chosen to balance both goals of objective function reduction and constraint satisfaction.

Interior-Point Methods
Interior-point methods, formerly discussed in Chapters 2 and 5, can be extended to general nonlinear programming under suitable differentiability conditions. The gist of the method is to solve the optimality conditions (20.3).

Algorithm 20.6 Sequential quadratic programming

1: choose $\mathbf{x}^0, \mathbf{y}^0, \mathbf{s}^0$
2: **for** $k = 0, 1, \ldots$ **do**
3:     solve the quadratic program (20.9) to find a search direction $(\Delta \mathbf{x}^k, \Delta \mathbf{y}^k, \Delta \mathbf{s}^k)$
4:     choose a step length $\alpha > 0$ and set $(\mathbf{x}^{k+1}, \mathbf{y}^{k+1}, \mathbf{s}^{k+1}) = (\mathbf{x}^k, \mathbf{y}^k, \mathbf{s}^k) + \alpha(\Delta \mathbf{x}^k, \Delta \mathbf{y}^k, \Delta \mathbf{s}^k)$
5: **end for**

Similar to the linear and quadratic programming cases, interior-point metho generate a sequence of iterates that satisfy some inequalities strictly and ea iteration of the algorithm aims to make progress towards satisfying the optimali conditions (20.3). The algorithm inevitably becomes a bit more elaborate f nonlinear programs because of the nonlinearities in the constraints.

As before we use the following notational convention: given a vector $\mathbf{s} \in \mathbb{R}$ let $\mathbf{S} \in \mathbb{R}^{p \times p}$ denote the diagonal matrix defined by $S_{ii} = s_i$, for $i = 1, \ldots,$ and let $\mathbf{1} \in \mathbb{R}^p$ denote the vector whose components are all 1s. The optimali conditions (20.3) can be restated as

$$\begin{bmatrix} \nabla f(\mathbf{x}) + \nabla g(\mathbf{x})\mathbf{y} + \nabla h(\mathbf{x})\mathbf{s} \\ g(\mathbf{x}) - \mathbf{b} \\ h(\mathbf{x}) + \mathbf{z} - \mathbf{d} \\ \mathbf{SZ1} \end{bmatrix} = \begin{bmatrix} 0 \\ 0 \\ 0 \\ 0 \end{bmatrix}, \quad \mathbf{s}, \mathbf{z} \geq 0.$$

Given $\mu > 0$, let $(\mathbf{x}(\mu), \mathbf{y}(\mu), \mathbf{z}(\mu), \mathbf{s}(\mu))$ be the solution to the followi perturbed version of the above optimality conditions:

$$\begin{bmatrix} \nabla f(\mathbf{x}) + \nabla g(\mathbf{x})\mathbf{y} + \nabla h(\mathbf{x})\mathbf{s} \\ g(\mathbf{x}) - \mathbf{b} \\ h(\mathbf{x}) + \mathbf{z} - \mathbf{d} \\ \mathbf{SZ1} \end{bmatrix} = \begin{bmatrix} 0 \\ 0 \\ 0 \\ \mu\mathbf{1} \end{bmatrix}, \quad \mathbf{s}, \mathbf{z} > 0.$$

The first condition above can be written as $\mathbf{r}_\mu(\mathbf{x}, \mathbf{y}, \mathbf{z}, \mathbf{s}) = 0$ for the *residu vector*:

$$\mathbf{r}_\mu(\mathbf{x}, \mathbf{y}, \mathbf{z}, \mathbf{s}) := \begin{bmatrix} \nabla f(\mathbf{x}) + \nabla g(\mathbf{x})\mathbf{y} + \nabla h(\mathbf{x})\mathbf{s} \\ g(\mathbf{x}) - \mathbf{b} \\ h(\mathbf{x}) + \mathbf{z} - \mathbf{d} \\ \mathbf{SZ1} - \mu\mathbf{1} \end{bmatrix}.$$

The *central path* is the set $\{(\mathbf{x}(\mu), \mathbf{y}(\mu), \mathbf{z}(\mu), \mathbf{s}(\mu)) : \mu > 0\}$. Under suitab assumptions $(\mathbf{x}(\mu), \mathbf{y}(\mu), \mathbf{z}(\mu), \mathbf{s}(\mu))$ converges to a local optimal solution (20.3). This suggests the following algorithmic strategy: Suppose $(\mathbf{x}, \mathbf{y}, \mathbf{z}, \mathbf{s})$ "near" $(\mathbf{x}(\mu), \mathbf{y}(\mu), \mathbf{z}(\mu), \mathbf{s}(\mu))$ for some $\mu > 0$. Use $(\mathbf{x}, \mathbf{y}, \mathbf{z}, \mathbf{s})$ to move to a bett point $(\mathbf{x}^+, \mathbf{y}^+, \mathbf{z}^+, \mathbf{s}^+)$ "near" $(\mathbf{x}(\mu^+), \mathbf{y}(\mu^+), \mathbf{z}(\mu^+), \mathbf{s}(\mu^+))$ for some $\mu^+ < \mu$.

It can be shown that if a point $(\mathbf{x}, \mathbf{y}, \mathbf{z}, \mathbf{s})$ is on the central path, then t corresponding value of μ satisfies $\mathbf{z}^\mathsf{T}\mathbf{s} = p\mu$. Likewise, given $\mathbf{z}, \mathbf{s} > \mathbf{0}$, define

$$\mu(\mathbf{z}, \mathbf{s}) := \frac{\mathbf{z}^\mathsf{T}\mathbf{s}}{p}.$$

To move from a current point $(\mathbf{x}, \mathbf{y}, \mathbf{z}, \mathbf{s})$ to a new point, we use the Newton step for the nonlinear system of equations $\mathbf{r}_\mu(\mathbf{x}, \mathbf{y}, \mathbf{z}, \mathbf{s}) = \mathbf{0}$; that is,

$$(\Delta\mathbf{x}, \Delta\mathbf{y}, \Delta\mathbf{z}, \Delta\mathbf{s}) = -\mathbf{r}'_\mu(\mathbf{x}, \mathbf{y}, \mathbf{z}, \mathbf{s})^{-1}\mathbf{r}_\mu(\mathbf{x}, \mathbf{y}, \mathbf{z}, \mathbf{s}). \qquad (20.10)$$

Algorithm 20.7 presents a template for an interior-point method.

Algorithm 20.7 Interior-point method for nonlinear programming

---
1: choose $\mathbf{x}^0, \mathbf{y}^0$ and $\mathbf{z}^0, \mathbf{s}^0 > \mathbf{0}$
2: **for** $k = 0, 1, \dots$ **do**
3:      solve the Newton system (20.10) for $(\mathbf{x}, \mathbf{y}, \mathbf{z}, \mathbf{s}) = (\mathbf{x}^k, \mathbf{y}^k, \mathbf{z}^k, \mathbf{s}^k)$ and $\mu :=$ $0.1\mu(\mathbf{z}^k, \mathbf{s}^k)$
4:      choose a step length $\alpha \in (0, 1]$ and set $(\mathbf{x}^{k+1}, \mathbf{y}^{k+1}, \mathbf{z}^{k+1}, \mathbf{s}^{k+1}) =$ $(\mathbf{x}^k, \mathbf{y}^k, \mathbf{z}^k, \mathbf{s}^k) + \alpha(\Delta\mathbf{x}, \Delta\mathbf{y}, \Delta\mathbf{z}, \Delta\mathbf{s})$
5: **end for**

The step length α in step 4 should be chosen via a backtracking procedure so that $\mathbf{z}^{k+1}, \mathbf{s}^{k+1} > \mathbf{0}$ and the size of $\mathbf{r}_\mu(\mathbf{x}^{k+1}, \mathbf{y}^{k+1}, \mathbf{z}^{k+1}, \mathbf{s}^{k+1})$ is sufficiently smaller than $\mathbf{r}_\mu(\mathbf{x}^k, \mathbf{y}^k, \mathbf{z}^k, \mathbf{s}^k)$.

20.5 Estimating a Volatility Surface

We conclude this chapter with a description of nonlinear programming to estimate the volatility surface. The discussion in this section is based on Coleman et al. (1999a,b).

The Black–Scholes–Merton (BSM) equation for pricing European options is based on a geometric Brownian motion model for the movements of the underlying security. Namely, one assumes that the underlying security price S_t at time t satisfies

$$\frac{dS_t}{S_t} = \mu dt + \sigma dW_t, \qquad (20.11)$$

where μ is the *drift*, σ is the (constant) volatility, and W_t is the standard Brownian motion. Using this equation and some standard assumptions about the absence of frictions and arbitrage opportunities, one can derive the BSM partial differential equation for the value of a European option on this underlying security. Using the boundary conditions resulting from the payoff structure of the particular option, one determines the value function for the option. For example, for the European call and put options with strike K and maturity T, we obtain the following formulas:

$$C(K,T) = S_0 \Phi(d_1) - Ke^{-rT}\Phi(d_2), \qquad (20.1$$
$$P(K,T) = Ke^{-rT}\Phi(-d_2) - S_0\Phi(-d_1), \qquad (20.1$$

where

$$d_1 = \frac{\log(S_0/K) + (r + \sigma^2/2)T}{\sigma\sqrt{T}},$$
$$d_2 = d_1 - \sigma\sqrt{T},$$

and $\Phi(\cdot)$ is the cumulative distribution function for the standard normal dist bution. In the formula r represents the continuously compounded risk-free a constant interest rate and σ is the volatility of the underlying security that assumed to be constant.

The risk-free interest rate r, or a reasonably close approximation to it, is oft available, for example from Treasury bill prices in US markets. Therefore, all o needs to determine the call or put price using these formulas is a reliable estima of the volatility parameter σ. Conversely, given the market price for a particul European call or put, one can uniquely determine the *implied volatility* of t underlying security (implied by this option price) by solving the equations abo with the unknown σ.

Empirical evidence against the appropriateness of (20.11) as a model for t movements of most securities is abundant. Most such studies refute the assum tion of a volatility that does not depend on time or underlying price level. Inde studying the prices of options with the same maturity but different strike researchers observed that the implied volatilities for such options exhibited "smile" structure, i.e., higher implied volatilities away from the money in bo directions, decreasing to a minimum level as one approaches the at-the-mon option from up or down. This is clearly in contrast with the constant (fla implied volatilities one would expect had (20.11) been an appropriate model f the underlying price process.

There are quite a few models that try to capture the volatility smile, includi stochastic volatility models, jump diffusions, etc. Since these models introdu non-traded sources of risk, perfect replication via dynamic hedging as in the BS approach becomes impossible and the pricing problem is more complicated. alternative that is explored in Coleman et al. (1999b) is the one-factor continuo diffusion model:

$$\frac{dS_t}{S_t} = \mu(S_t, t)dt + \sigma(S_t, t)dW_t, \quad t \in [0, T], \qquad (20.1$$

where the constant parameters μ and σ of (20.11) are replaced by continuo and differentiable functions $\mu(S_t, t)$ and $\sigma(S_t, t)$ of the underlying price S_t a time t. Here T denotes the end of the fixed time horizon. If the instantaneo risk-free interest rate r is assumed constant and the dividend rate is constar given a function $\sigma(S,t)$, a European call option with maturity T and strike K has a unique price. Let us denote this price with $C(\sigma(S,t), K, T)$.

While an explicit solution for the price function $C(\sigma(S,t), K, T)$ as in (20.12) is no longer possible, the resulting pricing problem can be solved efficiently via numerical techniques. Since $\mu(S,t)$ does not appear in the generalized BSM partial differential equation, all one needs is the specification of the function $\sigma(S,t)$ and a good numerical scheme to determine the option prices in this generalized framework.

So, how does one specify the function $\sigma(S,t)$? First of all, this function should be consistent with the observed prices of currently or recently traded options on the same underlying security. If we assume that we are given market prices of m call options with strikes K_j and maturities T_j in the form of bid–ask pairs (β_j, α_j) for $j = 1, \ldots, n$, it would be reasonable to require that the volatility function $\sigma(S,t)$ is chosen so that

$$\beta_j \leq C(\sigma(S,t), K_j, T_j) \leq \alpha_j, \quad j = 1, \ldots, n. \tag{20.15}$$

To ensure that (20.15) is satisfied as closely as possible, one strategy is to minimize the violations of the inequalities in (20.15):

$$\min_{\sigma(S,t)\in\mathcal{H}} \sum_{j=1}^{n} [\beta_j - C(\sigma(S,t), K_j, T_j)]^+ + [C(\sigma(S,t), K_j, T_j) - \alpha_j]^+. \tag{20.16}$$

Above, $\mathcal{H}$ denotes the space of measurable functions $\sigma(S,t)$ with domain $\mathbb{R}_+ \times [0,T]$ and $[u]^+ = \max\{u, 0\}$. Alternatively, using the closing prices C_j for the options under consideration, or choosing the mid-market prices $C_j = (\beta_j + \alpha_j)/2$, we can solve the following nonlinear least-squares problem:

$$\min_{\sigma(S,t)\in\mathcal{H}} \sum_{j=1}^{n} (C(\sigma(S,t), K_j, T_j) - C_j)^2. \tag{20.17}$$

This is a nonlinear least-squares problem since the function $C(\sigma(S,t), K_j, T_j)$ depends nonlinearly on the variables, namely the local volatility function $\sigma(S,t)$.

While the calibration of the local volatility function to the observed prices using the objective functions in (20.16) and (20.17) is important and desirable, there are additional properties that are desirable in the local volatility function. The most common feature sought in existing models is regularity or smoothness. For example, in Lagnado and Osher (1997) the authors try to achieve a smooth volatility function by modifying the objective function in (20.17) as follows:

$$\min_{\sigma(S,t)\in\mathcal{H}} \sum_{j=1}^{n} (C(\sigma(S,t), K_j, T_j) - C_j)^2 + \lambda \|\nabla\sigma(S,t)\|_2. \tag{20.18}$$

Here, λ is a positive tradeoff parameter and $\|\cdot\|_2$ represents the L^2-norm in $\mathcal{H}$. Large deviations in the volatility function would result in a high value for the norm of the gradient function, and by penalizing such occurences, the formulation above encourages a smoother solution to the problem. The most appropriate

value for the tradeoff parameter λ must be determined experimentally. To sol
the resulting problem numerically, one must discretize the volatility function
the underlying price and time grid. Even for a relatively coarse discretization
the S_t and t spaces, one can easily end up with an optimization problem wi
many variables.

An alternative strategy is to build the smoothness into the volatility functi
by modeling it with spline functions. The use of the spline functions not on
guarantees the smoothness of the resulting volatility function estimates b
also reduces the degrees of freedom in the problem. As a consequence, t
optimization problem to be solved has many fewer variables and is easier. Th
strategy is proposed in Coleman et al. (1999b) and we review it below.

We start by assuming that $\sigma(S, t)$ is a bi-cubic spline. While higher-ord
splines can also be used, cubic splines often offer a good balance between fle
ibility and complexity. Next we choose a set of spline knots at points $(\bar{S}_i,$
for $i = 1, \dots, k$. If the value of the volatility function at these points is giv
by $\bar{\sigma}_j := \sigma(\bar{S}_j, \bar{t}_j)$, the interpolating cubic spline that goes through these kno
and satisfies a particular end condition (such as the natural spline end conditi
of linearity at the boundary knots) is uniquely determined. In other words,
completely determine the volatility function as a natural bi-cubic spline (a
therefore to determine the resulting call option prices) we have k degrees
freedom represented with the choices $\bar{\sigma} = (\bar{\sigma}_1, \dots, \bar{\sigma}_k)$. Let $\Sigma(S, t, \bar{\sigma})$ be t
bi-cubic spline local volatility function obtained by setting $\sigma(\bar{S}_j, \bar{t}_j) := \bar{\sigma}_j$. L
$C(\Sigma(S, t, \bar{\sigma}), S, t)$ denote the resulting call price function. Then the analog of t
objective function (20.17) is

$$\min_{\bar{\sigma} \in \mathbb{R}^k} \sum_{j=1}^{n} (C(\Sigma(S, t, \bar{\sigma}), K_j, T_j) - C_j)^2. \qquad (20.1$$

One can introduce positive weights w_j for each of the terms in the objecti
function above to address different accuracies or confidence in the call pric
C_j. One can also introduce lower and upper bounds l_i and u_i for the volatiliti
at each knot to incorporate additional information that may be available fro
historical data, etc. This way, we form the following nonlinear least-squar
problem with k variables:

$$\min_{\bar{\sigma} \in \mathbb{R}^k} f(\sigma) := \sum_{j=1}^{n} w_j \left(C(\Sigma(S, t, \bar{\sigma}), K_j, T_j) - C_j \right)^2 \qquad (20.2$$
$$\text{s.t. } l \leq \bar{\sigma} \leq u.$$

It should be noted that the formulation above will not be appropriate if the
are many more knots than prices, that is, if k is much larger than n. In this cas
the problem will be underdetermined and solutions may exhibit "overfitting
There should be fewer knots than available option prices.

The problem (20.20) is a standard nonlinear optimization problem except th
the objective function $f(\bar{\sigma})$ and in particular the function $C(\Sigma(S, t, \bar{\sigma}), K_j, T$

depends on the decision variables $\bar{\sigma}$ in a complicated and non-explicit manner. Since most of the nonlinear optimization methods we discussed in the previous section require at least the gradient of the objective function (and sometimes its Hessian matrix as well), this may sound alarming. Without an explicit expression for f, its gradient must be estimated either using a finite difference scheme or using automatic differentiation. Coleman et al. (1999b) implement both alternatives and report that local volatility functions can be estimated very accurately using these strategies. They also test the hedging accuracy of different delta-hedging strategies, one using a constant volatility estimation and another using the local volatility function produced by the strategy above. These tests indicate that the hedges obtained from the local volatility function are significantly more accurate.

20.6 Exercises

Exercise 20.1 Suppose $f : \mathbb{R}^n \to \mathbb{R}$ is a differentiable function at the point $\mathbf{x} \in \mathbb{R}^n$. Consider the first-order Taylor approximation to f around $\mathbf{x}$:

$$\hat{f}(\mathbf{p}) := f(\mathbf{x}) + \nabla f(\mathbf{x})^\mathsf{T} \mathbf{p}.$$

Show that if $\nabla f(\mathbf{x}) \neq \mathbf{0}$ then the solution to the problem

$$\min_{\|\mathbf{p}\| \leq 1} \hat{f}(\mathbf{p})$$

is

$$\mathbf{p}^* = -\frac{1}{\|\nabla f(\mathbf{x})\|} \nabla f(\mathbf{x}).$$

In other words, it is the unitary vector in the direction of negative gradient.

Exercise 20.2

(a) Let $\mathbf{A} \in \mathbb{R}^{m \times n}$, $\mathbf{b} \in \mathbb{R}^m$, $\mathbf{c} \in \mathbb{R}^n$. Show that the mixed binary program

$$\begin{aligned}
\min \quad & \mathbf{c}^\mathsf{T} \mathbf{x} \\
\text{s.t.} \quad & \mathbf{A}\mathbf{x} \leq \mathbf{b} \\
& x_j \in \{0, 1\}, \; j \in J
\end{aligned}$$

is equivalent to the nonlinear program

$$\begin{aligned}
\min \quad & \mathbf{c}^\mathsf{T} \mathbf{x} \\
\text{s.t.} \quad & \mathbf{A}\mathbf{x} \leq \mathbf{b} \\
& x_j(1 - x_j) = 0, \; j \in J.
\end{aligned}$$

(b) Let $\mathbf{A} \in \mathbb{R}^{m \times n}$, $\mathbf{b} \in \mathbb{R}^m$, $\mathbf{c} \in \mathbb{R}^n$. Show that the mixed integer program

$$\begin{aligned}
\min \quad & \mathbf{c}^\mathsf{T} \mathbf{x} \\
\text{s.t.} \quad & \mathbf{A}\mathbf{x} \leq \mathbf{b} \\
& x_j \in \mathbb{Z}, \; j \in J
\end{aligned}$$

is equivalent to the nonlinear program

$$\min \quad \mathbf{c}^\mathsf{T}\mathbf{x}$$
$$\text{s.t.} \quad \mathbf{A}\mathbf{x} \le \mathbf{b}$$
$$\sin(\pi x_j) = 0, \ j \in J.$$

Exercise 20.3 Let n be a positive integer. Show that for suitable differential functions $f, \mathbf{g}, \mathbf{h}$ the statement

"There exist $x, y, z \in \mathbb{Z}$ all different such that $x^n + y^n = z^n$."

can be equivalently stated as

"The optimal value of

$$\min \quad f(\mathbf{x})$$
$$\text{s.t.} \quad \mathbf{g}(\mathbf{x}) \le \mathbf{0}$$
$$\mathbf{h}(\mathbf{x}) = \mathbf{0}$$

is zero."

What does that suggest about the difficulty of solving generic nonlinear programming problems?

Appendices

Appendix Basic Mathematical Facts

A.1 Matrices and Vectors

For two positive integers m and n, let $\mathbb{R}^{m \times n}$ denote the space of $m \times n$ matrices with real entries. The *transpose* of an $m \times n$ matrix

$$\mathbf{A} = \begin{bmatrix} a_{11} & a_{12} & \cdots & a_{1n} \\ a_{21} & a_{22} & \cdots & a_{2n} \\ \vdots & \vdots & \ddots & \vdots \\ a_{m1} & a_{m2} & \cdots & a_{mn} \end{bmatrix} \in \mathbb{R}^{m \times n}$$

is the $n \times m$ matrix

$$\mathbf{A}^\mathsf{T} = \begin{bmatrix} a_{11} & a_{21} & \cdots & a_{m1} \\ a_{12} & a_{22} & \cdots & a_{m2} \\ \vdots & \vdots & \ddots & \vdots \\ a_{1n} & a_{2n} & \cdots & a_{mn} \end{bmatrix} \in \mathbb{R}^{n \times m}.$$

A square matrix $\mathbf{A} \in \mathbb{R}^{n \times n}$ is *symmetric* if $\mathbf{A}^\mathsf{T} = \mathbf{A}$.

The product of two matrices $\mathbf{A} = (a_{ik}) \in \mathbb{R}^{m \times n}$ and $\mathbf{B} = (b_{kj}) \in \mathbb{R}^{n \times p}$ is the matrix $\mathbf{C} = \mathbf{AB} = (c_{ij}) \in \mathbb{R}^{m \times p}$ defined componentwise as follows:

$$c_{ij} = \sum_{k=1}^{n} a_{ik} b_{kj}, \quad i = 1, \ldots, m, \; j = 1, \ldots, p.$$

Observe that the matrix product $\mathbf{AB}$ is well defined if the number of columns of $\mathbf{A}$ and the number of rows of $\mathbf{B}$ match.

The identity matrix $\mathbf{I} \in \mathbb{R}^{n \times n}$ is the matrix with components equal to 1 on the diagonal and all other components equal to 0. Observe that for all $\mathbf{A} \in \mathbb{R}^{m \times n}$, $\mathbf{B} \in \mathbb{R}^{n \times p}$ we have $\mathbf{AI} = \mathbf{A}$ and $\mathbf{IB} = \mathbf{B}$. If $\mathbf{A}, \mathbf{B} \in \mathbb{R}^{n \times n}$ and $\mathbf{AB} = \mathbf{BA} = \mathbf{I}$, then we say that $\mathbf{B}$ is the *inverse* of $\mathbf{A}$ and write $\mathbf{B} = \mathbf{A}^{-1}$.

The following kinds of matrix–vector products arise often. Suppose

$$\mathbf{c} := \begin{bmatrix} c_1 \\ c_2 \\ \vdots \\ c_n \end{bmatrix}, \quad \mathbf{Q} := \begin{bmatrix} q_{11} & q_{12} & \cdots & q_{1n} \\ q_{12} & q_{22} & \cdots & q_{2n} \\ \vdots & \vdots & \ddots & \vdots \\ q_{1n} & q_{2n} & \cdots & q_{nn} \end{bmatrix}, \quad \mathbf{x} := \begin{bmatrix} x_1 \\ x_2 \\ \vdots \\ x_n \end{bmatrix}.$$

Then

$$\mathbf{c}^\mathsf{T}\mathbf{x} = \begin{bmatrix} c_1 & \cdots & c_n \end{bmatrix} \begin{bmatrix} x_1 \\ \vdots \\ x_n \end{bmatrix} = c_1 x_1 + \cdots + c_n x_n$$

and

$$\mathbf{Q}\mathbf{x} = \begin{bmatrix} q_{11} & \cdots & q_{1n} \\ \vdots & \ddots & \vdots \\ q_{1n} & \cdots & q_{nn} \end{bmatrix} \begin{bmatrix} x_1 \\ \vdots \\ x_3 \end{bmatrix} = \begin{bmatrix} q_{11}x_1 + \cdots + q_{1n}x_n \\ \vdots \\ q_{1n}x_1 + \cdots + q_{nn}x_n \end{bmatrix}.$$

So

$$\mathbf{x}^\mathsf{T}\mathbf{Q}\mathbf{x} = \begin{bmatrix} x_1 & \cdots & x_n \end{bmatrix} \begin{bmatrix} q_{11}x_1 + \cdots + q_{1n}x_n \\ \vdots \\ q_{1n}x_1 + \cdots + q_{nn}x_n \end{bmatrix}$$
$$= \sum_{i=1}^{n}\sum_{j=1}^{n} q_{ij}x_i x_j$$
$$= q_{11}x_1^2 + \cdots + q_{nn}x_n^2 + 2q_{12}x_1 x_2 + 2q_{23}x_2 x_3 + \cdots + 2q_{n-1,n}x_{n-1}x_n$$

A symmetric matrix $\mathbf{M} \in \mathbb{R}^{n\times n}$ is *positive semidefinite* if $\mathbf{x}^\mathsf{T}\mathbf{M}\mathbf{x} \geq 0$ for $\mathbf{x} \in \mathbb{R}^n$ and it is *positive definite* if it satisfies the stronger condition $\mathbf{x}^\mathsf{T}\mathbf{M}\mathbf{x} >$ for all non-zero $\mathbf{x} \in \mathbb{R}^n$.

A.2   Convex Sets and Convex Functions

A set $S \subseteq \mathbb{R}^n$ is *convex* if for all $\mathbf{x}, \mathbf{y} \in S$ the straight segment joining $\mathbf{x}$ and is contained in S; that is,

$$[\mathbf{x}, \mathbf{y}] := \{\lambda\mathbf{x} + (1-\lambda)\mathbf{y} : \lambda \in [0,1]\} \subseteq S.$$

The following are types of convex sets that appear often in optimization mode It is easy to verify that they are indeed convex sets.

Half-space: Given a non-zero $\mathbf{a} \in \mathbb{R}^n$ and $b \in \mathbb{R}$ the half-space

$$\{\mathbf{x} \in \mathbb{R}^n : \mathbf{a}^\mathsf{T}\mathbf{x} \leq b\}$$

is convex.

Intersections of convex sets: Given a collection $S_i \subseteq \mathbb{R}^n$, for $i \in I$, of conv sets, their intersection $\bigcap_{i \in I} S_i$ is a convex set.

Affine images and preimages: Given a convex set $S \subseteq \mathbb{R}^n$, matrices $\mathbf{A} \in \mathbb{R}^{m \times n}$, $\mathbf{B} \in \mathbb{R}^{n \times p}$ and vectors $\mathbf{a} \in \mathbb{R}^m$, $\mathbf{b} \in \mathbb{R}^n$, the sets

$$\mathbf{A}(S) + \mathbf{a} = \{\mathbf{A}\mathbf{x} + \mathbf{a} : \mathbf{x} \in S\} \subseteq \mathbb{R}^m$$

and

$$\mathbf{B}^{-1}(S + \mathbf{b}) := \{\mathbf{v} \in \mathbb{R}^p : \mathbf{B}\mathbf{v} - \mathbf{b} \in S\} \subseteq \mathbb{R}^p$$

are convex.

Suppose $S \subseteq \mathbb{R}^n$ is a convex set. A function $f : S \to \mathbb{R}$ is *convex* if, for all $\mathbf{x}, \mathbf{y} \in S$ and $\lambda \in [0, 1]$,

$$f(\lambda \mathbf{x} + (1 - \lambda)\mathbf{y}) \le \lambda f(\mathbf{x}) + (1 - \lambda)f(\mathbf{y}).$$

A common way of dealing with the domain of a function is to consider *extended valued* functions; that is, functions defined on the whole space $\mathbb{R}^n$ and allowed to take the value ∞. The *domain* of an extended valued function $f : \mathbb{R}^n \to \mathbb{R} \cup \{\infty\}$ is the set

$$\text{dom}(f) := \{\mathbf{x} \in \mathbb{R}^n : f(\mathbf{x}) < \infty\}.$$

An alternative and equivalent definition of convexity is the following. An extended valued function $f : \mathbb{R}^n \to \mathbb{R} \cup \{\infty\}$ is *convex* if the set

$$\text{epigraph}(f) := \{(\mathbf{x}, t) \in \mathbb{R}^{n+1} : f(\mathbf{x}) \le t\}$$

is convex. Observe that if f is convex, then its domain is a convex set. Furthermore, if f is convex then for all $\ell \in \mathbb{R}$ the *sublevel* set $\{\mathbf{x} \in \mathbb{R}^n : f(\mathbf{x}) \le \ell\}$ is a convex set.

The following relationship between differentiability and convexity is particularly useful to verify that functions are convex.

Theorem A.1 *Suppose $f : S \to \mathbb{R}$ is twice differentiable on the open set $S \subseteq \mathbb{R}^n$ and $C \subseteq S$ is a convex set. Then f is convex on C if and only if $\nabla^2 f(\mathbf{x})$ is positive semidefinite for all $\mathbf{x} \in C$.*

As an immediate consequence of Theorem A.1 it follows that every affine function $f(\mathbf{x}) = \mathbf{c}^\mathsf{T}\mathbf{x} + b$ is convex. It also follows that a quadratic function

$$f(\mathbf{x}) = \frac{1}{2}\mathbf{x}^\mathsf{T}\mathbf{Q}\mathbf{x} + \mathbf{c}^\mathsf{T}\mathbf{x} + b$$

is convex if and only if $\mathbf{Q}$ is positive semidefinite.

A.3   Calculus of Variations: the Euler Equation

The calculus of variations is the analog of calculus that works with functionals rather than functions. Functionals are often integrals of functions. Many problems in the calculus of variations arose from the need to find a function

that optimizes a given functional. The *Euler equation* for the minimization
a functional subject to boundary conditions is a kind of first-order optimal
condition for the problem

$$\min_{x} \int_0^T L(t, x(t), \dot{x}(t))dt, \ x(0) = x_0, \ x(T) = x_T.$$

The optimal solution $x^*(t)$ must satisfy the differential equation

$$L_x = \frac{d}{dt} L_{\dot{x}}. \tag{A}$$

Equation (A.1) is called the *Euler equation*. For a derivation of this optimal
condition as well as a detailed discussion on the interesting subject of calcu
of variations, see Fleming and Rishel (1975).

References

Alizadeh F. (1991). *Combinatorial Optimization with Interior Point Methods and Semi-definite Matrices*. PhD thesis, University of Minnesota.

Almgren R. and N. Chriss (2000). Optimal execution of portfolio transactions. *Journal of Risk*, 3:5–39.

Almgren R., C. Thum, E. Hauptmann, and H. Li (2005). Direct estimation of equity market impact. *Risk*, 18:58-62.

Andersson F., H. Mausser, D. Rosen, and S. Uryasev (2001). Credit risk optimization with conditional value-at-risk criterion. *Mathematical Programming*, 89:273-291.

Artzner P., F. Delbaen, J. Eber, and D. Heath (1999). Coherent measures of risk. *Mathematical Finance*, 9:203–228.

Back K. (2010). *Asset Pricing and Portfolio Choice Theory*. Oxford University Press.

Basel Committee on Banking Supervision (2011). Basel III: A Global Regulatory Framework for More Resilient Banks and Banking Systems. Technical Report, Bank for International Settlements.

Bawa V.S., S.J. Brown, and R.W. Klein (1979). *Estimation Risk and Optimal Portfolio Choice*. North-Holland.

Bellman R. (1954). The theory of dynamic programming. *Bulletin of the American Mathematical Society*, 60:503–515.

Bellman R. (1957). *Dynamic Programming*. Princeton University Press.

Ben-Tal A. and A. Nemirovski (1998). Robust convex optimization. *Mathematics of Operations Research*, 23(4):769–805.

Ben-Tal A. and A. Nemirovski (2002). Robust optimization – methodology and applications. *Mathematical Programming*, 92(3):453–480.

Ben-Tal A., L. El Ghaoui, and A. Nemirovski (2009). *Robust Optimization*. Princeton University Press.

Bertsekas D. (1999). *Nonlinear Programming*. Athena Scientific.

Bertsekas D. (2005). *Dynamic Programming and Optimal Control*. Athena Scientific.

Bertsimas D. and A. Lo (1998). Optimal control of execution costs. *Journal of Financial Markets*, 1:1–50.

Bertsimas D. and J. Tsitsiklis (1997). *Introduction to Linear Optimization*. Athena Scientific.

Bertsimas D., V. Gupta, and I.Ch. Paschalidis (2012). Inverse optimization: new perspective on the Black–Litterman model. *Operations Research*, 138 1403.

Birge J. and F. Louveaux (1997). *Introduction to Stochastic Programmii* Springer.

Black F. and R. Litterman (1992). Global portfolio optimization. *Financ Analysts Journal*, 48:28–43.

Black F. and M. Scholes (1973). The pricing of options and corporate liabiliti *Journal of Political Economy*, 81:637–659.

Blume M. (1975). Betas and the regression tendencies. *Journal of Finan* 30:785–795.

Boyd S. and L. Vandenberghe (2004). *Convex Optimization*. Cambridge Univ sity Press.

Brinson G., B. Singer, and G. Beebower (1991). Determinants of portfo performance. *Financial Analysts Journal*, 47:40–48.

Broadie M. (1993). Computing efficient frontiers using estimated paramete *Annals of Operations Research*, 45(1):21–58.

Campbell J., A. Lo, and A. MacKinlay (1997). *The Econometrics of Financ Markets*. Princeton University Press.

Cariño D., T. Kent, D. Myers, C. Stacy, M. Sylvanus, A. Turner, K. Watanal and W. Ziemba (1994). The Russell–Yasuda Kasai model: an asset/liabili model for a Japanese insurance company using multistage stochastic progra ming. *Interfaces*, 24(1):29–49.

Ceria S. and R. Stubbs (2006). Incorporating estimation errors into portfo selection: robust portfolio selection. *Journal of Asset Management*, 7:109–1:

Choueifaty Y. and Y. Coignard (2008). Toward maximum diversification. *Jour of Portfolio Management*, 40–51.

Chvátal V. (1983). *Linear Programming*. W.H. Freeman.

Coleman T.F., Y. Kim, Y. Li, and A. Verma (1999a). Dynamic Hedging in Volatile Market. Technical Report, Cornell Theory Center.

Coleman T.F., Y. Kim, Y. Li, and A. Verma (1999b). Reconstructing t unknown volatility function. *Journal of Computational Finance*, 2:77–102.

Conforti M., G. Cornuéjols, and G. Zambelli (2014). *Integer Programmir* Springer.

Connor G. (1995). The three types of factor models: a comparison of the explanatory power. *Financial Analysts Journal*, 51:42–46.

Constantinides G. (1983). Capital market equilibrium with personal tax. *Econ metrica*, 51:611–636.

Constantinides G. (1984). Optimal stock trading with personal taxes: imp cations for prices and the abnormal January returns. *Journal of Financ Economics*, 13:65–89.

Cornuéjols G., M. Fisher, and G. Nemhauser (1977). Location of bank accoun to minimize float: an analytical study of exact and approximate algorithn *Management Science*, 23:229–263.

Cox J., S. Ross, and M. Rubinstein (1979). Option pricing: a simplified approach. *Journal of Financial Economics*, 7:229–263.

Dammon R., C. Spatt, and H. Zhang (2001). Optimal consumption and investment with capital gains taxes. *Review of Financial. Studies*, 14:583–616.

Dammon R., C. Spatt, and H. Zhang (2004). Optimal asset location and allocation with taxable and tax-deferred investing. *Journal of Finance*, 59:999–1038.

Dantzig G. (1963). *Linear Programming and Extensions*. Princeton University Press.

Dantzig G. (1990). The diet problem. *Interfaces*, 20(4):43–47.

Dantzig G., R. Fulkerson, and S. Johnson (1954). Solution of a large-scale traveling-salesman problem. *Operations Research*, 2:393–410.

Davarnia D. and G. Cornuéjols (2017). From estimation to optimization via shrinkage. *Operations Research Letters*, 45:642–646.

De Vries S. and R. Vohra (2003). Combinatorial auctions: a survey. *INFORMS Journal on Computing*, 15(3):284–309.

Duffie D. (2001). *Dynamic Asset Pricing Theory*. Princeton University Press.

Efron B. and C. Morris (1977). Stein's paradox in statistics. *Scientific American*, 236:119–127.

El Ghaoui L. and H. Lebret (1997). Robust solutions to least-squares problems with uncertain data. *SIAM Journal on Matrix Analysis and Applications*, 18(4):1035–1064.

El Ghaoui L., F. Oustry, and H. Lebret (1998). Robust solutions to uncertain semidefinite programs. *SIAM Journal on Optimization*, 9(1):33–52.

Engle R. (1982). Autoregressive conditional heteroscedasticity with estimates of the variance of United Kingdom inflation. *Econometrica*, 50:987–1007.

Fabozzi F. (2004). *Bonds, Markets, Analysis and Strategies*, fifth edition. Prentice-Hall.

Fabozzi F., P. Kolm, D. Pachamanova, and S. Focardi (2007). *Robust Portfolio Optimization and Management*. Wiley.

Fama E. and K. French (1992). The cross-section of expected stock returns. *Journal of Finance*, 67:427–465.

Fleming W. and R. Rishel (1975). *Deterministic and Stochastic Optimal Control*. Springer.

Friedman J., T. Hastie, and R. Tibshirani (2001). *The Elements of Statistical Learning*, volume 1. Springer.

Gârleanu N. and L. Pedersen (2013). Dynamic trading with predictable returns and transaction costs. *Journal of Finance*, 68(6):2309–2340.

Goldfarb D. and G. Iyengar (2003). Robust portfolio selection problems. *Mathematics of Operations Research*, 28(1):1–38.

Gomory R.E. (1958). Outline of an algorithm for integer solutions to linear programs. *Bulletin of the American Mathematical Society*, 64:275–278.

Gomory R.E. (1960). An Algorithm for the Mixed Integer Problem. Technical Report RM-2597, The Rand Corporation.

Gondzio J. and R. Kouwenberg (2001). High performance for asset liabili management. *Operations Research*, 49:879–891.

Grinold R. and R. Kahn (1999). *Active Portfolio Management: A Quantitati Approach for Producing Superior Returns and Controlling Risk*, second editio McGraw-Hill.

Güler O. (2010). *Foundations of Optimization*. Springer.

Halldórsson B. and R. Tütüncü (2003). An interior-point method for a cla of saddle-point problems. *Journal of Optimization Theory and Applicatio* 116(3):559–590.

Harrison J. and D. Kreps (1979). Martingales and arbitrage in multiperi security markets. *Journal of Economic Theory*, 20:381–408.

Harrison J. and S. Pliska (1981). Martingales and stochastic integrals in t theory of continuous trading. *Stochastic Processes and their Applicatio* 11:215–260.

Heath D., R. Jarrow, and A. Morton (1992). Bond pricing and the ter structure of interest rates: a new methodology for contingent clai valuation. *Econometrica*, 60:77–105.

Herzel S. (2005). Arbitrage opportunities on derivatives: a linear programmi approach. *Dynamics of Continuous, Discrete and Impulsive Systems. Series Applications and Algorithms*, 12:589–606.

Hodges S. and S. Schaefer (1977). A model for bond portfolio improveme *Journal of Financial and Quantitative Analysis*, 12:243–260.

Hoyland K. and S.W. Wallace (2001). Generating scenario trees for multista decision problems. *Management Science*, 47:295–307.

Jorion P. (1986). Bayes–Stein estimation for portfolio analysis. *Journal Financial and Quantitative Analysis*, 21:279–292.

Jorion P. (1992). Portfolio optimization in practice. *Financial Analysts Journ* 48:68–74.

Jorion P. (2003). Portfolio optimization with tracking-error constraint *Financial Analysts Journal*, 59:70–82.

Karmarkar N. (1984). A new polynomial time algorithm for linear programmi *Combinatorica*, 4:373–395.

Kelly J.L. (1956). A new interpretation of information rate. *Bell Syste Technical Journal*, 35:917–926.

Klaassen P. (2002). Comment on "Generating scenario trees for multista decision problems". *Management Science*, 48:1512–1516.

Kocuk B. and G. Cornuéjols (2017). Incorporating Black–Litterman Vie in Portfolio Construction When Stock Returns Are a Mixture of Norma Technical Report, Carnegie–Mellon University, Pittsburgh.

Konno H. and H. Yamazaki (1991). Mean-absolute deviation portfolio optimiz tion model and its applications to Tokyo stock market. *Management Scien* 37(5):519–531.

Kouwenberg R. (2001). Scenario generation and stochastic programming models for asset liability management. *European Journal of Operational Research*, 134:279–292.

Kritzman M. (2002). *Puzzles of Finance: Six Practical Problems and their Remarkable Solutions*. Wiley.

Lagnado R. and S. Osher (1997). Reconciling differences. *Risk*, 10:79–83.

Land A.H. and A.G. Doig (1960). An automatic method of solving discrete programming problems. *Econometrica*, 28:497–520.

Ledoit O. and M. Wolf (2003). Improved estimation of the covariance matrix of stock returns with an application to portfolio selection. *Journal of Empirical Finance*, 10:602–621.

Ledoit O. and M. Wolf (2004). A well-conditioned estimator for large-dimensional covariance matrices. *Journal of Multivariate Analysis*, 88:365–411.

Lintner J. (1965). The valuation of risk assets and the selection of risky investments in stock portfolios and capital budgets. *Review of Economics and Statistics*, 47:13–37.

Litterman B. (2003). *Modern Investment Management: An Equilibrium Approach*. Wiley.

Markowitz H. (1952). Portfolio selection. *Journal of Finance*, 7:77–91.

Merton R. (1973). Theory of rational option pricing. *Bell Journal of Economics and Management Science*, 4:141–183.

Meucci A. (2005). *Risk and Asset Allocation*. Springer.

Meucci A. (2010). Return calculations for leveraged securities and portfolios. *GARP Risk Professional*, October:40–43.

Michaud R. and R. Michaud (2008). *Efficient Asset Management*. Oxford University Press.

Mossin J. (1966). Equilibrium in a capital asset market. *Econometrica*, 34:768–783.

Nesterov Y. (2004). *Introductory Lectures on Convex Optimization: A Basic Course*. Kluwer Academic.

Nesterov Y. and A. Nemirovskii (1994). *Interior-Point Polynomial Algorithms in Convex Programming*. SIAM.

Nesterov Y. and M. Todd (1997). Self-scaled barriers and interior-point methods for convex programming. *Mathematics of Operations Research*, 22:1–42.

Nesterov Y. and M. Todd (1998). Primal–dual interior-point methods for self-scaled cones. *SIAM Journal on Optimization*, 8:324–364.

Nocedal J. and S. Wright (2006). *Numerical Optimization*. Springer.

Padberg M. and G. Rinaldi (1987). Optimization of a 532-city symmetric traveling salesman problem by branch and cut. *Operations Research Letters*, 6:1–7.

Pérold A. (1988). The implementation shortfall: paper versus reality. *Journal of Portfolio Management*, 14:4–9.

Pokutta S. and C. Schmaltz (2012). Optimal bank planning under Basel III regulations. *Capco Institute Journal of Financial Transformation*, 34:165–174.

Porteus E. (2002). *Foundations of Stochastic Inventory Theory*. Stanford University Press.

Poundstone W. (2005). *Fortune's Formula: The Untold Story of the Scientific Betting System that Beat the Casinos and Wall Street*. Hill and Wang.

Ragsdale C. (2007). *Spreadsheet Modeling & Decision Analysis: A Practical Introduction to Management Science*, fifth edition. Thomson South-Western.

Renegar J. (2001). *A Mathematical View of Interior-Point Methods in Convex Optimization*. SIAM.

Rockafellar T. and S. Uryasev (2000). Optimization of conditional value-at-risk. *Journal of Risk*, 2:21–41.

Ronn E. I. (1987). A new linear programming approach to bond portfolio management. *Journal of Financial and Quantitative Analysis*, 22:439–466.

Roos C., T. Terlaky, and J.-Ph. Vial (2005). *Interior Point Methods for Linear Optimization*, second edition. Springer.

Rosenberg B. (1974). Extra-market components of covariance in security returns. *Journal of Financial and Quantitative Analysis*, 9(2):263–274.

Ross S. (1976). The arbitrage theory of capital asset pricing. *Journal of Economic. Theory*, 13:341–360.

Rustem B. and M. Howe (2002). *Algorithms for Worst-Case Design and Applications to Risk Management*. Princeton University Press.

Schaefer S.M. (1982). Tax induced clientele effects in the market for British government securities. *Journal of Financial Economics*, 10:121–159.

Scherer B. (2002). Portfolio resampling: review and critique. *Financial Analyst Journal*, 58:98–109.

Scherer B. (2007). Can robust portfolio optimization help to build better portfolios? *Journal of Asset Management*, 7:374–387.

Schmieta S. and F. Alizadeh (2001). Associative and Jordan algebras, and polynomial time interior-point algorithms for symmetric cones. *Mathematics of Operations Research*, 26(3):543–564.

Schmieta S. and F. Alizadeh (2003). Extension of primal–dual interior point algorithms to symmetric cones. *Mathematical Programming*, 96(3):409–438.

Shapiro A., D. Dentcheva, and A. Ruszczynski (2009). *Lectures on Stochastic Programming: Modeling and Theory*. SIAM.

Sharpe W. (1964). Capital asset prices: a theory of market equilibrium under conditions of risk. *Journal of Finance*, 19(3):425–442.

Sharpe W. (1992). Asset allocation: management style and performance measurement. *Journal of Portfolio Management*, 18(2):7–19.

Shleifer A. (2000). *Inefficient Markets*. Oxford University Press.

Shreve S. (2000). *Stochastic Calculus for Finance*, volumes I and II. Springer.

Sra S., S. Nowozin, and S. Wright (2012). *Optimization for Machine Learning*. MIT Press.

Stein C. (1956). Inadmissibility of the usual estimator for the mean of multivariate normal distribution. In *Proceedings of the Third Berkeley Symposium of Mathematical Statistics and Probability*, pp. 197–206.

Sturm J. (1999). Using SeDuMi 1.02, a Matlab toolbox for optimization over symmetric cones. *Optimization Methods and Software*, 11:625–653.

Tibshirani R. (1996). Regression shrinkage and selection via the lasso. *Journal of the Royal Statistical Society*, 58:267–288.

Tobin J. (1958). Liquidity preference as behavior towards risk. *Review of Economic Studies*, 25(2):65–86.

Toh K., M. Todd, and R. Tütüncü (1999). SDPT3 – a MATLAB software package for semidefinite programming. *Optimization Methods and Software*, 11:545–581.

Tuckman B. (2002). *Fixed Income Securities: Tools for Today's Markets*. Wiley.

Tütüncü R. and M. Koenig (2004). Robust asset allocation. *Annals of Operations Research*, 132:157–187.

Vapnik V. (2013). *The Nature of Statistical Learning Theory*. Springer.

Werner R. (2010). Costs and Benefits of Robust Optimization. Technical Report, Technical University of München.

Ye Y. (1997). *Interior-Point Algorithms: Theory and Analysis*. Wiley.

Zhao Y. and W.T. Ziemba (2001). A stochastic programming model using an endogenously determined worst case risk measure for dynamic asset allocation. *Mathematical Programming*, 89(2):293–309.

Index

accrued interest, 42
active constraint, *see* binding constraint
active return, 105
active risk, 105
active set, 313
active-set methods, 81
adaptive decision, 7
adjoint, 285
Almgren–Chriss model, 201
alpha
  Jensen, 113
  t-statistic, 113
alpha of a security, 104
APT, 111
arbitrage, 55
arbitrage pricing theory, *see* APT
Armijo–Goldstein condition, 309
asset allocation, 95
asset–liability management, 262
auction
  combinatorial, 161
autoregressive model, 256

backtracking, 309
Basel III, 15
basic feasible solution, 25
basis, 25
  optimal, 25
Bellman's optimality principle, 215, 216, 221
Benders decomposition, 255
Benders decomposition method, 177
bequest, 227
beta
  long–short threshold, 119
beta of a security, 104
bid, 161
bid–ask spread, 66
binary program, 140
binding constraint, 3
binomial lattice, 238
binomial lattice model, 238
binomial pricing model, 56
Black–Litterman model, 126
Black–Scholes–Merton equation, 245

Black–Scholes–Merton option pricing
  formula, 316
blocking constraint, 82
bond
  clean price, 42
  coupon rate, 42
  dirty price, 42
  maturity date, 42
  term to maturity, 42
  yield, 42
bond allocation, 12
bond portfolio
  dedicated, 35
boostrapping, 131
branch-and-bound method, 150, 151
branch-and-bound tree, 152
branch-and-cut method, 150
branching, 151
Brownian motion, 315
BSM formula, 316
bundle, 161

capital allocation line, 98
capital asset pricing model, *see* CAPM
CAPM, 98, 108, 111, 112
captured value, 202
cash flow problems, 44
central path, 29, 83
clientele effects, 63
clustering, 142
combinatorial auction problem, 161
complementary slackness conditions, 24
conditional value at risk, *see* CVaR
cone
  second-order, 277
  symmetric, 287
conic program, 277
  dual, 282
  primal, 282
constraint
  turnover, 101
constraint set, 3
consumption, 174
contingent claim, 55

convex function, 325
convex optimization, 4
convex set, 324
cutting plane, 154
cutting-plane method, 150, 154
CVaR, 181, 184

decision variables, 3
descent direction, 309
deterministic model, 4
diffusion model, 316
dispersion measure, 181
diversification, 134
  maximum, 135
drift, 244
dual cone, 282
dual problem, 21

efficient frontier, 93, 124
estimator
  inadmissable, 129
  James–Stein shrinkage, 129
  risk, 129
Euler equation, 206
event tree, 250
excess return, 102

factor exposure, *see* factor loading
factor loading, 107
factor model, 106
factor portfolio, 120
Farkas's lemma, 21, 34
feasibility cut, 178
feasible point, 3
feasible region, 3
feasible solution, *see* feasible point
Fisher–Weil convexity, 40
Fisher–Weil dollar convexity, 40
Fisher–Weil dollar duration, 39
Frobenius inner product, 281
fund allocation
  linear programming model, 12
fundamental theorem of asset pricing, 56

geometric Brownian motion model, 244
Gomory mixed integer cut, 155
Gordan's theorem, 22, 34
gradient descent method, 309

homogenization, 102
hyperplane separation theorem, 34

ice-cream cone, *see* cone, second-order
immunization, 39
immunized portfolio, 39
implementation shortfall, *see* total cost of
  trading
implied volatility, 273
index fund, 165
infeasible problem, 3
information ratio, 105

insurance company ALM problem, 263
interior-point method, 28
  infeasibility, 30
  nonlinear programming, 313
  quadratic program, 83

Jacobian, 310
Jensen's alpha, 112

Kelly criterion, 197

L-shaped method, 177
Lagrange multiplier, 75
Lagrangian dual, 147
Lagrangian function, 22, 77
lasso regression, 87
line search, 309
line-search procedure, 30
linear independence constraint qualification,
  307
linear optimization model, *see* linear
  programming model
linear program, 11
linear programming, 5
linear programming model, 11
  non-degenerate, 19
  standard form, 13
linear–quadratic regulator, 216
lockbox problem, 163
Lorenz cone, *see* cone, second-order
loss function, 193

MAD, 181
market completeness, 59
Markowitz mean–variance, 91
Markowitz mean–variance model, 8
master problem, 177
matrix
  inverse, 323
  positive semidefinite, 5, 280
mean absolute deviation, *see* MAD
mean–variance, 71, 296
  stochastic optimization, 175
mean–variance model
  basic, 94
  general, 100
mean–variance optimization model, 93
minimum position constraints, 168
mixed integer linear program, 140
mixed integer optimization, 4
mixed integer program, 140
mixed integer programming, 6
multi-period model, 4
multiple-factor risk model, 109

newsvendor problem, 173, 175, 177
Newton step, 29, 286
Newton's method, 310
nonlinear program, 305
NP-hardness, 150

objective function, 3
one-fund separation theorem, 97
optimal decision rule, 216
optimal policy, 216
optimal solution, 3
optimal value, 3
optimality conditions, 24
optimality cut, 178, 180
optimization
    robust, 289
option
    American, 241
    European, 239
option pricing, 238

par, *see* principal value
pension fund, 263
performance analysis, 112
portfolio
    benchmark, 103
    characteristic, 96
    efficient, 93
    equally weighted, 133
    factor mimicking, 111
    minimum risk, 95
    risk-parity, 134
    tangency, 97
    value-weighted, 133
portfolio management, 8
portfolio optimization
    dynamic, 198
positive linear pricing rule, 55
primal problem, 21
principal, *see* principal value
principal value, 42
program
    semidefinite, 280
pruning a node, 152
pure integer linear program, 140

quadratic program, 71
    dual, 76
    primal, 76
    standard form, 71
quadratic programming, 5
    sequential, 313
quadratic programming model, *see* quadratic
    program

random sampling, 257
    adjusted, 257
recourse, 7
recourse problem, 177
reduced cost, 25
reduced gradient
    generalized, 312
regret, 292
relaxation, 141, 145
    linear programming, 145

resampled efficiency, 131
residual vector, 83
reward-to-risk ratio, *see* Sharpe ratio
Ricatti equation, 220
ridge regression, 86
risk contribution, 134
    marginal, 134
risk management, 9
risk measure, 6, 175, 181
    coherent, 184
risk-neutral probability measure, 55
robust portfolio optimization, 299
robustness, 289
    constraint, 290
    objective, 291
    relative, 292
    sampling, 294

saddle-point problem, 292
scenario optimization, 176
scenario tree, 248
    arbitrage-free , 258
scenario trees
    construction, 256
second-order program, 277
security selection, 95, 103
selection return, 114
semidefinite program
    standard form, 281
sensitivity, 18, 38, 75
separation theorem, *see* hyperplane
sequential system, 214
shadow price, 17, 38
Sharpe ratio, 101, 116
shrinkage estimators, 129
shrinkage factor, 129
shrinkage procedure, 108
shrinkage target, 129
signal, 111
simplex method, 25
    dual, 25, 27
single-factor risk model, 107
slack variable, 14
Slater condition, 284
static model, 4
Stein paradox, 129
Stiemke's theorem, 22, 34
stochastic and dynamic optimization, 4
stochastic discount factor, 56
stochastic model, 4
stochastic optimization, 6, 173, 174
    two-stage with recourse, 6
    with recourse, 174
stochastic program
    linear two-stage, 175
stochastic programming, 248
    multi-stage, 248

stochastic sequential decision problem, 221
stochastic sequential system, 221
strong duality theorem, 21
style analysis, 113
subdifferential, 311
subgradient, 311
subgradient method, 311
support vector machine, 85
surplus variable, 14
synthetic option, 270
synthetic option strategy, 270

tangent subspace, 308
term structure, 39
  implied, 38
total cost of trading, 203
tracking error, *see* active risk
tractability, 4
trade list, 202
trading strategy
  execution, 202
trading trajectory, 202
treasury yield curve, 42
tree fitting, 257

two-fund separation theorem, 96
two-fund theorem, 115

unbounded problem, 4
uncertainty set, 289
  ellipsoidal, 290
urgency, 206
utility, 173
  logarithmic, 199
  power, 199
  quadratic, 93

value at risk, *see* VaR
  conditional, *see* CVaR
value-to-go function, 215
VaR, 183
  α, 183
volatility, 244
volatility smile, 316
volume-weighted average price (VWAP),
  204

weak duality theorem, 21
winner selection problem, *see* combinatorial
  auction problem

Printed in the United States
by Baker & Taylor Publisher Services

Printed in the United States
by Baker & Taylor Publisher Services